Two and Three Dimensional Calculus

Two and Three Dimensional Calculus

With Applications in Science and Engineering

Phil Dyke
University of Plymouth
UK

This edition first published 2018
© 2018 John Wiley & Sons Ltd

The right of Phil Dyke to be identified as the author of this work has been asserted in accordance with law.

Registered Offices
John Wiley & Sons, Inc., 111 River Street, Hoboken, NJ 07030, USA
John Wiley & Sons Ltd, The Atrium, Southern Gate, Chichester, West Sussex, PO19 8SQ, UK

Editorial Office
9600 Garsington Road, Oxford, OX4 2DQ, UK

For details of our global editorial offices, customer services, and more information about Wiley products visit us at www.wiley.com.

Wiley also publishes its books in a variety of electronic formats and by print-on-demand. Some content that appears in standard print versions of this book may not be available in other formats.

Library of Congress Cataloging-in-Publication Data

Names: Dyke, P. P. G., author.
Title: Two and three dimensional calculus : with applications in science and
   engineering / by Phil Dyke.
Description: Hoboken, NJ : John Wiley & Sons, 2018. | Includes
   bibliographical references and index. |
Identifiers: LCCN 2017055996 (print) | LCCN 2017061105 (ebook) | ISBN
   9781119221791 (pdf) | ISBN 9781119221807 (epub) | ISBN 9781119221784
   (cloth)
Subjects: LCSH: Calculus of variations.
Classification: LCC QA315 (ebook) | LCC QA315 .D95 2018 (print) | DDC
   515/.63–dc23
LC record available at https://lccn.loc.gov/2017055996

Cover Design: Wiley
Cover Images: (Background) © sorendls/Gettyimages;
(inset image) - Courtesy of Phil Dyke

Set in 10/12pt WarnockPro by SPi Global, Chennai, India

Printed in Great Britain by TJ International Ltd, Padstow, Corwall

10 9 8 7 6 5 4 3 2 1

To My Son Adrian

Contents

Preface

Calculus, first developed by Newton, or perhaps Leibniz, in the seventeenth century, forms the bedrock of a large swathe of physics and engineering. In the intervening few hundred years, the mathematical foundation underlying calculus has developed into first analysis and now it has become embedded in advanced set theory. Meanwhile, classical differential and integral calculus continues to be applied not only to problems of physics and engineering but also to economics, biology, medicine and almost every other discipline where variables are continuously changing. The book has originated from teaching this material first to engineers, throughout the 1990s and again for the past 10 years to second-year undergraduate mathematics students. Therefore, the author has experienced first-hand the difficulty many students have both with some aspects of the computational side and with trying to make sense of the theoretical aspects. This text is written mainly from the practical viewpoint of getting answers to problems rather than from the viewpoint of putting advanced calculus in a pure mathematical setting. For students of mathematics, therefore, this book is a useful guide to calculation but is certainly not the whole story. Placing calculus in mathematical constructs such as manifolds and the like is not done in this book.

In this text, it is assumed that the reader is already familiar with single-variable calculus, and Chapter 1 is given as a refresher. After this, partial derivatives are introduced followed by their application to finding extreme values. Throughout these chapters, the tuition is done via explanation and examples. It is the firm view of the author that the best way to learn mathematics is to do it, and for calculus this means using calculus to solve practical problems. The decision has been made not to emphasise the use of either computer algebra or numerical mathematics, so direct reference to code or software is absent. These are mature subjects in their own right and although they are both used to help when necessary, to allow them to dominate would be a distraction. Therefore, most of the problems solved here are there to help in the understanding of two- and three-dimensional problems that involve continuously varying quantities. The fourth chapter is not calculus but introduces vectors. This might be revision too, but is necessary to the following chapters on differentiating vectors. In this part of the book, curvilinear co-ordinate systems are introduced, and these are applied to vector quantities with applications in geometry and mechanics. The last chapters of the book concern integration. Theorems and results on integration are introduced as well as applications of integration in electromagnetism, fluid mechanics, elasticity and thermodynamics. There is a nod in the direction of tensors that are an n-dimensional generalisation of vectors.

The style adopted is a conversational one where the reader is taken through two- and three-dimensional calculus in a student-friendly way. This book can be used as an accompanying text to university courses in mathematics, physics, engineering and other quantitative disciplines. It can also be used to help anyone who wants to study the subject themselves using YouTube and other online material. This text is also available as an ebook that has at least two advantages. First, it is cheaper as an ebook and second, the diagrams in the ebook are all in colour. This is particularly useful for the three-dimensional graphs drawn with MAPLE that illustrate surfaces in finding maxima and minima (Chapter 3), exploring surface integrals (Chapter 10) and finding multiple integrals (Chapter 9).

It is a pleasure to thank all the students who through the years have helped shape the way this material is presented through their feedback and through road-testing many of the examples and exercises. However, all mistakes that have crept through are securely my own fault. Special mention is due here to my son Adrian who has helped me on the software and networking side of writing. The text has been created using LaTeX.

Phil Dyke
Professor of Applied Mathematics
University of Plymouth
May 2017

1

Revision of One-Dimensional Calculus

In this chapter, there will be a brief run through of those bits of mathematics that will be required in subsequent chapters. Such a run through cannot, of course, be exhaustive or even particularly thorough. Therefore, if you should find any material in this chapter that is completely new, then you should revisit fundamental reading material on single-variable calculus. When reading this chapter, you may find either familiar content in which you may be rusty or a few unfamiliar nuances here and there. To understand the concepts of differentiation and integration, let us introduce what is meant by a *limit* and in the process also introduce *convergence*.

1.1 Limits and Convergence

The standard notation for a function $f(x)$ dates back to Leonhard Euler in the eighteenth century. The notation implies that one simply inserts the value x into the definition of the function $f(x)$ in order to calculate the corresponding value of the function itself. Therefore, for example, if $f(x) = x^2 - 3x + 4$ its value where $x = 1$ would simply be $1^2 - (3 \times 1) + 4 = 2$ and so at $x = 1$ the value of the function is indisputably 2. When there is no doubt that $f(x)$, when $x = 1$, has the value 2, it is usually written neatly as $f(1) = 2$. Such certainty is however sometimes not the case. Take the function

$$g(x) = \frac{x^2 + 5x + 4}{x^2 - 5x + 4}$$

at the point where $x = 1$. We see here that the numerator is 10 but the denominator is 0 so $g(1)$ does not have a value. We could say 'it is infinite' and move on rather than worry too much. Mathematically, it is better to say that $g(x)$ increases without limit as x approaches the value 1. Here, the concept of a limit appears, and the limit is written as

$$x \to 1 \quad g(x) \to \infty$$

Two and Three Dimensional Calculus: With Applications in Science and Engineering, First Edition. Phil Dyke.
© 2018 John Wiley & Sons Ltd. Published 2018 by John Wiley & Sons Ltd.

or perhaps

$$\lim_{x \to 1} g(x) = \infty$$

which is not a good use of the mathematical equals sign as neither the limit nor ∞ is a number; it's an abstract concept that at the time worried most mathematicians and was in fact responsible for the subsequent mental breakdowns of several nineteenth-century pioneers of number theory. Those interested in learning more should explore the definitions of the transfinite numbers $\aleph_0, \aleph_1, \ldots$ (the use of the Hebrew letter aleph is standard) but be assured that it has no connection with madness these days. Note that x can approach 1 from the right $(x > 1)$ or the left $(x < 1)$. If $x \to 1$ from the left, then the notation $x \to 1-$ is used, if the approach is from the right, then the notation $x \to 1+$ applies. Sometimes, the minus or plus symbols are written as suffices or superfixes thus $1_-, 1_+$ or $1^-, 1^+$. Examination of the function $g(x)$ shows that

$$\lim_{x \to 1-} g(x) = +\infty \quad \text{but that} \quad \lim_{x \to 1+} g(x) = -\infty.$$

Another reason for having to use limits is if the function takes an indeterminate form. Perhaps the limit

$$\lim_{x \to 0} \frac{\sin x}{x}$$

is familiar. Its value is 1 even though both numerator and denominator tend to 0 as x approaches 0 and from either side of 0. The evaluation of such indeterminate forms can be done from first principles. For example, in the case of $\sin x / x$, simply expand $\sin x$ as a series in powers of x called a McLaurin series:

$$\sin x = x - \frac{x^3}{3!} + \frac{x^5}{5!} + \cdots$$

so that

$$\frac{\sin x}{x} = 1 - \frac{x^2}{3!} + \frac{x^4}{5!} + \cdots$$

and letting $x \to 0$, all the terms on the right vanish apart from the 1 which is the value of the limit. Of course, we must state that the series is certainly convergent for small values of x so as $x \to 0$ convergence of the right hand side to 1 is assured. The use of the term *convergence* should be noted. There is a departure from standard texts on pure mathematics that legitimately make a great deal of what is meant by convergence and limits and give the precise definitions of both. This is not done here; not just for reasons of space, but for reasons of emphasis. In this text, the emphasis is on mathematical methods and practicalities. Books on real analysis need to be consulted for the theorems and proofs. Convergence of any series of real numbers $\sum_{n=0}^{\infty} a_n$, is assured provided

$$\lim_{n \to \infty} \left| \frac{a_{n+1}}{a_n} \right| < 1.$$

This is the ratio test that is not infallible in the sense that this ratio could tend to unity yet the series still be convergent, but it is this test that comes to our aid here. The series for $\sin x / x$ can be written as

$$\frac{\sin x}{x} = \sum_{n=0}^{\infty} \frac{(-1)^n x^{2n}}{(2n+1)!}$$

and the absolute value of the ratio of successive terms is

$$\left| \frac{a_{n+1}}{a_n} \right| = \frac{x^2}{(2n+3)(2n+2)},$$

which always tends to zero as $n \to \infty$ no matter what the value of x. Therefore, by the ratio test, absolute convergence of the series is assured.

There will be more on limits later, but the definition of derivative needs to be given now.

1.2 Differentiation

If a function is smooth, that is it has a tangent at a point that changes smoothly as the point moves along the graph of the curve $y = f(x)$ in the x, y plane, then the limit

$$\lim_{\Delta x \to 0} \left\{ \frac{f(x_0 + \Delta x) - f(x_0)}{\Delta x} \right\}$$

exists and is unique and is called the *derivative* of $f(x)$ at the point x_0. Again this will do for our purposes, but pure mathematics involving ϵ and δ (due to Karl Weierstrass (1815–1897)) is required for rigour and clarity as to what *smooth* actually means here, and this can be found in books on real analysis. Without mathematics, it means sharp corners and breaks in the graph of $y = f(x)$ are disallowed. At a corner, there are two tangents; only one can be permitted, otherwise it is not unique and so there is no unique value to the above limit and hence no unique derivative. Derivatives really do come in very handy for calculations in a variety of applied areas, so the evaluation of this limit has received a great deal of attention since calculus was first proposed by Isaac Newton (1642–1727) in 1665 in England, and Gottfried Leibniz (1646–1716) a little later, around 1675, in Germany. Leibniz' approach wins here and it is his notation that is now followed; Newton used pure geometry and only the dot to denote differentiation with respect to time in some areas of mechanics survives today from his pioneering research. His *fluxions* are fascinating to study, but now are only of interest to historians. To do a simple example straight from the limit definition, let us find the derivative of the function $f(x) = x^2$ at an arbitrary point $x = x_0$. The numerator is $(x_0 + \Delta x)^2 - x_0^2$ and this simplifies to $x_0^2 + 2x_0 \Delta x + (\Delta x)^2 - x_0^2 = 2x_0 \Delta x + (\Delta x)^2$. The denominator is just Δx and this is a factor of the numerator that can be cancelled to leave the quotient

$$\frac{(x_0 + \Delta x)^2 - x_0^2}{\Delta x} = 2x_0 + \Delta x,$$

the right hand side of which tends to $2x_0$ as $\Delta x \to 0$. Thus, the derivative of x^2 at the point $x = x_0$ is equal to $2x_0$, or more succinctly the derivative of x^2 is $2x$ for any value x. Notice the cancellation of Δx occurs *before* it is allowed to tend to zero, so this is legal. Do not let it tend to zero first, *then* cancel, that is a mathematical sin; cancelling zeros usually leads to nonsense. Thus, if the limit exists, and in this text it usually does, then write

$$\frac{df}{dx} = \lim_{\Delta x \to 0} \left\{ \frac{f(x + \Delta x) - f(x)}{\Delta x} \right\} \qquad (1.1)$$

and Equation 1.1 is the derivative of the function $f(x)$ with respect to x. There are many standard results for differentiation (finding derivatives), and all can be derived by finding this limit using various mathematical expansion techniques. The derivative of x^2 was found above and this generalises to

$$\frac{d}{dx}(x^n) = nx^{n-1},$$

where n is any real number. In addition, the addition formulas for trigonometric functions can be used to derive

$$\frac{d}{dx}(\sin x) = \cos x, \qquad \frac{d}{dx}(\cos x) = -\sin x, \quad \text{etc.}$$

When finding derivatives, however, one has to be careful not to use formulas that themselves depend on differentiation. In particular, Taylor's theorem and the special case McLaurin's theorem come later in this chapter; they are expansions of functions in terms of polynomials that are obtained through differentiation. Let's take a look at the exponential function e^x where $e = 2.71828\ldots$ is the base for natural logarithms. There are many ways to define e mathematically: it is that value of a such that the function a^x differentiates to itself. Equivalently, it is the value of a such that the slope of the function $y = a^x$ has unit slope at $x = 0$. This can be taken as the definition of the number e: that is, if

$$\frac{d}{dx}(a^x) = a^x$$

for some real number a, then $a = e$. Many mathematicians prefer the limit definition

$$\lim_{n \to \infty} \left(1 + \frac{1}{n}\right)^{1/n} = e$$

as it is clean and isolated, but then one would have to *prove* the first definition. It is more natural here to use the functional definition in terms of a^x. The inverse of differentiation is integration, so writing $y = e^x$ the above equation is

$$\frac{dy}{dx} = y$$

from which we get

$$\frac{dy}{y} = dx$$

and integrating both sides gives

$$x = \int \frac{dy}{y}$$

but since the inverse of $y = e^x$ is $x = \ln y$, it must be the case that

$$\int \frac{dy}{y} = \ln y$$

always allowing for an arbitrary constant of course. The last equation can be the definition of the natural logarithm, in which case, the derivative of $\ln y$ with respect to y is $1/y$ and so

$$\frac{dx}{dy} = \frac{1}{y}$$

and so

$$\frac{dy}{dx} = y$$

is derived, in which case, the function y treated as dependent upon x differentiates to itself. This is, therefore, the exponential function. As long as one treats either this property as a definition of the exponential function, or the integration of $1/y$ as the definition of the logarithm to the base e (natural logarithm), one can derive the other. One certainly cannot derive both out of thin air, and it is rather messy to start with the limit definition of e as one has to use obscure properties of limits as well as differentiation rules yet to be introduced.

1.2.1   Rules for Differentiation

There are rules most will be familiar with for differentiating functions of functions, products and quotients. Here they are:

$$\frac{d}{dt}\{f(x(t))\} = \frac{df}{dx}\frac{dx}{dt},$$

$$\frac{d}{dx}\{u(x)v(x)\} = v\frac{du}{dx} + u\frac{dv}{dx},$$

$$\frac{d}{dx}\left(\frac{u}{v}\right) = \left(v\frac{du}{dx} - u\frac{dv}{dx}\right)/v^2.$$

These can all be proved using the limit definition of derivative; here is a proof of the product rule.

Example 1.1 Prove that

$$\frac{d}{dx}\{u(x)v(x)\} = v\frac{du}{dx} + u\frac{dv}{dx}$$

for suitably well-behaved functions $u = u(x)$ and $v = v(x)$.

Solution: From the limit definition of derivative we have

$$\frac{d}{dx}\{u(x)v(x)\} = \lim_{\Delta x \to 0}\left\{\frac{u(x+\Delta x)v(x+\Delta x) - u(x)v(x)}{\Delta x}\right\},$$

so the right-hand side becomes

$$\lim_{\Delta x \to 0}\left\{\frac{u(x+\Delta x)v(x+\Delta x) - u(x)v(x+\Delta x) + u(x)v(x+\Delta x) - u(x)v(x)}{\Delta x}\right\}$$

upon subtracting then adding $u(x)v(x+\Delta x)$ in the numerator. Hence, grouping the numerator gives

$$\lim_{\Delta x \to 0} v(x+\Delta x)\left\{\frac{u(x+\Delta x) - u(x)}{\Delta x}\right\} + \lim_{\Delta x \to 0} u(x)\left\{\frac{v(x+\Delta x) - v(x)}{\Delta x}\right\}$$

and letting the limit $\Delta x \to 0$ establishes the result. □

The quotient rule follows from applying the product rule to $u(x) \times 1/v(x)$ and the function of a function rule to $1/v(x)$. Note that the more mature name for the 'function of a function' rule is the 'chain rule'. This gets a lot of attention later in this book starting with the next chapter. Another topic worthy of mention that follows directly from these rules is *implicit differentiation*, and this is best introduced with an example:

Example 1.2 Determine dy/dx given that

$$xy - y^4 = 3x + 5y.$$

Solution: First, note that it is not possible to get y on one side of the equation and differentiate, so the differentiation has to be done implicitly, that is on the expression as given. xy can be treated as a product so

$$\frac{d}{dx}(xy) = y + x\frac{dy}{dx}$$

by direct application of the product rule. In addition, the term y^4 can be differentiated through applying the chain rule, thus

$$\frac{d}{dx}y^4 = 4y^3\frac{dy}{dx}.$$

Thus, differentiating the given equation leads to

$$y + x\frac{dy}{dx} + 4y^3\frac{dy}{dx} = 3 + 5\frac{dy}{dx},$$

which is a simple equation for the derivative dy/dx. Solving thus gives the required answer

$$\frac{dy}{dx} = \frac{3 - y}{4y^3 + x - 5}.$$

$\square$

Finally, it should be mentioned that if $y = f(x)$, the derivative df/dx can also be written $f'(x)$; a notation due to Joseph-Louis Lagrange (1736–1813).

1.2.2 Mean Value Theorem

This section will be more about finding limits and differentiation; but let's start with the mean value theorem.

Theorem 1.1 (The First Mean Value Theorem) If the function $f(x)$ is continuous on the closed domain $a \le x \le b$ written $[a, b]$ and differentiable in the open domain $a < x < b$ written (a, b), then there exists at least one point $c \in (a, b)$ such that

$$\frac{f(b) - f(a)}{b - a} = f'(c),$$

where the dash denotes differentiation with respect to x.

The proof of this belongs squarely in real analysis texts, but to see what it means take a look at Figure 1.1. This figure shows the graph of $y = f(x)$ that passes through the points $(a, f(a))$ and $(b, f(b))$. The mean value theorem is explained in words as follows: If a function $f(x)$ is continuous and differentiable in between two values $x = a$ and $x = b$, which means that the points $(a, f(a))$ and $(b, f(b))$ are connected by an unbroken piece of string without kinks, then the mean value theorem says that the slope of the straight line joining the end points of the string equals the slope of the tangent to the string at least once at

Figure 1.1 The mean value theorem.

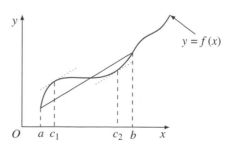

a point along its length. In the example shown in Figure 1.1, there are in fact two such points at $x = c_1$ and $x = c_2$; this is because the graph crosses the line between the end points at $x = a$ and $x = b$. If the graph was entirely above or below this line, then there would be just one. The important thing is that there is at least one. The proof of this is outside this text as it is definitely analysis. Here is a special case of this theorem called Rolle's Theorem.

Theorem 1.2 (Rolle's Theorem) If the function $f(x)$ is continuous on the closed domain $a \leq x \leq b$ written $[a, b]$ and differentiable in the open domain $a < x < b$ written (a, b) and in addition $f(a) = f(b)$, then there exists at least one point $c \in (a, b)$ such that

$$f'(c) = 0$$

where the dash continues to denote differentiation with respect to x.

The proof of this theorem is easier than the earlier one, and it forms part of the end-of-chapter exercises. The third theorem in this short series is more practical and is used on and off throughout the rest of this text. Therefore, it gets a section of its own.

1.2.3 Taylor's Series

Sometimes called Taylor's theorem, the basic idea of a Taylor's series is to express the value of a function $f(x)$ at the point $x = a + h$ in terms of its value and that of its derivatives at the point $x = a$. The value of $f(a + h)$ is expressed as a power series in h which is considered small. The theorem is stated as follows.

Theorem 1.3 If $f(x)$ is a single-valued continuous function of x, then the following series is valid

$$f(x) = f(a) + hf'(a) + \frac{h^2}{2!}f''(a) + \frac{h^3}{3!}f'''(a) + \cdots + \frac{h^n}{n!}f^n(a) + R_{n+1}(x),$$

where $h = x - a$ and $R_n(x)$ is a remainder. The commonest form of the remainder is due to Lagrange and is given by

$$R_n(x) = \frac{h^n}{n!}f^{(n)}(a + \theta h), \quad \text{where} \quad 0 < \theta < 1.$$

It is assumed that all the derivatives exist in all the right intervals. In applied science and engineering textbooks, *Taylor's series* is written:

$$f(a + h) = f(a) + hf'(a) + \frac{h^2}{2!}f''(a) + \frac{h^3}{3!}f'''(a) + \cdots + \frac{h^n}{n!}f^{(n)}(a) + \cdots$$

with convergence assumed. If $n = 1$ in the statement of Taylor's theorem, then

$$f(a + h) = f(a) + R_1(x)$$

and

$$R_1(x) = hf'(a + \theta h),$$

which is a restatement of the mean value theorem with $c = a + \theta h$ and $b - a = h$. The proof of this theorem is not difficult in the sense that it is not too analytical and it is instructive; so here it is:

Proof: Define the remainder $R_{n+1}(x)$ as

$$R_{n+1}(x) = f(x) - f(a) - (x - a)f'(a) - \frac{f''(a)}{2!}(x - a)^2 - \cdots$$
$$- \frac{f^{(n)}(a)}{n!}(x - a)^n.$$

Now, the technique takes advantage of this being true for any x and a, so in particular choose t to be between a and x and define

$$F(t) = f(x) - f(t) - (x - t)f'(t) - \frac{f''(t)}{2!}(x - t)^2 - \cdots - \frac{f^{(n)}(t)}{n!}(x - t)^n$$

(1.2)

so that $F(a) = R_{n+1}(x)$. Consider both x and a as fixed and differentiate this last expression with respect to t so that, using the product rule,

$$F'(t) = -f'(t) - (x - t)f''(t) + f'(t)$$
$$- \frac{f'''(t)}{2!}(x - t)^2 + f''(t)(x - t) - \cdots - \frac{f^{(n+1)}(t)}{n!}(x - t)^n$$
$$+ \frac{f^{(n)}(t)}{(n - 1)!}(x - t)^{(n-1)},$$

where all but the second to last term cancels, so

$$F'(t) = -\frac{f^{(n+1)}(t)}{n!}(x - t)^n$$

is obtained. The proof is completed by defining a new function

$$G(t) = F(t) - \left(\frac{x - t}{x - a}\right)^{n+1} F(a).$$

Features of this new function $G(t)$ include $G(a) = 0$ and $G(x) = F(x) = 0$. Therefore, the function $G(t)$ satisfies the conditions of Rolle's theorem, in which case, there exists a value c such that $a < c < x$ where $G'(c) = 0$ so

$$0 = G'(c) = F'(c) + (n + 1)\frac{(x - c)^n}{(x - a)^{(n+1)}}F(a)$$
$$= -\frac{f^{(n+1)}(c)}{n!}(x - c)^n + (n + 1)\frac{(x - c)^n}{(x - a)^{(n+1)}}F(a).$$

Therefore,

$$R_{n+1}(x) = F(a) = \frac{f^{(n+1)}(c)}{(n+1)!}(x-a)^{(n+1)}.$$

From Equation 1.2,

$$F(a) = f(x) - f(a) - (x-a)f'(a) - \frac{f''(a)}{2!}(x-a)^2 - \cdots - \frac{f^{(n)}(a)}{n!}(x-a)^n$$

so

$$f(x) = f(a) + hf'(a) + \frac{h^2}{2!}f''(a) + \frac{h^3}{3!}f'''(a) + \cdots$$

$$+ \frac{h^{(n-1)}}{(n-1)!}\,f^{(n-1)}(a) + R_{n+1}(x),$$

which with $x = a + h$ is Taylor's theorem. $\qquad\qquad\square$

The special case $a = 0$ is particularly useful and commonplace. It is called Maclaurin's series or theorem,

$$f(h) = f(0) + hf'(0) + \frac{h^2}{2!}f''(0) + \frac{h^3}{3!}f'''(0) + \cdots$$

$$+ \frac{h^{(n-1)}}{(n-1)!}f^{(n-1)}(0) + R_{n+1}(h),$$

and it gives a power series expansion in x of the function $f(x)$.

One of the direct applications of Taylor's series relates to the evaluation of limits. The following is a useful result stated in the form of a theorem.

Theorem 1.4 **(L'Hôpital's Rule)** If $f(x)$ and $g(x)$ are two functions that are differentiable in the vicinity of the point $x = c$ and both take the value zero at the same point $x = c$ where c can be 0, any finite value or $\pm\infty$, then

$$\lim_{x\to c}\left\{\frac{f(x)}{g(x)}\right\} = \lim_{x\to c}\left\{\frac{f'(x)}{g'(x)}\right\}.$$

Proof: This proof is for the cases where c is finite. The infinite case is left for Exercise 1.5 at the end of this chapter. There are many proofs, most of which can be found on the internet these days, but here is one that uses Taylor's theorem. Expanding both $f(x)$ and $g(x)$ about the point $x = c$ gives

$$f(x) = f(c) + (x-c)f'(c) + \cdots,$$

$$g(x) = g(c) + (x-c)g'(c) + \cdots,$$

which since $f(c) = g(c) = 0$ reduce to lowest order $f(x) \approx (x-c)f'(c)$ and $g(x) \approx (x-c)g'(c)$. As $x \to c$ the approximation tends to equality. Therefore,

$$\lim_{x\to c}\left\{\frac{f(x)}{g(x)}\right\} = \lim_{x\to c}\left\{\frac{(x-c)f'(c)}{(x-c)g'(c)}\right\} = \frac{f'(c)}{g'(c)} = \lim_{x\to c}\left\{\frac{f'(x)}{g'(x)}\right\}$$

as required. This completes the proof provided $g'(c) \neq 0$ and c is finite. If $g'(c) = 0$ and $f'(c) \neq 0$, then the limit does not exist (is infinite). If both $f'(c)$ and $g'(c)$ are 0, then the earliest non-zero terms in both Taylor's series are the $f''(c)$ and $g''(c)$ square terms and the limit is the ratio of these two second derivatives. If they too are zero, we carry on until non-zero terms are reached. □

The most frequent use of this result is for the case $c = 0$. Here are some examples.

Example 1.3   Evaluate the following limits using L'Hôpital's Rule;

1) $\sin(x)/(x - \pi)$ as $x \to \pi$;
2) $\tan(x)/x$ as $x \to 0$;
3) $(1 - \cos(5t))/(\cos(7t) - 1)$ as $t \to 0$;
4) $(\sin^2\theta - \sin\theta^2)/\theta^4$ as $\theta \to 0$.

Solution: These limits are done reasonably straightforwardly:

1) Differentiating both numerator and denominator once gives

$$\lim_{x \to \pi} \left\{ \frac{\sin(x)}{x - \pi} \right\} = \lim_{x \to \pi} \left\{ \frac{\cos(x)}{1} \right\} = -1$$

so the answer is -1.

2)

$$\lim_{x \to 0} \left\{ \frac{\tan(x)}{x} \right\} = \lim_{x \to 0} \left\{ \frac{\sec^2 x}{1} \right\} = 1.$$

3)

$$\lim_{t \to 0} \left\{ \frac{1 - \cos(5t)}{\cos(7t) - 1} \right\} = \lim_{t \to 0} \left\{ \frac{5 \sin(5t)}{-7 \sin(7t)} \right\}$$

and this limit is still indeterminate of the form $0/0$ so we re-apply the theorem, differentiating again gives

$$\lim_{t \to 0} \left\{ \frac{5 \sin(5t)}{-7 \sin(7t)} \right\} = \lim_{t \to 0} \left\{ \frac{25 \cos(5t)}{-49 \cos(7t)} \right\} = -\frac{25}{49}.$$

4)

$$\lim_{\theta \to 0} \left\{ \frac{(\sin^2\theta - \sin\theta^2)}{\theta^4} \right\} = \lim_{\theta \to 0} \left\{ \frac{(2 \sin\theta \cos\theta - 2\theta \cos\theta^2)}{4\theta^3} \right\}.$$

Cancelling 2 and differentiating again gives:

$$\lim_{\theta \to 0} \left\{ \frac{(\sin\theta \cos\theta - \theta \cos\theta^2)}{2\theta^3} \right\}$$

$$= \lim_{\theta \to 0} \left\{ \frac{(\cos^2\theta - \sin^2\theta - \cos\theta^2 + 2\theta^2 \sin\theta^2)}{6\theta^2} \right\}.$$

Again, this is indeterminate. Therefore, differentiating again:

$$= \lim_{\theta \to 0} \left\{ \frac{(-4\sin\theta\cos\theta - 2\theta\sin\theta^2 + 4\theta\sin\theta^2 + 2\theta^4\cos\theta^2)}{12\theta} \right\}$$

and only the first term contributes. The answer is $-1/3$.   □

As an aside, the last part of the last example is probably easier done by expanding the numerator using the standard Maclaurin series for $\sin\theta$. Only the first two terms are required $\sin\theta \approx \theta - \theta^3/6$. Other exercises in the use of L'Hôpital's Rule are met in the end-of-chapter exercises. Do have a go at them, and remember that all solutions are found at the end of the book.

1.2.4   Maxima and Minima

Perhaps the most obvious application of differentiation that is generalised in later chapters is finding the location of maxima and minima. For a function of a single variable $y = f(x)$ that is differentiable everywhere, no gaps or spikes are allowed, one simply solves

$$\frac{dy}{dx} = 0.$$

Geometrically, this equation finds where the slope of the tangent to the curve $y = f(x)$ is parallel to the x-axis. Where this happens, the function $y = f(x)$ either has a minimum value, a maximum value or possibly a point of inflexion, all supposing that the function is differentiable throughout of course. All of these are shown in Figure 1.2 prefixed by the word 'local' to distinguish them from global extrema that might occur at end points of a range. It is assumed that most of you are familiar with the determination of maxima and minima using calculus, but here is a reminder in the form of an example:

Example 1.4   Find the turning points of the function $f(x) = x^4 - 8x^2 - 3$ and classify them.

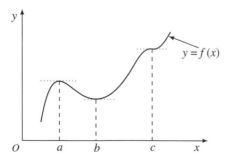

Figure 1.2 A function with a local maximum at $x = a$, a local minimum at $x = b$ and a local point of inflexion at $x = c$.

Figure 1.3 The graph of the function $f(x) = x^4 - 8x^2 - 3$.

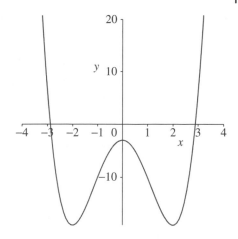

Solution: Solving

$$\frac{df}{dx} = f'(x) = 4x^3 - 16x = 0$$

gives the roots $x = 0, -2$ and 2. The second derivative is $f''(x) = 12x^2 - 16$ which is positive where $x = -2$ and 2 whence these are minima, and at $x = 0, f'' = -16 < 0$ giving a maximum. As an alternative to using the second derivatives, it is possible to determine the species of turning point by examining the sign of the derivative of $f(x)$ over a range that includes the turning points, here a range might be $-4 \leq x \leq 4$ as shown in Figure 1.3. Between -4 and $-2, f'(x) < 0$, between -2 and $0, f'(x) > 0$. Between 0 and $2, f'(x) < 0$ and finally for $x > 2, f'(x) > 0$ once more. As the slope changes sign, $f(x)$ passes through a point where $f'(x) = 0$, that is a turning point and the species is dictated by the direction of the change of sign; negative to positive denotes a minimum, positive to negative denotes a maximum as seen in the figure. If the sign of $f'(x)$ does not change yet still passes through a zero, then this zero is a point of inflexion. There are none in this example. □

Although there is, in Chapter 3, an analogy to this second derivative criteria for classifying turning points for functions of two variables $f(x, y)$, the most useful part of the solution to the above example is probably the last part. Both technical and numerical ways of optimising functions, that is finding where they attain maximum or minimum values, often mean determining directions of greatest or least slope. This example is an introduction to this kind of technique.

1.2.5 Numerical Differentiation

The use of numerical techniques to solve problems is quite old but has really come into its own with high-powered computing. Finite difference and other numerical techniques are now routinely used to solve realistic problems throughout science, engineering and medicine. They remain a side issue for this text, but this subsection introduces a one-dimensional numerical method that features and is generalised later in the section of Chapter 3 on optimisation. This is the Newton–Raphson method, and it is introduced not through the formality of a theorem but via the following example.

Example 1.5 (Newton–Raphson Method) Show that if $x = x_n$ is an approximate root of the equation $f(x) = 0$, then the expression

$$x_{n+1} = x_n - \frac{f(x_n)}{f'(x_n)}$$

is usually a better approximation, but not always.

Solution: Using Taylor's series but retaining just the first two terms leads to

$$f(x + h) \approx f(x) + hf'(x).$$

Let $x = x_n$ and $x + h = x_{n+1}$ then if x_n is an approximation to the exact root, we have

$$f(x_{n+1}) \approx 0 \approx f(x_n) + hf'(x_n)$$

so

$$h \approx -\frac{f(x_n)}{f'(x_n)}$$

and

$$x_{n+1} = x_n + h \quad \text{becomes} \quad x_{n+1} \approx x_n - \frac{f(x_n)}{f'(x_n)}.$$

which is the iterative Newton–Raphson method. The value x_{n+1} is hopefully a closer approximation to the solution of the equation $f(x) = 0$. Sadly though this method often works well, it is not foolproof. To see how it works geometrically, examine Figure 1.4. The first guess $x = x_n$ is considered close to the real root, $f'(x_n)$ is the slope of the curve $y = f(x)$ at the guess and the figure shows the adjustment h. Here is the detail: Figure 1.4 shows a function $y = f(x)$ that rather idealistically crosses the x-axis just once. This is the solution of the equation $f(x) = 0$. The point x_n represents the initial guess; this looks rather a poor guess but this is done for visualisation purposes. The tangent has a positive slope $f'(x_n) = a/h$ and the Newton–Raphson formula gives

$$x_{n+1} \approx x_n - \frac{f(x_n)}{f'(x_n)} = x_n - \frac{-a}{a/h} = x_n + h$$

hence arriving at the point marked x_{n+1} a second, usually improved approximation to the actual root. The method can fail if the initial guess is very poor

Figure 1.4 A graphical representation of the Newton–Raphson method.

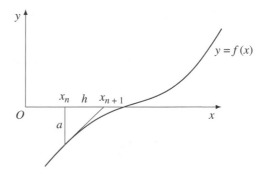

or $f(x)$ is oscillatory near the root. In either case, the slope $f'(x_n)$ could be such that x_{n+1} is actually further away from the root than x_n. □

The Newton–Raphson technique can be applied iteratively to get better and better approximations to the root. Here is a practical example.

Example 1.6 Determine the real root of the cubic equation $f(x) = x^3 - 2x^2 - x - 1$ using four iterations of the Newton–Raphson technique with an initial guess of $x = 3$.

Solution: Differentiating the cubic gives $f'(x) = 3x^2 - 4x - 1$ and applying the Newton–Raphson iteration:

$$x_{n+1} = x_n - \frac{x_n^3 - 2x_n^2 - x_n - 1}{3x_n^2 - 4x_n - 1}$$

gives the following results, starting with $n = 0, x_0 = 3$. The value under x_{n+1} is taken as the next guess.

n	x_{n+1}	f(x)
0	3	5
1	2.64	0.847
2	2.553	0.0476
3	2.5468	0.0002

After four iterations, the answer is accurate to four places of decimals. The function is drawn in Figure 1.5. If the initial guess was $x = 1$, admittedly a poor guess, then 43 iterations would be required for similar accuracy as the tangent to $f(x)$ at $x = 1$ can be seen to take the next guess further from the real answer. The guesses then oscillate before eventually the correct solution is approached

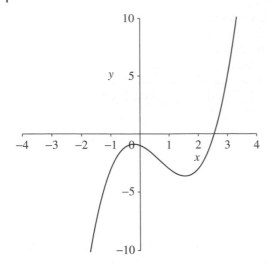

Figure 1.5 The graph of the function $f(x) = x^3 - 2x^2 - x - 1$.

quickly once the values of guesses pass above the minimum value of $f(x)$ at 1.55 changing the direction of the tangent. The advice is be careful and perhaps do some research on $f(x)$ before deciding on a first guess. □

1.3 Integration

The reverse of differentiation is integration, in much the same way as subtraction is the reverse of addition and division the reverse of multiplication. It is usual that the reverse operations are technically more difficult, and in the case of differentiation the reverse, integration, may not be possible to carry out at all analytically. Of course, once differentiation has occurred, the reverse can be achieved theoretically. It is just that, technically, it may be beyond us apart from using the integral to define a new function; these are termed transcendental or special functions. The function that describes the area under a normal distribution between two variates in statistics is one example of a special function, called the error function. There are many techniques–tricks–one can try for evaluating integrals. It is not really the point of this revision chapter to spend a good deal of time running through these, but they do come in handy later on in the text. Three common techniques are substitution, integration by parts and for rational functions the use of partial fractions. It is true these days that computer algebra systems can come to our aid, and these are used in this text. The upshot is that here two examples will be done to revise these techniques. After this, those content to rely on computer algebra are free to do so. The view of the author is that an ability to do detailed algebra does help to boost the confidence and self esteem of the student, factors often sadly neglected. The confident student is usually the successful one.

Integration between limits can be shown to be the area under the curve provided the curve does not cross the x-axis. This is defined as the limit of a sum. It is not obvious that this is the same as the reverse or inverse of differentiation, but it is and a look at the detail helps show this and the result is called the *Fundamental Theorem of the Calculus*.

Theorem 1.5 **(Fundamental Theorem of the Calculus)** If $f(t)$ is a function whose integral exists in the interval $a \leq t \leq b$ and $F(x)$ is defined as

$$F(x) = \int_a^x f(t)dt, \quad \text{where} \quad a < x \leq b,$$

then $F'(x) = f(x)$, where the dash denotes the derivative of a function with respect to x.

Proof: This result as the name indicates is fundamental and links differentiation to integration in a formal way. The proof is reasonably mechanistic, though books on analysis will present all the rigour. Figure 1.6 shows the function $f(t)$; the use of t instead of the usual x is explained because x is needed later. First of all, let us write the derivative of $F(x)$ in terms of its definition

$$F'(x) = \lim_{\Delta x \to 0} \left\{ \frac{F(x + \Delta x) - F(x)}{\Delta x} \right\}.$$

Graphically, this is the area under $f(t)$ between the lines $t = x$ and $t = x + \Delta x$ in Figure 1.6. In the interval $[x, x + \Delta x]$, there will be a value ξ that makes the area under $f(x)$ in this interval the same as the rectangle Δx wide and $f(\xi)$ tall. That is there exists ξ such that $F(x + \Delta x) - F(x) = f(\xi)\Delta x$, or

$$\frac{F(x + \Delta x) - F(x)}{\Delta x} = f(\xi), \quad \text{with} \quad x \leq \xi \leq x + \Delta x.$$

This is shown in Figure 1.6. It resembles the mean value theorem, see Figure 1.1 in Section 1.2.2 and is in fact a form of what is called the mean value theorem for

Figure 1.6 The function $y = f(t)$ is integrable in $[a, b]$.

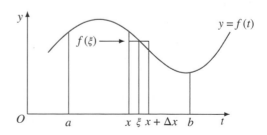

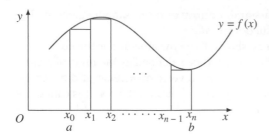

Figure 1.7 An approximation based on the minimum value of $f(x^*)$ for integration between limits.

integrals. As $\Delta x \to 0$, ξ gets squeezed between x and $x + \Delta x$ and in the limit it must be the case that $\xi \to x$. Therefore, from the above definition of the derivative of $F(x)$, it is seen that $F'(x) = f(\xi) = f(x)$. □

Using the fundamental theorem just proved, it is therefore legitimate for an integral as the inverse or reverse of differentiation to be defined in terms of evaluating the area under a graph, see Figure 1.7. Those familiar with approximate integration would have seen this kind of diagram but instead of rectangles, for the trapezoidal rule the top is a straight line approximating $f(x)$ so that each strip is a much more accurate trapezium. For other numerical integration rules, an even better approximation for $f(x)$ is used. However, for the formal definition of what is called Riemann integration as the width of each strip is made smaller, the approximation gets better

$$\int_a^b f(x)dx \approx \sum_{i=1}^n (x_i - x_{i-1})f(x^*),$$

where $x_i \leq x^* \leq x_{i+1}$. The analysis does this formally via lower and upper bounds. If $f(x)$ increases in the interval $[x_i, x_{i+1}]$, then it attains its maximum value at x_{i+1} and minimum value at x_i; if on the other hand $f(x)$ decreases in this interval, then the opposite is true. If it attains a maximum or minimum in the interval, then neither is true, but a value of $x \in [x_i, x_{i+1}]$ can always be chosen so that $f(x)$ is either a maximum or a minimum. Therefore, if the value of $f(x)$ in this interval is replaced by the maximum value in this and every other interval, then the sum on the right will be an overestimate for the integral. Similarly, if the minimum value of x in each interval is chosen, as in Figure 1.7, then it will be an underestimate. The definite integral can thus be squeezed between two approximations, one definitely less than the other which is definitely greater than the definite integral. As the number of intervals is allowed to increase indefinitely and the interval length decreases, if these approximations approach the same limit, then the definite integral also tends to this unique limit, and the function $f(x)$ is termed integrable. The condition of integrability turns out to be mild in mathematical terms. For example, all continuous functions are integrable, and some that aren't that contain gaps can also be integrated. This is in stark contrast to being differentiable which demands smoothness. Later in the book, a different kind of integration called the line integral will be introduced. This is all that will be said about the formal definition; let us turn to practicalities. Here is a short table of a few standard integrals; it is by no means exhaustive:

$f(x)$	$\int f(x)dx$
x^n	$\dfrac{1}{n+1}x^{n+1}$, where $n \neq -1$
$\dfrac{f'(x)}{f(x)}$	$\ln f(x)$ for any function $f(x)$ that has a derivative $f'(x)$
e^{ax}	$\dfrac{1}{a}e^{ax}$
$\sin(x)$	$-\cos(x)$
$\cos(x)$	$\sin(x)$
$\tan(x)$	$-\ln\cos(x)$
$\cot(x)$	$\ln\sin(x)$
$\sec(x)$	$\ln[\sec(x) + \tan(x)]$
$\mathrm{cosec}(x)$	$\ln[\mathrm{cosec}(x) - \cot(x)]$
$\dfrac{1}{\sqrt{a^2 - x^2}}$	$\arcsin\dfrac{x}{a}$
$\dfrac{1}{a^2 + x^2}$	$\dfrac{1}{a}\arctan\dfrac{x}{a}$

and of course a constant C called *the constant of integration* has to be added to the right-hand side. This is because the differential of a constant is always zero, so when integrating a function, one has to include such a constant. Therefore, here are four integrals to find, the last of which is not particularly straightforward:

Example 1.7 Evaluate the following indefinite integrals:

1) $\displaystyle\int x \arctan x\, dx$,

2) $\displaystyle\int \frac{dx}{x(\ln x)^2}$,

3) $\displaystyle\int \sin^3 4x \cos 4x\, dx$,

4) $\displaystyle\int \frac{dx}{1 + x^4}$.

Solution:

1) This first integral is solved by integrating by parts. The formula is

$$\int u\,dv = uv - \int v\,du$$

and is the integral of the formula for differentiating a product. Identify $u = \arctan(x)$ and $dv = x\,dx$ as we can integrate $dv = x\,dx$ but not $u = \arctan(x)$. This gives $du = \dfrac{1}{1+x^2}dx$ and $v = \dfrac{1}{2}x^2$. Therefore, applying the formula yields

$$\int x \arctan x\,dx = \frac{1}{2}x^2 \arctan x - \frac{1}{2}\int \frac{x^2 dx}{(1+x^2)}.$$

The second integral can be evaluated easily by writing the numerator as $1 + x^2 - 1$ whence

$$\int x \arctan x\,dx = \frac{1}{2}x^2 \arctan x - \frac{1}{2}(x - \arctan x) + C$$

$$= \frac{1}{2}x^2 \arctan x + \frac{1}{2}\arctan x - \frac{1}{2}x + C,$$

where C is an arbitrary constant.

2) This is an example of substitution whereby letting $u = \ln x$ gives $du = dx/x$ rendering the integral in the simpler form

$$\int \frac{dx}{x(\ln x)^2} = \int \frac{du}{u^2} = -\frac{1}{u} + C,$$

giving the answer

$$-\frac{1}{\ln(x)} + C.$$

3) This is another substitution, but this time trigonometric. Put $u = \sin 4x$ giving $du = 4\cos 4x\,dx$ so the integral becomes

$$\int \sin^3 4x \cos 4x\,dx = \frac{1}{4}\int u^3\,du = \frac{1}{12}u^4 + C,$$

hence the answer is

$$\frac{1}{12}\sin^4 4x + C.$$

4) This is the traditional 'last example'; trickier than the rest to stop the best students finishing too early. Here's the solution without all the intermediate algebra steps: First of all note that

$$1 + x^4 = 1 + 2x^2 + x^4 - 2x^2 = (1 + x^2)^2 - 2x^2$$

$$= (1 + x\sqrt{2} + x^2)(1 - x\sqrt{2} + x^2).$$

This enables the integrand $1/(1 + x^4)$ to be split into partial fractions as follows:

$$\frac{1}{1 + x^4} = \frac{1}{(1 + x\sqrt{2} + x^2)(1 - x\sqrt{2} + x^2)}$$

$$= \frac{ax + b}{1 + x\sqrt{2} + x^2} + \frac{cx + d}{1 - x\sqrt{2} + x^2}.$$

The constants a, b, c and d are found by equating coefficients of powers of x in the identity

$$1 \equiv (ax + b)(1 - x\sqrt{2} + x^2) + (cx + d)(1 + x\sqrt{2} + x^2)$$

giving $b = d = 1/2$ and $a = -c = 1/\sqrt{2}$. Thus,

$$\int \frac{dx}{1 + x^4}$$

$$= \frac{1}{4\sqrt{2}} \left\{ \int \frac{2x + \sqrt{2} + \sqrt{2}}{1 + x^2 + x\sqrt{2}} dx - \int \frac{2x - \sqrt{2} - \sqrt{2}}{1 + x^2 - x\sqrt{2}} dx \right\}.$$

The first two terms in each numerator on the right is the derivative of the denominator leading to ln terms, and the remaining term in each integral can be re-arranged into an arctan integral:

$$\int \frac{dx}{1 + x^4} = \frac{1}{4\sqrt{2}} \ln \left\{ \frac{1 + x^2 + x\sqrt{2}}{1 + x^2 - x\sqrt{2}} \right\}$$

$$+ \frac{1}{4} \int \frac{dx}{(x + \frac{1}{\sqrt{2}})^2 + \frac{1}{2}} + \frac{1}{4} \int \frac{dx}{(x - \frac{1}{\sqrt{2}})^2 + \frac{1}{2}}.$$

Therefore,

$$\int \frac{dx}{1 + x^4} = \frac{1}{4\sqrt{2}} \ln \left\{ \frac{1 + x^2 + x\sqrt{2}}{1 + x^2 - x\sqrt{2}} \right\}$$

$$+ \frac{\sqrt{2}}{4} [\arctan(x\sqrt{2} + 1) + \arctan(x\sqrt{2} - 1)] + C$$

is the final answer. □

Here is an example involving integration from first principles. It also brings together other ideas introduced in this chapter, namely the evaluation of limits.

Example 1.8 Using the result

$$\sum_{k=1}^{n} k^2 = \frac{1}{6} n(n + 1)(2n + 1),$$

evaluate the integral $\int_0^1 x^2 dx$ from first principles.

Solution: The evaluation of definite integrals from first principles is contrived and depends on 'guessing' the correct form for the discrete integral then being able to go to the limit successfully. Here, a convenient discrete form of $f(x) = x^2$ is

$$f\left(\frac{k}{n}\right) = \frac{k^2}{n^2}$$

as this is zero at $k = 0$ and 1 at $k = n$. As n gets larger, there are more and more values, and eventually as $n \to \infty$, it becomes the continuous $f(x) = x^2$. Therefore,

$$\int_0^1 x^2 dx = \lim_{n\to\infty} \frac{1}{n} \sum_{k=0}^{n} \frac{k^2}{n^2}$$

and the given value of the sum to infinity of the square numbers, this gives

$$\int_0^1 x^2 dx = \lim_{n\to\infty} \frac{1}{n} \frac{1}{6n^2} n(n+1)(2n+1)$$

$$= \lim_{n\to\infty} \left(\frac{1}{6}\left(1+\frac{1}{n}\right)\left(2+\frac{1}{n}\right)\right) = \frac{1}{3},$$

which is consistent with using standard integration. □

It remains the case that integration from first principles is more difficult than differentiation from first principles.

Exercises

1.1 Use first principles to find the derivatives of the following functions:
a) x^3,
b) $\sin x$,
c) $\ln x$.

1.2 Find the first four terms of the Taylor series for each of the following functions about the value of x indicated:
a) $\cos x, x = \pi/3$;
b) $\ln x, x = 1$;
c) $x^4, x = 3$.

1.3 Prove Rolle's theorem, Theorem 1.2 in the text.

1.4 Prove the following theorem called the *Extended Mean Value Theorem*:
Let the functions f and g be differentiable on (a, b) and continuous on

$[a, b]$. If $g(x) \neq 0$ for all $x \in (a, b)$, then there is at least one point $c \in (a, b)$ such that

$$\frac{f'(c)}{g'(c)} = \frac{f(b) - f(a)}{g(b) - g(a)}.$$

1.5 Use the previous question to prove the special case of L'Hôpital's Rule: If $h(x)$ and $k(x)$ are functions differentiable on the half line $x \geq 0$. Additionally suppose that $k'(x) > 0$ for $x \geq 0$ and that $h(x) \to \infty$ and $k(x) \to \infty$ with

$$\frac{h'(x)}{k'(x)} \to \lambda \quad \text{for some real} \quad \lambda \text{ as } x \to \infty.$$

Prove that

$$\frac{h(x)}{k(x)} \to \lambda \quad \text{as} \quad x \to \infty.$$

1.6 Evaluate the following limits:

a) $\lim\limits_{x \to 3} \left(\dfrac{x^3 - 2x^2 - x - 6}{x - 3} \right)$,

b) $\lim\limits_{x \to 0} \left(\dfrac{1 - \cos x}{x^2} \right)$,

c) $\lim\limits_{x \to 0} \left(\dfrac{\sqrt{x + 1} - 1}{x} \right)$,

d) $\lim\limits_{x \to 0} \dfrac{|x|}{x}$,

e) $\lim\limits_{x \to 0} \left(\dfrac{\sin x - x}{x - \tan x} \right)$.

1.7 Evaluate the following limits:

a) $\lim\limits_{x \to \infty} (\ln x)^{1/x}$,

b) $\lim\limits_{x \to \infty} (3^x + 5^x)^{1/x}$,

c) $\lim\limits_{x \to 0} \left(\dfrac{\sin(\tan x) - \tan(\sin x)}{x^7} \right)$ [Hint: Use Maclaurin series, and computer algebra].

1.8 Find and classify the extrema of the following functions:

a) $y = x^4 - 12x^3 + 48x^2 - 64x$,

b) $y = 2x(x^2 + 2)^{-1/2}$,

c) $y = 2x^{5/3} - 5x^{4/3}$.

[Hint: These days, it is useful to draw the functions using MAPLE, MATLAB or similar software.]

1.9 Apply the Newton–Raphson method to the equation $x^2 = a$ to derive the iteration formula:

$$x_{n+1} = \frac{1}{2}\left(x_n + \frac{a}{x_n}\right).$$

Hence, find the value of $\sqrt{5}$ to seven significant figures given the approximation $\sqrt{5} \approx 2.24$.

1.10 Evaluate the following indefinite integrals:

a) $\displaystyle\int x \ln x\, dx.$

b) $\displaystyle\int \frac{x e^x dx}{(x+1)^2}.$

c) Show that if $I_n = \displaystyle\int \sin^n x\, dx$, then $nI_n = (n-1)I_{n-2} - \cos x \sin^{n-1} x$.
Hence, evaluate I_4 and I_5.

d) $\displaystyle\int \frac{1}{\sqrt{3 + 2x - x^2}} dx.$

e) $\displaystyle\int \frac{1}{1 + x^6} dx$ [Hint: $1 + x^6 = (1 + x^2)(1 - x^2 + x^4)$].

1.11 Using the substitution $t = \tan(x/2)$, prove that

$$\int_0^{\pi/2} \frac{\cos x}{\alpha \cos x + \sin x} dx = \frac{\alpha \pi}{2(\alpha^2 + 1)} - \frac{\ln \alpha}{\alpha^2 + 1},$$

where α is a constant.

2

Partial Differentiation

2.1 Introduction

In the first chapter a revision of what might be called straightforward differentiation took place. It is assumed that this was revision; and it reacquainted you with calculus as a vehicle for analysing how functions that depend on only a single variable change. This change might be with time, t, or with space in one dimension, x. This is in fact the starting point here; attention is still on how a function changes with time and space but now they both can change and change independently. There are three different spatial directions too. It is possible to cater for all these variations and more, but let us start simply. If only one variable is permitted to change and the rest are held constant, then this leads to what is called *partial differentiation*. To be specific, suppose a function $f(x, y, z)$ varies with the three variables x, y and z in a way that enables us to define derivatives with respect to any of the three variables, then start by differentiating the function $f(x, y, z)$ with respect to x holding y and z constant. This derivative is called the partial derivative of f with respect to x and is denoted by a curly ∂ as follows:

$$\frac{\partial f}{\partial x}.$$

In terms of a limiting process, the following limit is assumed to exist:

$$\lim_{\Delta x \to 0} \left\{ \frac{f(x + \Delta x, y, z) - f(x, y, z)}{\Delta x} \right\} = \frac{\partial f}{\partial x}$$

and it is seen right away that only x is changing. The variables y and z do not change at all, in fact if they were not mentioned, we get the definition

$$\lim_{\Delta x \to 0} \left\{ \frac{f(x + \Delta x) - f(x)}{\Delta x} \right\} = \frac{df}{dx},$$

which is in Chapter 1. However, this is really missing the point that it is possible to use differentiation on functions of many variables. Similarly we could find the

Two and Three Dimensional Calculus: With Applications in Science and Engineering, First Edition. Phil Dyke.
© 2018 John Wiley & Sons Ltd. Published 2018 by John Wiley & Sons Ltd.

derivative with respect to y as follows

$$\lim_{\Delta y \to 0} \left\{ \frac{f(x, y + \Delta y, z) - f(x, y, z)}{\Delta y} \right\} = \frac{\partial f}{\partial y}$$

or z

$$\lim_{\Delta z \to 0} \left\{ \frac{f(x, y, z + \Delta z) - f(x, y, z)}{\Delta z} \right\} = \frac{\partial f}{\partial z}.$$

Therefore, there are three partial derivatives, each representing the rate of change of f in the three co-ordinate directions. Some of you might, at this juncture, be wondering if they can be combined to get some kind of 'overall change' using vectors. Indeed, this is done and for those who already know vector notation, the quantity

$$\nabla f = \mathbf{i} \frac{\partial f}{\partial x} + \mathbf{j} \frac{\partial f}{\partial y} + \mathbf{k} \frac{\partial f}{\partial z}$$

does the job. It will be introduced properly later. In order to consolidate this new idea of partial differentiation, let us consider an example.

Example 2.1 Find all the partial derivatives of the function $f(x, y, z) = xy^2 z^3$.

Solution: Differentiating with respect to x and holding y and z constant gives

$$\frac{\partial f}{\partial x} = y^2 z^3$$

as the derivative of x with respect to x is 1, and $y^2 z^3$ is simply a constant multiplier here. Differentiating with respect to y this time holding both x and z constant gives

$$\frac{\partial f}{\partial y} = 2xyz^3$$

as the derivative of y^2 with respect to y is $2y$. Finally, differentiating f with respect to z holding both x and y constant gives

$$\frac{\partial f}{\partial z} = 3xy^2 z^2$$

as the derivative of z^3 with respect to z is $3z^2$. □

If we consider only the functions having only two independent variables $f(x, y)$, then there is a useful interpretation of the partial derivative. Writing $z = h(x, y)$ as the function, this can be the height above sea-level on a map, x denoting East and y North. In the old days, most were familiar with maps and the keen walker still may carry Ordnance Survey maps of an area when hiking. Though Google maps or satnav on a smartphone might have replaced these paper maps now, the principle still holds. A map can show the height above sea level in the form of contours (lines joining points of equal $h(x, y)$). A tangent line to a contour line in the direction of x (eastwards) will be a line along which y does not change, on the other hand in the three dimensions of the rolling hills, it will slope either upwards or downwards. This is the gradient of $h(x, y)$ in the direction of x holding y constant, and so it is

$$\frac{\partial h}{\partial x}.$$

Similarly, a line tangential to a contour but now in the y direction will have a gradient

$$\frac{\partial h}{\partial y}.$$

Figure 2.1 shows the paraboloid with equation $z = x^2 + y^2$. It is drawn with the z-axis down so the nose of the paraboloid is upward. A paraboloid is the shape of a satellite dish; this one is more pointy and is more like the nose of a submarine. The cross-section is circular and it is in fact generated by revolving the parabola $z = x^2 (y = 0)$ about its axis of symmetry. The partial derivatives are

$$\frac{\partial}{\partial x}(z - x^2 - y^2) = -2x; \quad \frac{\partial}{\partial y}(z - x^2 - y^2) = -2y; \quad \text{and}$$

$$\frac{\partial}{\partial z}(z - x^2 - y^2) = 1.$$

The first two of these are important as they represent the two tangents to the surface $z = x^2 + y^2$. In the figure on the lower left, the surface is drawn and there is a cross located at the point $(1, 1, 2)$ on the surface of the paraboloid. This cross is tangential to the surface and represents a tangent plane. The cross itself is composed of the two tangent lines that represent lines along which the partial derivatives $\frac{\partial z}{\partial x}$ and $\frac{\partial z}{\partial y}$ are constant. This information will be used in later chapters along with vectors to find out about the directions of tangents to curves and surfaces, for now the figure on the lower right is the plan view of the paraboloid and shows that these lines are indeed parallel to the x- and y-axes as viewed from above. The three-dimensional figure on the lower left however shows that they slope in the z direction. A fact confirmed by the two upper pictures that show two-dimensional cross-sections through each tangent. If the point $(0, 0, 0)$ were chosen then these partial derivatives would both be zero, and when this occurs the point itself is termed a turning point (maximum, minimum or one of a variety of saddle points). This topic will also be discussed in a later chapter. Here let us get to more technical algebraic matters; starting with the *differential*. In one dimension, the following can be written at once, trivially

$$df = \frac{df}{dx}dx.$$

In higher dimensions, it gets more subtle.

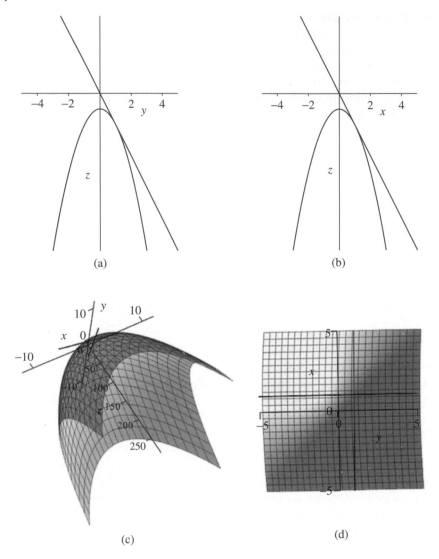

Figure 2.1 The paraboloid $z = x^2 + y^2$ with a tangent cross at $(1, 1, 2)$ showing the lines of the cross: $z = 2x$ with $y = 1$, and $z = 2y$ with $x = 1$. (a) Cross-section through the plane $x = 1$; (b) Cross-section through the plane $y = 1$; (c) Isometric view, (z downward); (d) Plan view.

2.2   Differentials

Suppose there is a function f that varies with respect to the three variables x, y and z, three mutually perpendicular axes that are used to define points through-out three-dimensional space. The quantity df is called the *differential* of f and has nothing to do with x, y or z that might be used to describe how f varies. It was precisely this kind of quantity that upset the mathematical and scientific establishment back in the seventeenth century as Newton developed the calculus. df is actually zero, in the sense that if you select any small value, then df is less than the value you have chosen, no matter how small it was. This seemed a silly notion 350 or so years ago, but it is in fact the basis of mathematical analysis that began in the 1680s and reached its peak 200 years later in the hands of the Germans, notably Karl Weierstrass. It can be defined mathematically as follows.

Definition 2.1   The *differential df* of a function of three variables $f = f(x, y, z)$ is defined by

$$df = \frac{\partial f}{\partial x}dx + \frac{\partial f}{\partial y}dy + \frac{\partial f}{\partial z}dz. \tag{2.1}$$

This definition is easily extended to a function of any number of variables

$$df = \sum_{n=1}^{N} \frac{\partial f}{\partial x_n}dx_n.$$

In many dimensions, it is as if one is adding up all the contributions from all the x_n that make the total variation of f. One does not prove definitions, but some mathematics does make the definition seem more reasonable. Let us concentrate only on three variables to make things a bit easier. A function that varies with three variables will have a domain of definition; for example the temperature in a lecture theatre. At any point in the theatre the temperature can be measured and its value at that instant is a number (in degrees Kelvin let us say). It will be higher near a radiator, lower perhaps near a leaky window especially in a British winter. If two points are very close together, then one would expect the temperature at each of these points to be correspondingly close. It would be bizarre if the temperature were not a smooth function. There would be sharp gradients around radiators and that feverish sweaty student at the back who should really be in bed. What is true of the temperature in the lecture theatre is assumed also to be true for all reasonable functions of the three variables $f(x, y, z)$. It is, therefore, reasonable to write

$$f(x + \Delta x, y + \Delta y, z + \Delta z) - f(x, y, z) = \Delta f = a\Delta x + b\Delta y + c\Delta z + \epsilon\rho, \tag{2.2}$$

where $\epsilon \to 0$ as $\rho = \sqrt{(\Delta x)^2 + (\Delta y)^2 + (\Delta z)^2} \to 0$. This ensures that as the small distances Δx, Δy and Δz get smaller so does Δf and the two values $f(x + \Delta x, y + \Delta y, z + \Delta z)$ and $f(x, y, z)$ get ever closer themselves. One captures this rigorously by saying that the constants a, b and c are independent of the small distances Δx, Δy and Δz, respectively. Suppose in Equation (2.2) one assumes that f and $f + \Delta f$ are at points beside each other along the x-axis. In this case, Δy and Δz are both zero and Equation (2.2) becomes

$$f(x + \Delta x, y, z) - f(x, y, z) = \Delta f = a\Delta x + \epsilon \rho \tag{2.3}$$

so that dividing by Δx yields

$$\frac{f(x + \Delta x, y, z) - f(x, y, z)}{\Delta x} = \frac{\Delta f}{\Delta x} = a + \frac{\epsilon \rho}{\Delta x}$$

and letting $\Delta x \to 0$ gives

$$\frac{\partial f}{\partial x} = a$$

since $\frac{\rho}{\Delta x} \to 1$ but $\epsilon \to 0$. Similarly by considering the values of f at adjacent points on the y- and z-axes, it is seen that

$$\frac{\partial f}{\partial y} = b \quad \text{and} \quad \frac{\partial f}{\partial z} = c.$$

This proves Equation (2.1). The crucial part of this is the role played by ϵ. This quantity will not tend to zero for jumpy or jerky functions as the small values Δx, Δy and Δz tend to zero and it is this seemingly small point that is behind the precision of mathematical analysis. If we cannot find any suitable ϵ, then the whole edifice collapses, there are no partial derivatives and calculations become very different, possibly they become impossible. There are special points called singularities where things *do* break down, the most notorious being at the centre of a black hole. These are however exotic and rare. Having made differentials sound very theoretical, there is a good practical application, and we will look at this now.

2.2.1 Small Errors

If we allow the differentials to be very small without being exactly zero, then Equation (2.1) could be replaced by the approximation

$$\Delta f \approx \frac{\partial f}{\partial x}\Delta x + \frac{\partial f}{\partial y}\Delta y + \frac{\partial f}{\partial z}\Delta z. \tag{2.4}$$

This approximation can be expressed in words as the small error in f is composed of the sums of the errors in x, y and z. For example, when an item is being manufactured, then the overall error can be calculated from taking into account the places where each process can produce an error then adding them to get the total error. Here is a simple example.

Example 2.2 Suppose a cylindrically shaped solid object was being manufactured. Its radius is r, and height is h. Calculate the error ΔV in the volume $V = \pi r^2 h$ in terms of the errors in radius Δr and height Δh. If $\Delta r = 0.1$ cm and $\Delta h = 0.05$ cm then calculate the error in the volume given $r = 5$ cm and $h = 4$ cm.

Solution: Given $V = \pi r^2 h$, Equation (2.4) gives

$$\Delta V \approx 2\pi r h \Delta r + \pi r^2 \Delta h. \tag{2.5}$$

Inserting the values given in this example results in

$$\Delta V \approx 2 \times \pi \times 4 \times 5 \times 0.1 + \pi \times (4)^2 \times 0.05 = 16.49 \text{ c.c.},$$

where (with apologies) old units have been used: centimetres *cm* and cubic centimetres *c.c.* The volume is $100\pi = 314.16$ c.c. so the error can be expressed as a percentage 5.2%. There were apologies for using non-SI units, but in fact mathematicians as a breed really do not like units, they are seen as either an annoyance or even an irrelevance. Something engineers and applied scientists have to bother with. In this problem, the units are practical, but the answer (16.49 c.c.) only means anything if you know what the original volume was, so the percentage answer is in this sense better. 16.49 is the *absolute error* whereas 5.2% is the *relative error*. To find the relative error for this example, simply divide Equation (2.5) by V on the left and $\pi r^2 h$ on the right to get

$$\frac{\Delta V}{V} \approx 2\frac{\Delta r}{r} + \frac{\Delta h}{h}$$

and express the answer as a percentage by multiplying by 100. As this is mathematics textbook, let us find the *exact* error in this case and see how good this approximation by partial differentiation is. The actual dimensions of the cylinder are $r = 5.1$ cm and $h = 4.05$ cm so the actual volume is $\pi(5.1)^2(4.05) = 330.94$ c.c. giving an error of $330.94 - 314.16 = 16.78$ c.c. Compare this with the above result 16.49 c.c. and the 'error in the error' is $16.78 - 16.49 = 0.29$ c.c. and has occurred due solely to the r^2 in the formula for the volume of a cylinder. Remember too that by maximising the radius (to 5.1 cm) and the height (to 4.05 cm) we also maximised the volume. This is because volume is an increasing function of both radius and height, the larger the radius or the larger the height, the larger the volume. Next, we shall do calculations where this is not the case so we need to use the absolute value when necessary to prevent the cancellation of errors and the false impression of greater accuracy. □

Example 2.3 A torque T is applied to a uniform circular rod of rigidity G, radius R and length L about its axis. The angle of twist θ of the rod about its axis is given by the formula

$$\theta = \frac{2LT}{\pi R^4 G}.$$

Using partial differentiation, determine the maximum approximate % error in θ given that L is in error by, $\pm 1\%$, R is in error by $\pm 2\%$ and G is in error by $\pm 3\%$.

Solution: Partially differentiating the equation for θ gives

$$d\theta = \frac{2T}{\pi R^4 G} dL + \frac{2L}{\pi R^4 G} dT - 4\frac{2LT}{\pi R^5 G} dR - \frac{2LT}{\pi R^4 G^2} dG$$

and dividing by θ on the left and $\frac{2LT}{\pi R^4 G}$ on the right gives

$$\frac{d\theta}{\theta} = \frac{dL}{L} + \frac{dT}{T} - 4\frac{dR}{R} - \frac{dG}{G}$$

and this gives the appropriate expression for calculating the error in θ. First of course the exact d is replaced by the approximation Δ:

$$\frac{\Delta\theta}{\theta} \approx \frac{\Delta L}{L} + \frac{\Delta T}{T} - 4\frac{\Delta R}{R} - \frac{\Delta G}{G}. \tag{2.6}$$

The data in the question gives the following

$$\frac{\Delta L}{L} = \pm 0.01; \quad \frac{\Delta T}{T} = 0; \quad \frac{\Delta R}{R} = \pm 0.02; \quad \frac{\Delta G}{G} = \pm 0.03.$$

Note that there is no error in T as this is the applied torque that is entirely under our control and is not being measured. Note also that each (non-zero) error is prefixed by $\pm$, and the plus or minus is chosen so as to maximise the total error. Therefore, the result is

$$\frac{\Delta\theta}{\theta} \approx 0.01 + 0 + 4 \times 0.02 + 0.03 = 0.12 = 12\%.$$

$\square$

There is a general and convenient way of getting to the relative error using logarithmic differentiation; relying on the knowledge that

$$\frac{d}{dx}[\ln(f)] = \frac{1}{f}\frac{df}{dx}$$

generalised to partial differentiation. Using the previous example taking logs first,

$$\ln(\theta) = \ln\left\{\frac{2LT}{\pi R^4 G}\right\} = \ln 2 + \ln L + \ln T - \ln \pi - 4\ln R - \ln G,$$

then differentiating

$$\frac{1}{\theta}\frac{\partial\theta}{\partial x} = 0 + \frac{1}{L}\frac{\partial L}{\partial x} + \frac{1}{T}\frac{\partial T}{\partial x} - 0 - 4\frac{1}{R}\frac{\partial R}{\partial x} - \frac{1}{G}\frac{\partial G}{\partial x},$$

where x is a dummy variable upon which all variables depend. The torque T does not vary so its derivative is zero. Replace all derivatives $\frac{\partial}{\partial x}$ by the approximation $\frac{\Delta}{\Delta x}$ cancelling $\frac{1}{\Delta x}$, Equation (2.6) is regained and the solution to the example proceeds as before. Which of the two derivations is preferable is a question of personal taste.

This method touches upon and leads naturally to the total derivative, the subject of the next section.

2.3 Total Derivative

First of all consider a generalisation of the *function of a function* rule. Put $u = f(x, y)$ and let $z = z(u)$, then try to differentiate z with respect to x. There is no doubt that

$$\frac{\partial u}{\partial x} = \lim_{\Delta x \to 0} \left\{ \frac{u(x + \Delta x, y) - u(x, y)}{\Delta x} \right\}$$

$$= \lim_{\Delta x \to 0} \left\{ \frac{f(x + \Delta x, y) - f(x, y)}{\Delta x} \right\}.$$

It is equally true that the total derivative

$$\frac{dz}{du} = \lim_{\Delta u \to 0} \left\{ \frac{z(u + \Delta u) - z(u)}{\Delta u} \right\}$$

from the definition in Chapter 1. Hence, we can write

$$\frac{\partial z}{\partial x} = \lim_{\Delta u \to 0} \left\{ \frac{z(u + \Delta u) - z(u)}{\Delta u} \right\} \lim_{\Delta x \to 0} \left\{ \frac{u(x + \Delta x, y) - u(x, y)}{\Delta x} \right\},$$

where it is understood that the Δu in the denominator of the first limit only involves changes in x and not y; so $\Delta u = \Delta f = f(x + \Delta x, y) - f(x, y)$. Thus,

$$\frac{\partial z}{\partial x} = \frac{dz}{du} \frac{\partial u}{\partial x}.$$

Similarly

$$\frac{\partial z}{\partial y} = \frac{dz}{du} \frac{\partial u}{\partial y}.$$

The following example will help cement these relationships.

Example 2.4   If u is an arbitrary function of y/x viz.

$$u = f\left(\frac{y}{x}\right)$$

show that

$$\frac{\partial u}{\partial x} = -\frac{y}{x^2} f' \quad \text{and} \quad \frac{\partial u}{\partial y} = \frac{1}{x} f',$$

where the prime symbol denotes differentiation. Hence or otherwise show that

$$x\frac{\partial u}{\partial x} + y\frac{\partial u}{\partial y} = 0.$$

Solution: Put $t = y/x$ so that $u = f(t)$ and we use

$$\frac{\partial u}{\partial x} = \frac{df}{dt}\frac{\partial t}{\partial x} = f'\frac{(-y)}{x^2}$$

and

$$\frac{\partial u}{\partial y} = \frac{df}{dt}\frac{\partial t}{\partial y} = f'\frac{1}{x}$$

as required. Thus,

$$x\frac{\partial u}{\partial x} + y\frac{\partial u}{\partial y} = -x.f'.\frac{y}{x^2} + y.f'.\frac{1}{x} = 0.$$

$\square$

Now, consider the reverse situation: a function of two variables $f(x, y)$ where both x and y are themselves functions of the same single variable t. One then asks what is the relationship between the rates of change of f with respect to x, with respect to y and with respect to t. The rate of change of f with respect to t will be a straight derivative, not partial. This exact derivative will have the limit definition seen in Chapter 1

$$\frac{df}{dt} = \lim_{\Delta t \to 0}\left\{\frac{f(t + \Delta t) - f(t)}{\Delta t}\right\}$$

$$= \lim_{\Delta t \to 0}\left\{\frac{f(x(t + \Delta t), y(t + \Delta t)) - f(x, y)}{\Delta t}\right\}.$$

This could be written as the pair

$$\frac{df}{dt} = \lim_{\Delta t \to 0}\left\{\left[\frac{f(x(t + \Delta t), y(t + \Delta t)) - f(x, y)}{x(t + \Delta t) - x(t)}\right]\left[\frac{x(t + \Delta t) - x(t)}{\Delta t}\right]\right\}$$

provided the limits exist and are unique, or indeed the pair

$$\frac{df}{dt} = \lim_{\Delta t \to 0}\left\{\left[\frac{f(x(t + \Delta t), y(t + \Delta t)) - f(x, y)}{y(t + \Delta t) - y(t)}\right]\left[\frac{y(t + \Delta t) - y(t)}{\Delta t}\right]\right\}$$

with a similar proviso. The first is simply

$$\frac{df}{dt} = \frac{\partial f}{\partial x}\frac{dx}{dt},$$

the second is simply

$$\frac{df}{dt} = \frac{\partial f}{\partial y}\frac{dy}{dt},$$

or is it? There is a subtlety here some of you may have noticed. The two alternate pairs really should be written

$$\frac{df}{dt} = \lim_{\Delta t \to 0} \left\{ \left[\frac{f(x(t + \Delta t), y(t)) - f(x, y)}{x(t + \Delta t) - x(t)} \right] \left[\frac{x(t + \Delta t) - x(t)}{\Delta t} \right] \right\}$$

and

$$\frac{df}{dt} = \lim_{\Delta t \to 0} \left\{ \left[\frac{f(x(t), y(t + \Delta t)) - f(x, y)}{y(t + \Delta t) - y(t)} \right] \left[\frac{y(t + \Delta t) - y(t)}{\Delta t} \right] \right\}$$

the first has only x varying and not y, the second has y varying and not x. As all are functions of t this is only possible if in the first expression $f(x(t), y)$ and y does not depend on t, and in the second expression $f(x, y(t))$ and this time x does not depend on t. As t varies in these expressions, first y does not vary, and then x does not vary. These expressions are therefore valid *partial* derivatives. Later when differential geometry is discussed, it will be seen that t is a parameter that traces curves on the surface $z = f(x, y)$ as it varies. There are lots of choices of course as the number of scribbles one can make on a given surface is infinite. The two choices just made for our particular $f(x, y)$ are quite valid and correspond to the partial derivatives in the x- and y-direction. This will lead to directional derivatives, but that will have to wait until vectors have been covered (but see Exercise 2.3). Combining these two results by simply adding them defines what is called a *total derivative*

$$\frac{df}{dt} = \frac{\partial f}{\partial x} \frac{dx}{dt} + \frac{\partial f}{\partial y} \frac{dy}{dt}. \tag{2.7}$$

Adding a third variable $z = z(t)$ would give Equation (2.1) divided by dt

$$\frac{df}{dt} = \frac{\partial f}{\partial x} \frac{dx}{dt} + \frac{\partial f}{\partial y} \frac{dy}{dt} + \frac{\partial f}{\partial z} \frac{dz}{dt} \tag{2.8}$$

and a generalisation to N variables is possible:

$$\frac{df}{dt} = \sum_{n=1}^{N} \frac{\partial f}{\partial x_n} \frac{dx_n}{dt},$$

where $f = f(x_1(t), x_2(t), \ldots, x_N(t))$, but $N = 3$ will do most of the time in this text. First of all the above expression is easy to remember. It is as if each term captures the variation of f due to each component $\frac{\partial f}{\partial x_n} \frac{dx_n}{dt}$ then the total variation of f is obtained by simply adding together all these contributions. This adding up of individual contribution implies linearity and is valid due to a limiting processes of the type involving ϵ and ρ met in the last section; any non-linearity in f is smoothed and ultimately everything is linearised.

If t is time, then Equation (2.8) will have applications in the physical world, notably mechanics where the time derivatives of x, y and z are written using the more Newtonian inspired notation as $\dot{x}, \dot{y}$ and $\dot{z}$ and they denote the speed in each of the co-ordinate directions. Equation (2.8) can be extended to the case where $f = f(x, y, z, t)$, that is t is also, in addition, an explicit variable for the function f. In this case

$$\frac{df}{dt} = \frac{\partial f}{\partial x}\frac{dx}{dt} + \frac{\partial f}{\partial y}\frac{dy}{dt} + \frac{\partial f}{\partial z}\frac{dz}{dt} + \frac{\partial f}{\partial t}, \tag{2.9}$$

and this has applications in fluid mechanics.

Another useful notation introduced in Chapter 1 and reused in Example 2.4 is

$$f'(t) = \frac{df}{dt},$$

where the prime symbol denotes differentiation with respect to, in this instance, t, but f' means differentiation with respect to the argument of the function f whatever that argument might be. For example if $t = xy$, then

$$f'(t) = \frac{df}{d(xy)}$$

so using the above results in

$$f'(t) = \frac{df}{dt} = \frac{\partial f}{\partial x}\frac{dx}{dt} + \frac{\partial f}{\partial y}\frac{dy}{dt} = \frac{\partial f}{\partial x}y + \frac{\partial f}{\partial y}x$$

so

$$f'(t) = y\frac{\partial f}{\partial x} + x\frac{\partial f}{\partial y}$$

in this particular case. This leads nicely to the changing of variables between (x, y) to (u, v) and vice versa.

2.4 Chain Rule

In one dimension there is the 'function-of-a-function' rule:

$$\frac{dy}{dt} = \frac{dy}{dx}\frac{dx}{dt}.$$

This has been extended in Equation (2.7) but let us now extend it further to the pair of equations:

$$\frac{\partial f}{\partial u} = \frac{\partial f}{\partial x}\frac{\partial x}{\partial u} + \frac{\partial f}{\partial y}\frac{\partial y}{\partial u},$$

$$\frac{\partial f}{\partial v} = \frac{\partial f}{\partial x}\frac{\partial x}{\partial v} + \frac{\partial f}{\partial y}\frac{\partial y}{\partial v}.$$

This pair is useful if the relationships $x = x(u, v)$ and $y = y(u, v)$ are known. However, if we only know $u = u(x, y)$ and $v = v(x, y)$ then the pair

$$\frac{\partial f}{\partial x} = \frac{\partial f}{\partial u}\frac{\partial u}{\partial x} + \frac{\partial f}{\partial v}\frac{\partial v}{\partial x},$$
$$\frac{\partial f}{\partial y} = \frac{\partial f}{\partial u}\frac{\partial u}{\partial y} + \frac{\partial f}{\partial v}\frac{\partial v}{\partial y}$$

is the more useful. These changes of variable relations is known as the *chain rule* and generalisations will follow; they are straightforward. If we know $u = u(x, y)$ and $v = v(x, y)$ but require $\frac{\partial f}{\partial u}$ and $\frac{\partial f}{\partial v}$ then either reverse the equations for u and v into equations for x and y or solve the last pair of equations simultaneously, whichever is more convenient. The proof of this and similar chain rules follow the same kind of arguments met in demonstrating the definition of differential after Equation (2.1) so they will not be presented here. They would be a little out of place, belonging as they do in books on real analysis. Here is an example which comes in useful later:

Example 2.5 Find the relationships between partial derivatives in Cartesian co-ordinates (x, y) and plane polar co-ordinates (r, θ) given $x = r \cos \theta$ and $y = r \sin \theta$

Solution: Since we have been given $x = x(r, \theta)$ and $y = y(r, \theta)$ we need the following chain rule pair

$$\frac{\partial f}{\partial r} = \frac{\partial f}{\partial x}\frac{\partial x}{\partial r} + \frac{\partial f}{\partial y}\frac{\partial y}{\partial r},$$
$$\frac{\partial f}{\partial \theta} = \frac{\partial f}{\partial x}\frac{\partial x}{\partial \theta} + \frac{\partial f}{\partial y}\frac{\partial y}{\partial \theta}.$$

Using the four partial derivatives:

$$\frac{\partial x}{\partial r} = \cos \theta; \quad \frac{\partial y}{\partial r} = \sin \theta; \quad \frac{\partial x}{\partial \theta} = -r \sin \theta; \quad \text{and} \quad \frac{\partial y}{\partial \theta} = r \cos \theta.$$

Hence,

$$\frac{\partial f}{\partial r} = \frac{\partial f}{\partial x} \cos \theta + \frac{\partial f}{\partial y} \sin \theta,$$
$$\frac{\partial f}{\partial \theta} = -\frac{\partial f}{\partial x} r \sin \theta + \frac{\partial f}{\partial y} r \cos \theta.$$

One now has the choice of solving this last pair or reversing the relationships $x = r \cos \theta$ and $y = r \sin \theta$. We shall do both. Reversing first, we square and add to get $r^2 = x^2 + y^2$ so $r = \sqrt{x^2 + y^2}$, and we divide to get

$$\frac{y}{x} = \tan\theta, \quad \text{so} \quad \theta = \arctan\left\{\frac{y}{x}\right\}.$$

Therefore, we have achieved $r = r(x, y)$ and $\theta = \theta(x, y)$ and we need the first chain rule pair

$$\frac{\partial f}{\partial x} = \frac{\partial f}{\partial r}\frac{\partial r}{\partial x} + \frac{\partial f}{\partial \theta}\frac{\partial \theta}{\partial x}$$

$$\frac{\partial f}{\partial y} = \frac{\partial f}{\partial r}\frac{\partial r}{\partial y} + \frac{\partial f}{\partial \theta}\frac{\partial \theta}{\partial y}.$$

The derivatives are a little more trickier to find; here they are:

$$\frac{\partial r}{\partial x} = \frac{x}{\sqrt{x^2 + y^2}} = \frac{x}{r} = \cos\theta;$$

$$\frac{\partial r}{\partial y} = \frac{y}{\sqrt{x^2 + y^2}} = \frac{y}{r} = \sin\theta;$$

$$\frac{\partial \theta}{\partial x} = \frac{1}{1 + y^2/x^2}\cdot-\frac{y}{x^2} = -\frac{y}{r^2} = -\frac{\sin\theta}{r}; \quad \text{and}$$

$$\frac{\partial \theta}{\partial y} = \frac{1}{1 + y^2/x^2}\cdot\frac{1}{x} = \frac{x}{r^2} = \frac{\cos\theta}{r}.$$

Hence, we obtain the relationships

$$\frac{\partial f}{\partial x} = \cos\theta\frac{\partial f}{\partial r} - \frac{\sin\theta}{r}\frac{\partial f}{\partial \theta},$$

$$\frac{\partial f}{\partial y} = \sin\theta\frac{\partial f}{\partial r} + \frac{\cos\theta}{r}\frac{\partial f}{\partial \theta}.$$

Perhaps an easier alternative in this case would be to start from the already derived pair

$$\frac{\partial f}{\partial r} = \frac{\partial f}{\partial x}\cos\theta + \frac{\partial f}{\partial y}\sin\theta,$$

$$\frac{\partial f}{\partial \theta} = -\frac{\partial f}{\partial x}r\sin\theta + \frac{\partial f}{\partial y}r\cos\theta.$$

Simply take the first multiplied by $\cos\theta$ and subtract the second multiplied by $\sin\theta/r$ to get

$$\cos\theta\frac{\partial f}{\partial r} - \frac{\sin\theta}{r}\frac{\partial f}{\partial \theta} = \frac{\partial f}{\partial x}.$$

Now take the first multiplied by $\sin\theta$ added to the second multiplied by $\cos\theta/r$ to get

$$\sin\theta\frac{\partial f}{\partial r} + \frac{\cos\theta}{r}\frac{\partial f}{\partial \theta} = \frac{\partial f}{\partial y}.$$

These are the same relationships as we obtained above. For notational convenience the derivative of f with respect to x is often written f_x and is called the *suffix derivative* notation. The above pairs written this way would be:

$$f_r = \cos \theta f_x + \sin \theta f_y,$$
$$f_\theta = r \sin \theta f_x + r \cos \theta f_y,$$

and

$$f_x = \cos \theta f_r - \frac{\sin \theta}{r} f_\theta,$$
$$f_y = \sin \theta f_r + \frac{\cos \theta}{r} f_\theta.$$

□

This notation will be used increasingly from now. While on notation, sometimes it is necessary to be specific about what is being held constant when differentiating partially, so the notation

$$\left. \frac{\partial f}{\partial x} \right|_y \quad \text{or alternatively} \quad \left(\frac{\partial f}{\partial x} \right)_y$$

is adopted to emphasise that y is being held constant. Those of you that have to study thermodynamics will surely meet this. In this book it is very seldom used as it is obvious what is being held constant.

2.4.1 Leibniz Rule

An application of the chain rule, Leibniz rule is sometimes called 'differentiation under the integral sign'. Here is the rule in the form of a theorem:

Theorem 2.1 If $F(\alpha) = \displaystyle\int_{a(\alpha)}^{b(\alpha)} f(x, \alpha)dx$ where $f(x, \alpha)$ is integrable in x and differentiable in the variable α, and $a(\alpha), b(\alpha)$ are differentiable functions, then

$$\frac{dF}{d\alpha} = \int_{a(\alpha)}^{b(\alpha)} \frac{\partial f(x, \alpha)}{\partial \alpha} dx + f(b, \alpha) \frac{db}{d\alpha} - f(a, \alpha) \frac{da}{d\alpha}.$$

Proof: There is a longer proof from first principles that uses the mean value theorem, this proof uses the chain rule. Assuming that $F = F(a, b, \alpha)$ where a and b are themselves functions of α the chain rule can take the form:

$$\frac{dF}{d\alpha} = \frac{\partial F}{\partial \alpha} + \frac{\partial F}{\partial b} \frac{db}{d\alpha} + \frac{\partial F}{\partial a} \frac{da}{d\alpha}, \tag{2.10}$$

where the explicit dependence of F on α is restricted to $f(x, \alpha)$ and does not include the dependence of a and b on α. This means that

$$\frac{\partial F}{\partial \alpha} = \frac{\partial}{\partial \alpha} \left(\int_{a(\alpha)}^{b(\alpha)} f(x, \alpha)dx \right) = \int_{a(\alpha)}^{b(\alpha)} \frac{\partial f}{\partial \alpha} dx.$$

Using the fundamental theorem of the calculus:

$$\frac{d}{dx}\int^x f(t)dt = f(x),$$

which essentially is a statement that differentiation and integration are inverse operations, it can be stated that

$$\frac{dF}{db} = \frac{d}{db}\int_a^b f(x,\alpha)dx = f(b,\alpha) \quad \text{and}$$

$$\frac{dF}{da} = \frac{d}{da}\int_a^b f(x,\alpha)dx = -f(a,\alpha).$$

Inserting these results into Equation (2.10) gives the required result

$$\frac{dF}{d\alpha} = \int_{a(\alpha)}^{b(\alpha)} \frac{\partial f(x,\alpha)}{\partial\alpha}dx + f(b,\alpha)\frac{db}{d\alpha} - f(a,\alpha)\frac{da}{d\alpha}. \qquad \square$$

Here is an example of its use.

Example 2.6 Given the result

$$\int_0^{\pi/2}\frac{\cos x}{\alpha\cos x + \sin x}dx = \frac{\alpha\pi}{2(\alpha^2+1)} - \frac{\ln\alpha}{\alpha^2+1}$$

find

$$\int_0^{\pi/2}\frac{\cos^2 x}{(2\cos x + \sin x)^2}dx.$$

Solution: The integral given is by no means obvious, but can be proved by the substitution $t = \tan(x/2)$. See Exercise 1.9 at the end of the last chapter. Differentiating both sides with respect to α to give

$$\int_0^{\pi/2}\frac{-\cos^2 x}{(\alpha\cos x + \sin x)^2}dx = \frac{d}{d\alpha}\left\{\frac{\alpha\pi}{2(\alpha^2+1)} - \frac{\ln\alpha}{(\alpha^2+1)}\right\}$$

$$= \frac{2(\alpha^2+1)\pi - 4\pi\alpha^2}{4(\alpha^2+1)^2} - \frac{(\alpha^2+1) - 2\alpha^2\ln\alpha}{\alpha(\alpha^2+1)^2}$$

$$= \frac{(1-\alpha^2)\pi}{2(\alpha^2+1)^2} - \frac{1+\alpha^2 - 2\alpha^2\ln\alpha}{\alpha(\alpha^2+1)^2}.$$

Thus,

$$\int_0^{\pi/2}\frac{\cos^2 x}{(\alpha\cos x + \sin x)^2}dx = -\frac{(1-\alpha^2)\pi}{2(\alpha^2+1)^2} + \frac{1+\alpha^2 - 2\alpha^2\ln\alpha}{\alpha(\alpha^2+1)^2}.$$

Substituting $\alpha = 2$ in this equation gives the result:

$$\int_0^{\pi/2}\frac{\cos^2 x}{(2\cos x + \sin x)^2}dx = \frac{3\pi + 5 - 8\ln 2}{50}$$

a result quite difficult to derive any other way. $\qquad \square$

2.4.2 Chain Rule in *n* Dimensions

The two-dimensional version of the chain rule is readily generalised, first to three dimensions

$$\frac{\partial f}{\partial u} = \frac{\partial f}{\partial x}\frac{\partial x}{\partial u} + \frac{\partial f}{\partial y}\frac{\partial y}{\partial u} + \frac{\partial f}{\partial z}\frac{\partial z}{\partial u}$$

$$\frac{\partial f}{\partial v} = \frac{\partial f}{\partial x}\frac{\partial x}{\partial v} + \frac{\partial f}{\partial y}\frac{\partial y}{\partial v} + \frac{\partial f}{\partial z}\frac{\partial z}{\partial v}$$

$$\frac{\partial f}{\partial w} = \frac{\partial f}{\partial x}\frac{\partial x}{\partial w} + \frac{\partial f}{\partial y}\frac{\partial y}{\partial w} + \frac{\partial f}{\partial z}\frac{\partial z}{\partial w}$$

and now to N dimensions where the co-ordinates are $(x_1, x_2, \ldots, x_N)$ and $(u_1, u_2, \ldots, u_N)$.

$$\frac{\partial f}{\partial u_1} = \frac{\partial f}{\partial x_1}\frac{\partial x_1}{\partial u_1} + \frac{\partial f}{\partial x_2}\frac{\partial x_2}{\partial u_1} + \cdots + \frac{\partial f}{\partial x_N}\frac{\partial x_N}{\partial u_1},$$

$$\frac{\partial f}{\partial u_2} = \frac{\partial f}{\partial x_1}\frac{\partial x_1}{\partial u_2} + \frac{\partial f}{\partial x_2}\frac{\partial x_2}{\partial u_2} + \cdots + \frac{\partial f}{\partial x_N}\frac{\partial x_N}{\partial u_2},$$

$$\vdots \quad = \quad \vdots \qquad \vdots \qquad\qquad \vdots$$

$$\frac{\partial f}{\partial u_N} = \frac{\partial f}{\partial x_1}\frac{\partial x_1}{\partial u_N} + \frac{\partial f}{\partial x_2}\frac{\partial x_2}{\partial u_N} + \cdots + \frac{\partial f}{\partial x_N}\frac{\partial x_N}{\partial u_N}.$$

This is here for reference as it is not implemented in this text. $N = 3$ is the largest dimension encountered here. Here is another constructive example.

Example 2.7   Starting with the two-dimensional chain rule:

$$\frac{\partial f}{\partial x} = \frac{\partial f}{\partial u}\frac{\partial u}{\partial x} + \frac{\partial f}{\partial v}\frac{\partial v}{\partial x}$$

$$\frac{\partial f}{\partial y} = \frac{\partial f}{\partial u}\frac{\partial u}{\partial y} + \frac{\partial f}{\partial v}\frac{\partial v}{\partial y}.$$

determine the version in the special case $u = u(x), v = v(x)$ and $f = u/v$.

Solution: In these circumstances, the second equation is identically zero of course, and the first is

$$\frac{df}{dx} = \frac{\partial f}{\partial u}\frac{du}{dx} + \frac{\partial f}{\partial v}\frac{dv}{dx},$$

which is similar to the total derivative, see Equation (2.7) but with notational differences. With

$$f(x) = \frac{u(x)}{v(x)} \quad \text{we have} \quad \frac{\partial f}{\partial u} = \frac{1}{v} \quad \text{and} \quad \frac{\partial f}{\partial v} = -\frac{u}{v^2}$$

and so the above equation which is

$$\frac{df}{dx} = \frac{\partial f}{\partial u}u' + \frac{\partial f}{\partial v}v'$$

now becomes

$$\frac{d}{dx}\left(\frac{u}{v}\right) = \frac{1}{v}u' - \frac{u}{v^2}v' = \frac{vu' - uv'}{v^2}$$

the familiar quotient rule for the derivative in a single variable. □

2.4.3 Implicit Functions

If there is a relationship $f(x, y) = 0$, then we say that there is an *implicit* relation between y and x. In practical terms it is assumed that y cannot be written in terms of x so derivatives cannot be found directly. However, for such implicit functions partial differentiation comes to our aid. We know that $f(x, y) = 0$, so in theory $y = y(x)$ we just do not know what the function is. Therefore, the derivative $\frac{dy}{dx}$ exists, and to find it we use

$$df = f_x\, dx + f_y\, dy,$$

which is the two-dimensional version of Equation (2.1) written using the suffix derivative notation. As $f(x, y) = 0$ so $df = 0$ too, and we mean identically zero not just infinitesimally small. Hence,

$$0 = f_x\, dx + f_y\, dy \quad \text{or} \quad \frac{dy}{dx} = -\frac{f_x}{f_y}.$$

Here are two examples

Example 2.8   Find $\frac{dy}{dx}$ if

$$\frac{x^2}{a^2} + \frac{y^2}{b^2} = 1.$$

Solution: Here it is possible to find $y = y(x)$ but easier not to.

$$f_x = \frac{2x}{a^2} \quad \text{and} \quad f_y = \frac{2y}{b^2}$$

so by direct partial differentiation

$$\frac{dy}{dx} = -\frac{b^2 x}{a^2 y}.$$

□

Here is a second example where implicit differentiation is essential

Example 2.9   Determine $\frac{dy}{dx}$ if $(\cos x)^y = (\sin y)^x$.

Solution: Some ingenuity is required for this example. Write

$$g(x, y) = (\cos x)^y \quad \text{and} \quad h(x, y) = (\sin y)^x$$

then take logs and use logarithmic differentiation:

$$\frac{\partial}{\partial x}(\ln g) = \frac{g_x}{g} = \frac{\partial}{\partial x}(y \ln \cos x) = -y \tan x, \quad \text{so} \quad g_x = -yg \tan x$$

$$\frac{\partial}{\partial y}(\ln g) = \frac{g_y}{g} = \frac{\partial}{\partial y}(y \ln \cos x) = \ln \cos x, \quad \text{so} \quad g_y = g \ln \cos x$$

$$\frac{\partial}{\partial x}(\ln h) = \frac{h_x}{h} = \frac{\partial}{\partial x}(x \ln \sin y) = \ln \sin y, \quad \text{so} \quad h_x = h \ln \sin y$$

and

$$\frac{\partial}{\partial y}(\ln h) = \frac{h_y}{h} = \frac{1}{h}\frac{\partial}{\partial y}(x \ln \sin y) = x \cot y \quad \text{so} \quad h_y = xh \cot y.$$

Now $f(x, y) = g(x, y) - h(x, y)$ so

$$f_x = g_x - h_x = -gy \tan x - h \ln \sin x, \quad \text{and}$$

$$f_y = g_y - h_y = g \ln \cos x - hx \cot y$$

but of course as $f = 0, g = h$ these cancel from the quotient $-f_x/f_y$ so this gives

$$\frac{dy}{dx} = -\frac{f_x}{f_y} = \frac{yg \tan x + h \ln \sin x}{g \ln \cos x - xh \cot y} = \frac{y \tan x + \ln \sin x}{\ln \cos x - x \cot y}.$$

This quotient is the rather unlikely answer.   □

2.5   Jacobian

Differentiation comes before integration just as addition comes before subtraction and multiplication before division, however there is a quantity that will be needed when doing integration involving several variables and that quantity is the Jacobian. It is named after Carl Gustav Jacob Jacobi (1804–1851) a German mathematician who worked himself too hard and eventually succumbed to smallpox, but not before contributing to the theory of equations, differential equations and elliptic functions.

Definition 2.2   In two dimensions the *Jacobian* of two functions of two variables $u = u(x, y)$ and $v = v(x, y)$ is the determinant

$$\frac{\partial(u, v)}{\partial(x, y)} = \begin{vmatrix} \dfrac{\partial u}{\partial x} & \dfrac{\partial u}{\partial y} \\[2mm] \dfrac{\partial v}{\partial x} & \dfrac{\partial v}{\partial y} \end{vmatrix} = \frac{\partial u}{\partial x}\frac{\partial v}{\partial y} - \frac{\partial v}{\partial x}\frac{\partial u}{\partial y}$$

and the extension to higher dimensions is straightforward: for $u = u(x, y, z)$, $v = v(x, y, z)$ and $w = w(x, y, z)$ the Jacobian is

$$\frac{\partial(u, v, w)}{\partial(x, y, z)} = \begin{vmatrix} \frac{\partial u}{\partial x} & \frac{\partial u}{\partial y} & \frac{\partial u}{\partial z} \\[2mm] \frac{\partial v}{\partial x} & \frac{\partial v}{\partial y} & \frac{\partial v}{\partial z} \\[2mm] \frac{\partial w}{\partial x} & \frac{\partial w}{\partial y} & \frac{\partial w}{\partial z} \end{vmatrix}$$

this time not multiplying it out in full! In complete generality,

$$\frac{\partial(u_1, u_2, \ldots, u_N)}{\partial(x_1, x_2, \ldots, x_N)} = \begin{vmatrix} \frac{\partial u_1}{\partial x_1} & \frac{\partial u_1}{\partial x_2} & \cdots & \frac{\partial u_1}{\partial x_N} \\[2mm] \frac{\partial u_2}{\partial x_1} & \frac{\partial u_2}{\partial x_2} & \cdots & \frac{\partial u_2}{\partial x_N} \\[2mm] \vdots & \vdots & \ddots & \vdots \\[2mm] \frac{\partial u_N}{\partial x_1} & \frac{\partial u_N}{\partial x_2} & \cdots & \frac{\partial u_N}{\partial x_N} \end{vmatrix}.$$

You might wonder what is the point of defining these. It turns out they are quite useful in many respects. They will be met in multiple integration when they occur naturally when changing variables, but they have other interesting properties. In the last section, we saw that if the function $f(x, y) = 0$, then as $df = 0$ identically in this case we have that

$$0 = f_x \, dx + f_y \, dy \quad \text{so} \quad \frac{dy}{dx} = -\frac{f_x}{f_y} \quad \text{and also} \quad \frac{dx}{dy} = -\frac{f_y}{f_x}$$

this means that

$$\frac{dy}{dx} \cdot \frac{dx}{dy} = 1,$$

which may seem obvious. However, Jacobians help us to generalise this and interpret the results. Therefore, let us start with

$$df = f_x \, dx + f_y \, dy = 0,$$

which if x and y are independent variables, then it implies $f_x = 0$ and $f_y = 0$, of course. Rewriting this as the pair of chain rule relationships,

$$\frac{\partial f}{\partial x} = \frac{\partial f}{\partial u} \frac{\partial u}{\partial x} + \frac{\partial f}{\partial v} \frac{\partial v}{\partial x} = 0,$$

$$\frac{\partial f}{\partial y} = \frac{\partial f}{\partial u} \frac{\partial u}{\partial y} + \frac{\partial f}{\partial v} \frac{\partial v}{\partial y} = 0,$$

where we have assumed that $f = f(u, v)$ and $u = u(x, y)$ and $v = v(x, y)$. From the elementary linear algebra of a pair of equations, the condition that not both

$f_u = 0$ and $f_v = 0$ is that the determinant

$$\begin{vmatrix} \dfrac{\partial u}{\partial x} & \dfrac{\partial u}{\partial y} \\[2mm] \dfrac{\partial v}{\partial x} & \dfrac{\partial v}{\partial y} \end{vmatrix} = 0$$

or

$$\frac{\partial(u, v)}{\partial(x, y)} = 0.$$

This proves that if there is a functional relationship between $u = u(x, y)$ and $v = v(x, y)$ expressed above in the form $f(x, y) = 0$ then it is necessary that the Jacobian $\frac{\partial(u,v)}{\partial(x,y)} = 0$. It is also sufficient. To see this, start with the zero Jacobian

$$\frac{\partial u}{\partial x}\frac{\partial v}{\partial y} - \frac{\partial v}{\partial x}\frac{\partial u}{\partial y} = 0$$

or $u_y/u_x = v_y/v_x$. Since $u = u(x, y)$ in theory this can be solved so that $x = F(u, y)$ so $u = u(F(u, y), y)$ in which case

$$du = u_x\, dx + u_y\, dy = u_x\left(\frac{\partial F}{\partial y}dy + \frac{\partial F}{\partial u}du\right) + u_y\, dy.$$

So

$$du = u_x F_u\, du + (u_x F_y + u_y)dy,$$

which gives the two relationships: $1 = u_x F_u$ and $F_y = -u_y/u_x$. Proceeding similarly with $v = v(x, y)$ leads to $v = v(F(u, y), y)$ so

$$dv = v_x\, dx + v_y\, dy = v_x\left(\frac{\partial F}{\partial y}dy + \frac{\partial F}{\partial u}du\right) + v_y\, dy.$$

Substituting for F_y gives

$$dv = \left(v_y - \frac{u_y v_x}{v_x}\right)dy + \frac{v_x}{u_x}du$$

and the expression in front of dy is zero due to the zero Jacobian. However since

$$dv = v_y\, dy + v_u\, du$$

this means that

$$v_y = 0$$

so

$$v = v(u)$$

and v and u are connected by a functional relationship; call it $G(u, v) = 0$. Thus, we have proved the following theorem.

Theorem 2.2   If $u = u(x, y)$ and $v = v(x, y)$ are suitably well behaved functions (continuously differentiable) then there is a functional relationship of the kind $G(u, v) = 0$ if and only if

$$\frac{\partial(u, v)}{\partial(x, y)} = 0.$$

2.6   Higher Derivatives

In one dimension, derivatives were extended to second and higher order. The same is done for partial derivatives. For a function of a single variable, $f = f(x)$ say, the second derivative is the derivative of its derivative, so

$$\frac{d^2f}{dx^2} = \frac{d}{dx}\left(\frac{df}{dx}\right).$$

Similarly if $f = f(x, y)$ then

$$\frac{\partial^2 f}{\partial x^2} = \frac{\partial}{\partial x}\left(\frac{\partial f}{\partial x}\right), \quad \text{and} \quad \frac{\partial^2 f}{\partial y^2} = \frac{\partial}{\partial y}\left(\frac{\partial f}{\partial y}\right).$$

Additionally, there are mixed derivatives:

$$\frac{\partial^2 f}{\partial x \partial y} = \frac{\partial}{\partial x}\left(\frac{\partial f}{\partial y}\right), \quad \text{and} \quad \frac{\partial^2 f}{\partial y \partial x} = \frac{\partial}{\partial y}\left(\frac{\partial f}{\partial x}\right).$$

These second-order derivatives are clumsily written in this way and almost always these four derivatives are written

$$f_{xx}, f_{yy}, f_{xy} \quad \text{and} \quad f_{yx},$$

respectively, in the suffix derivative notation which comes into its own now. The two mixed derivatives f_{xy} and f_{yx} are in fact almost always equal. The exceptions are where (x, y) are points where the function $f(x, y)$ misbehaves and either jumps or jerks. These are points where either the derivatives do not exist of if they do, they are not continuous. The proof of this is quite straightforward and simply means exchanging the order of the limiting processes in the definitions of the two derivatives. So

$$f_{xy} = \lim_{\Delta x \to 0} \left\{ \frac{\lim_{\Delta y \to 0}\frac{f(x+\Delta x, y+\Delta y)-f(x+\Delta x, y)}{\Delta y} - \lim_{\Delta y \to 0}\frac{f(x, y+\Delta y)-f(x, y)}{\Delta y}}{\Delta x} \right\},$$

whereas

$$f_{yx} = \lim_{\Delta y \to 0} \left\{ \frac{\lim_{\Delta x \to 0}\frac{f(x+\Delta x, y+\Delta y)-f(x, y+\Delta y)}{\Delta x} - \lim_{\Delta x \to 0}\frac{f(x+\Delta x, y)-f(x, y)}{\Delta x}}{\Delta y} \right\}.$$

Of course, the ability to exchange the order of differentiation lies in the existence of ϵ the all important small quantity met after the definition of differential. In almost all cases it will be assumed that $f_{xy} = f_{yx}$. It is also true for higher mixed derivatives, for example $f_{xxy} = f_{xyx} = f_{yxx}$ provided the derivatives are continuous. We will not make a song and dance about such properties, but before leaving these finer points of analysis, take a look at the function

$$f(x, y) = \frac{xy}{x^2 + y^2}$$

and try to find its value at the origin $(0, 0)$. The answer is that it has not got one. Ask a serial old fashioned piece of software, and it will put $x = 0$ first, deduce that the numerator is zero and as y is unassigned the denominator is 'most likely' not zero, so the answer is zero; and this is incorrect. In one dimension, letting $x \to 0$ can only be done from the left $(0-)$ or the right $(0+)$, whereas in two dimensions the origin can be approached along an infinite variety of curves, so this is why insisting that limits exist is very demanding in more than one dimension. In this example, approaching the origin along the straight line with gradient m so that $y = mx$ gives

$$f(x, y) = f(x, mx) = \frac{mx^2}{x^2 + m^2 x^2} = \frac{m}{1 + m^2}.$$

Cancelling x^2 being perfectly valid as we are not (yet) at $x = 0$. Therefore, it seems that at any point on this line f has the value

$$f = \frac{m}{1 + m^2},$$

which is completely independent of both x and y; this is not an issue except extremely close to the origin where all the lines $y = mx$ meet and where f can assume all kinds of different values. At the origin, therefore, f does not have a defined single value. It is possible to assign the value of zero to $f(0, 0)$ but this is putting a sticking plaster over an open wound and does not really help. Those of you who have studied analysis would have met *indeterminate forms* and '0/0' is one of these. Therefore $x = 0, y = 0$ is a singular point of this function and no amount of fiddling around can give a value there. However, being an indeterminate form is no guarantee that there is a singularity. For example, consider the function

$$f(x, y) = \begin{cases} \dfrac{x^3 y^3}{x^2 + y^2} & (x, y) \neq (0, 0), \\ 0 & (x, y) = (0, 0). \end{cases}$$

This behaves perfectly at the origin despite it being of the indeterminate form 0/0. In fact it has the value of 0 at the origin and is both continuous and twice differentiable there. Each case has to be examined from first principles, so let us leave matters there. This last example is completely free of such problems.

Example 2.10 If for a function $f(x, y)$

$$x\frac{\partial f}{\partial x} + y\frac{\partial f}{\partial y} = 0$$

show that

$$x^2\frac{\partial^2 f}{\partial x^2} + 2xy\frac{\partial^2 f}{\partial x \partial y} + y^2\frac{\partial^2 f}{\partial y^2} = 0.$$

Solution: Starting with the given equation, operate as follows:

$$\left(x\frac{\partial}{\partial x} + y\frac{\partial}{\partial y}\right)\left(x\frac{\partial f}{\partial x} + y\frac{\partial f}{\partial y}\right) = 0.$$

Expanding gives:

$$x\left(\frac{\partial}{\partial x}\left(x\frac{\partial f}{\partial x}\right)\right) + 2xy\frac{\partial^2 f}{\partial x \partial y} + y\left(\frac{\partial}{\partial y}\left(y\frac{\partial f}{\partial y}\right)\right) = 0$$

and using the product rule:

$$x^2\frac{\partial^2 f}{\partial x^2} + 2xy\frac{\partial^2 f}{\partial x \partial y} + y^2\frac{\partial^2 f}{\partial y^2} + x\frac{\partial u}{\partial x} + y\frac{\partial f}{\partial y} = 0,$$

where

$$x^2\frac{\partial^2 f}{\partial x^2} + 2xy\frac{\partial^2 f}{\partial x \partial y} + y^2\frac{\partial^2 f}{\partial y^2} = 0.$$

□

The equation

$$\frac{\partial^2 \phi}{\partial x^2} + \frac{\partial^2 \phi}{\partial y^2} = \phi_{xx} + \phi_{yy} = 0$$

is written

$$\nabla^2 \phi = 0$$

and is called Laplace's equation. The same equation but in three dimensions is

$$\frac{\partial^2 \phi}{\partial x^2} + \frac{\partial^2 \phi}{\partial y^2} + \frac{\partial^2 \phi}{\partial z^2} = \phi_{xx} + \phi_{yy} + \phi_{zz} = 0$$

and is also written identically as

$$\nabla^2 \phi = 0$$

the context telling whether two or three dimensions are intended. It has wide application both in two and in three dimensions throughout all physical sciences.

Example 2.11 Transform the two-dimensional version of $\nabla^2 \phi = 0$ from cartesian (x, y) to plane polar (r, θ) co-ordinates.

Solution: We have derived that

$$f_x = \cos\theta f_r - \frac{\sin\theta}{r} f_\theta$$

$$f_y = \sin\theta f_r + \frac{\cos\theta}{r} f_\theta,$$

so

$$f_{xx} = \left\{\cos\theta\frac{\partial}{\partial r} - \frac{\sin\theta}{r}\frac{\partial}{\partial\theta}\right\}\left\{\cos\theta f_r - \frac{\sin\theta}{r} f_\theta\right\}$$

$$f_{yy} = \left\{\sin\theta\frac{\partial}{\partial r} + \frac{\cos\theta}{r}\frac{\partial}{\partial\theta}\right\}\left\{\sin\theta f_r + \frac{\cos\theta}{r} f_\theta\right\},$$

or

$$f_{xx} = \cos\theta\frac{\partial}{\partial r}\left\{\cos\theta f_r - \frac{\sin\theta}{r} f_\theta\right\} - \frac{\sin\theta}{r}\frac{\partial}{\partial\theta}\left\{\cos\theta f_r - \frac{\sin\theta}{r} f_\theta\right\}$$

$$f_{yy} = \sin\theta\frac{\partial}{\partial r}\left\{\sin\theta f_r + \frac{\cos\theta}{r} f_\theta\right\} + \frac{\cos\theta}{r}\frac{\partial}{\partial\theta}\left\{\sin\theta f_r + \frac{\cos\theta}{r} f_\theta\right\}.$$

The expressions on the right are now carefully expanded using the product rule wherever appropriate. First, using the suffix derivative notation throughout, f_{xx} is given by

$$f_{xx} = \cos^2\theta f_{rr} - 2\frac{\cos\theta\sin\theta}{r}f_{r\theta} + \frac{\sin\theta\cos\theta}{r^2}f_\theta + \frac{\sin^2\theta}{r}f_r + \frac{\sin^2\theta}{r^2}f_{\theta\theta}$$

and f_{yy} is given by

$$f_{yy} = \sin^2\theta f_{rr} + 2\frac{\cos\theta\sin\theta}{r}f_{r\theta} - \frac{\sin\theta\cos\theta}{r^2}f_\theta + \frac{\cos^2\theta}{r}f_r + \frac{\cos^2\theta}{r^2}f_{\theta\theta}$$

so adding them gives

$$f_{xx} + f_{yy} = f_{rr} + \frac{1}{r^2}f_{\theta\theta} + \frac{1}{r}f_r.$$

Thus,

$$\nabla^2\phi = \phi_{rr} + \frac{1}{r}\phi_r + \frac{1}{r^2}\phi_{\theta\theta} = \frac{1}{r}\frac{\partial}{\partial r}\left\{r\frac{\partial\phi}{\partial r}\right\} + \frac{1}{r^2}\frac{\partial^2\phi}{\partial\theta^2}.$$

$\square$

2.6.1   Higher Differentials

Starting with the definition of the differential found in Equation (2.1), simply substitute the differential df for f to get

$$d^2f = \frac{\partial}{\partial x}(df)dx + \frac{\partial}{\partial y}(df)dy + \frac{\partial}{\partial z}(df)dz,$$

and we are left with sorting out the symbols on the right-hand side. We do this by substituting the definition of the differential df to obtain

$$d^2f = \frac{\partial}{\partial x}\left(\frac{\partial f}{\partial x}dx + \frac{\partial f}{\partial y}dy + \frac{\partial f}{\partial z}dz\right)dx$$

$$+ \frac{\partial}{\partial y} \left(\frac{\partial f}{\partial x} dx + \frac{\partial f}{\partial y} dy + \frac{\partial f}{\partial z} dz \right) dy$$

$$+ \frac{\partial}{\partial z} \left(\frac{\partial f}{\partial x} dx + \frac{\partial f}{\partial y} dy + \frac{\partial f}{\partial z} dz \right) dz.$$

Gathering like terms then simplifies this to

$$d^2 f = \frac{\partial^2 f}{\partial x^2} d^2 x + \frac{\partial^2 f}{\partial y^2} d^2 y + \frac{\partial^2 f}{\partial z^2} d^2 z + 2 \frac{\partial^2 f}{\partial x \partial y} dx \, dy + 2 \frac{\partial^2 f}{\partial y \partial z} dy dz$$

$$+ 2 \frac{\partial^2 f}{\partial z \partial x} dz dx.$$

This extends to higher dimensions, but the practical applications like those of single differentials to errors are lacking. They do have a part to play in the areas of mathematics inhabited by tensor products and Lie algebras, but these are securely for those who wish to study more advanced mathematics than is to be found in this introductory textbook.

2.7   Taylor's Theorem

In Chapter 1 Taylor's theorem was introduced in the form

$$f(a + h) = f(a) + hf'(a) + \frac{h^2}{2!} f''(a) + \cdots + \frac{h^n}{n!} f^{(n)}(a) + \cdots$$

as was the special case $a = 0$ called MacLaurin's theorem

$$f(h) = f(0) + hf'(0) + \frac{h^2}{2!} f''(0) + \cdots + \frac{h^n}{n!} f^{(n)}(0) + \cdots$$

As this is not a text on analysis, the remainder term for each of these series is not considered here, however a two-dimensional version is required for later chapters. MacLaurin's theorem is useful for this purpose. Let us start by restating the chain rule in the form

$$\frac{dF}{dt} = \frac{\partial f}{\partial u} \frac{du}{dt} + \frac{\partial f}{\partial v} \frac{dv}{dt},$$

where $F(t) = f(x + ht, y + kt)$. In this particular instance, let $u = x + ht$ and $v = y + kt$. Taylor's theorem concerns expressing the value of a function at one point in terms of its value and that of its derivatives at an adjacent point, so this linear form is appropriate and represents a first approximation of u and v as power series in t. This being so, we thus have

$$\frac{du}{dt} = h \quad \text{and} \quad \frac{dv}{dt} = k$$

so

$$\frac{dF}{dt} = h\frac{\partial f}{\partial u} + k\frac{\partial f}{\partial v}$$

or in terms of operators

$$\frac{d}{dt} \equiv h\frac{\partial}{\partial u} + k\frac{\partial}{\partial v}.$$

Reapplying this gives

$$\frac{d^2}{dt^2} \equiv \left(h\frac{\partial}{\partial u} + k\frac{\partial}{\partial v}\right)^2$$

and repeated reapplication n times gives

$$\frac{d^n}{dt^n} \equiv \left(h\frac{\partial}{\partial u} + k\frac{\partial}{\partial v}\right)^n.$$

Now write MacLaurin's theorem for $F(t)$

$$F(t) = F(0) + tF'(0) + \frac{t^2}{2!}F''(0) + \cdots + \frac{t^n}{n!}F^{(n)}(0) + \cdots \tag{2.11}$$

and substitute the values $u = x + ht, v = y + kt$ to the derivatives on the right. With $t = 0, u = x$ and $v = y$ so

$$F(0) = f(x,y); \quad F'(0) = h\frac{\partial f}{\partial x} + k\frac{\partial f}{\partial y}; \quad F''(0) = \left(h\frac{\partial f}{\partial x} + k\frac{\partial f}{\partial y}\right)^2; \quad \text{etc.}$$

Therefore, in terms of f Equation (2.11) becomes

$$f(x + ht, y + kt) = f(x,y) + t\left(h\frac{\partial}{\partial x} + k\frac{\partial}{\partial y}\right)f(x,y)$$

$$+ \frac{t^2}{2!}\left(h\frac{\partial}{\partial x} + k\frac{\partial}{\partial y}\right)^2 f(x,y) + \cdots + \frac{t^n}{n!}\left(h\frac{\partial}{\partial x} + k\frac{\partial}{\partial y}\right)^n f(x,y) + \cdots$$

This is true for any value of t, but if we put $t = 1$ Taylor's theorem in two dimensions results

$$f(x + h, y + k) = f(x,y) + \left(h\frac{\partial}{\partial x} + k\frac{\partial}{\partial y}\right)f(x,y)$$

$$+ \frac{1}{2!}\left(h\frac{\partial}{\partial x} + k\frac{\partial}{\partial y}\right)^2 f(x,y) + \cdots + \frac{1}{n!}\left(h\frac{\partial}{\partial x} + k\frac{\partial}{\partial y}\right)^n f(x,y) + \cdots$$

where the values of h and k are small and so the point $(x + h, y + k)$ is close to the point (x, y). This being so, only a few terms of the expansion in the right will give a good approximation to $f(x + h, y + k)$. The generalisation to dimensions higher than two is straightforward, but this is postponed until later. In one dimension, the Taylor's series for a particular function could be used to approximate the function, called the Taylor polynomial around a given value. The same is true in two dimensions, except that the calculation is rather cumbersome. However, this does need to be done, but not in this chapter. It belongs later, in the chapter that contains numerical optimisation. Therefore, this is the latest facet of this chapter and its role in preparing for what is to come; directional derivatives, Jacobians and now Taylor polynomials in two dimensions all feature in future chapters and all rely on material first covered here.

2.8 Conjugate Functions

Definition 2.3   If two functions of two variables $\phi(x, y)$ and $\psi(x, y)$ obey the two equations

$$\frac{\partial \phi}{\partial x} = \frac{\partial \psi}{\partial y} \quad \text{and} \quad \frac{\partial \phi}{\partial y} = -\frac{\partial \psi}{\partial x} \tag{2.12}$$

then they are termed *conjugate functions.*

Equations (2.12) are called the *Cauchy–Riemann equations*. The curves $\phi(x, y) = $ const. and $\psi(x, y) = $ const. intersect at right angles provided the functions are conjugate, but the proof of this also has to wait until the gradient functions has been introduced properly (but see Exercise 2.10). One result will be proved here in the form of an example, but others will wait until after vector calculus has been introduced.

Example 2.12   Show that conjugate functions must obey Laplace's equation.

Solution: Take the first of the Cauchy–Riemann equations and differentiate it with respect to x to obtain

$$\frac{\partial^2 \phi}{\partial x^2} = \frac{\partial^2 \psi}{\partial x \partial y}$$

and the second Cauchy–Riemann equation differentiated with respect to y gives

$$\frac{\partial^2 \phi}{\partial y^2} = -\frac{\partial^2 \psi}{\partial y \partial x}.$$

Adding these gives

$$\frac{\partial^2 \phi}{\partial x^2} + \frac{\partial^2 \phi}{\partial y^2} = 0,$$

which is Laplace's equation $\nabla^2 \phi = 0$. Performing the similar operations: differentiating the first with respect to y then subtracting the second differentiated with respect to x gives

$$\frac{\partial^2 \psi}{\partial x^2} + \frac{\partial^2 \psi}{\partial y^2} = 0,$$

both dependent upon the conjugate functions possessing continuous second order partial derivatives. □

In two dimensions functions that obey Laplace's equation are called *Harmonic*, so we have proved the theorem.

Theorem 2.3 All conjugate functions are harmonic.

For those who are familiar with complex analysis where $z = x + iy, i = \sqrt{-1}$ conjugate harmonic functions are generated by

$$f(z) = \phi(x, y) + i\psi(x, y).$$

Here is a simple example.

Example 2.13 Put $f(z) = z^3$ determine the functions $\phi(x, y)$ and $\psi(x, y)$ and show that they are harmonic.

Solution:

$$z^3 = (x + iy)^3 = x^3 + 3ix^2y - 3xy^2 - iy^3$$

so

$$\phi(x, y) = x^3 - 3xy^2 \quad \text{and} \quad \psi(x, y) = 3x^2y - y^3.$$

Direct differentiation yields

$$\phi_x = 3x^2 - 3y^2, \phi_{xx} = 6x, \phi_y = -6xy, \phi_{yy} = -6x$$

and so

$$\phi_{xx} + \phi_{yy} = \nabla^2 \phi = 0,$$

also

$$\psi_x = 6xy, \psi_{xx} = 6y, \phi_y = -3x^2 - 3y^2, \psi_{yy} = -6y$$

and so

$$\psi_{xx} + \psi_{yy} = \nabla^2 \psi = 0.$$

□

This very short foray into complex variables stops now as no prerequisites in this have been assumed. The next chapter is devoted to the application of partial differentiation to finding extreme values.

2.9   Case Study: Thermodynamics

This section can be treated as optional, but if physics or mechanical engineering is your main interest, then thermodynamics is a very good vehicle to demonstrate the use and power of partial differentiation. Of course, the main purpose here is to showcase partial differentiation and not to present a textbook on the principles of thermodynamics, so if you know no thermodynamics at all some of what follows may be difficult. That is why this section is optional. Do not give up immediately though, even though my own description of thermodynamics, stated in my early 20s fresh from gaining a mathematics degree when I first encountered it in an MSc Meteorology course was 'the science of defining as many heat related variables as possible and finding every possible connection between them'. The person teaching was the estimable Professor R C Sutcliffe FRS (1904–1991) who smiled amiably and more or less agreed with me.

Heat is a form of energy. Those of us familiar with energy in the mechanical sense, kinetic or potential energy, may wonder that it has the same name. It is the same stuff with the same units of measurements, and converting mechanical (kinetic or potential) energy into heat is common. Most machines do it, as they run they heat up. Conversion of mechanical (potential) energy into heat happens as a body (a meteor, for example) falls through the atmosphere. Now it does not happen the other way round very easily, but that is a different story. If two bodies at different temperatures are placed next to each other, then eventually they both end up at the same temperature somewhere between the two starting temperatures. Energy has transferred from the hotter body to the colder until they become the same. This is a consequence of the *second law of thermodynamics* usually stated that heat cannot travel from a body to one that is warmer. If this is the second law, one might wonder what the first law is; it is our old friend the energy equation but stated generally. Energy cannot be created or destroyed, it can only be converted from one form to another. This can be put in terms of infinitesimal quantities, for example

$$\Delta E = \Delta E_k + \Delta E_p + \Delta U,$$

where E is the total energy of a system, E_k is the kinetic energy, E_p the potential energy and U the internal heat energy. The prefix Δ denoting the change in these quantities. Sometimes this is written (wrongly, in the strict mathematical sense)

$$dE = dE_k + dE_p + dU.$$

U is actually the energy of all the molecules bouncing around, and in a gas is captured through the kinetic theory of gases. It is at this stage we commit ourselves to concentrate on thermodynamics and assume that the kinetic and potential energy of the system does not change, so the change in internal energy ΔU must be $Q - W$ where Q is the external heat added to the system and W the work done by the system. If $\Delta U = -W$ then the system is termed *adiabatic*. For an adiabatic system, the internal energy moves from one equilibrium state to another without any external heat, and logically this must be independent of the path, just dependent on the end states. It is at this stage the reader should be warned that many books on thermodynamics play fast and loose with notation, using 'd' or 'Δ' seemingly at a whim. The *heat capacity* is defined by

$$c = \frac{dQ}{dT},$$

where T is the temperature. Therefore, it is the rate of change of the total heat with temperature. This is not a measurable quantity as there is too much else going on. Instead the following two quantities are defined

$$c_p = \left(\frac{dQ}{dT}\right)_p \quad \text{and} \quad c_V = \left(\frac{dQ}{dT}\right)_V.$$

The first, c_p is the heat capacity, often called the specific heat, at constant pressure, the second is c_V the heat capacity at constant temperature and they are well defined in terms of physics. In mathematical terms the suffix says what is being held constant, so we could write

$$c_p = \left(\frac{\partial Q}{\partial T}\right)_p \quad \text{and} \quad c_V = \left(\frac{\partial Q}{\partial T}\right)_V,$$

where the suffix tells us what is being held constant. This *has* to be done in thermodynamics and is not an optional extra as in most mathematics. We are now in the thermodynamics of gases. In an ideal gas

$$dU = c_V dT$$

because both U and c_V are functions of temperature only, there are no partial derivatives. The *enthalpy H* is the total heat in a system and is defined by

$$H = U + pV.$$

Here is a simple example that uses this concept.

Example 2.14 Show that for an ideal gas $c_p = c_V + R$ where R is the gas constant.

Solution: By definition $H = U + pV$ and all variables depend upon temperature T, so differentiating with respect to T gives

$$\frac{dH}{dT} = \frac{dU}{dT} + \frac{d}{dT}(pV).$$

Now

$$\frac{dH}{dT} = c_p \quad \text{if pressure is kept constant, and}$$

$$\frac{dU}{dT} = c_V \quad \text{if the volume is kept constant, also}$$

$$\frac{d}{dT}(pV) = \frac{d}{dT}(RT) = R.$$

Hence for an ideal gas

$$c_p = c_V + R$$

so c_p is also only a function of the temperature. $\qquad\qquad\square$

Let us re-examine this in a little more generality. In terms of enthalpy, the first law of thermodynamics is:

$$dH = \delta q + Vdp.$$

The specific heat at constant pressure c_p and the specific heat at constant volume c_V have the definitions:

$$c_p = \frac{\delta q_p}{dT} \quad \text{and} \quad c_V = \frac{\delta q_V}{dT},$$

respectively. Most will be concerned about the notation that looks rather offensive to mathematicians with the use of δq but this is because this is neither the differential dq nor the usual incremental change in q. It represents that in the transition between two states, q can change by gaining or losing bits of both heat and mechanical energy no matter whether p, V or neither is being held constant. This is not true for V or U that are state variables and so change independently and so can be treated as standard mathematical variables. This kind of thing happens when the precise language of mathematics is used in a science, even if the science is physics. Fortunately in this text where the physics takes a back seat this notation problem quickly goes away.

In the context of the chain rule, the following can be written

$$dU = \left(\frac{\partial U}{\partial V}\right)_T dV + \left(\frac{\partial U}{\partial T}\right)_V dT,$$

where the suffix denotes, for emphasis, what has been held constant. Therefore, the first law of thermodynamics, expressed as

$$dU = \delta q - pdV$$

implies

$$\delta q = \left[\left(\frac{\partial U}{\partial V}\right)_T + p\right] dV + \left(\frac{\partial U}{\partial T}\right)_V dT.$$

For a process in a constant volume (isochoric, $dV = 0$) this implies

$$\left(\frac{\partial U}{\partial T}\right)_V = \frac{\delta q}{dT} = c_V.$$

For a process in a constant pressure (isobaric, $dp = 0$) we have

$$\left[\left(\frac{\partial H}{\partial T}\right)_T dV - V\right] + \left(\frac{\partial H}{\partial T}\right)_p = \delta q,$$

where

$$\left(\frac{\partial H}{\partial T}\right)_p = \frac{\delta q}{dT} = c_p.$$

Both c_p and c_V are state variables, but in some circumstances for example in the atmosphere the range of the other variables is such that both c_p and c_V are constant.

Example 2.15 There are two cells joined by a diaphragm as shown in Figure 2.2 in both cells the pressure and volume are constant with values p_1, V_1 and p_2, V_2. The diaphragm is then ruptured. This is a classic experiment first undertaken by James Prescott Joule (1818–1883) in 1845. Analyse the relationships between c_p and c_V.

Solution: In this situation, in each chamber $U = U(T)$ and $H = H(T)$. The convention is that both of these variables are zero at absolute zero, so it must be the case that $U = c_V T$ and $H = c_p T$. For the case of an ideal gas statistical mechanics applied to a monatomic gas (the Sackur–Tetrode equation) gives $c_V = (3/2)R$ and as $c_p - c_V = R$ this implies $c_p = (5/2)R$ it also can be found by considering the degrees of freedom of a molecule. The ratio of the specific heats is γ and

$$\gamma = \frac{c_p}{c_V} = \frac{5}{3}.$$

For a diatomic gas molecule, the figures are $c_V = (5/2)R$ and $c_p = (7/2)R$ so $\gamma = 1.4$ this is virtually the value for dry air. In general,

$$\gamma = 1 + \frac{2}{f},$$

where f is the number of degrees of freedom. In air with moisture it is more complex (see Adrian Gill's book *Atmosphere-Ocean Dynamics*, 1982, p. 43). □

Figure 2.2 Two ideal gases in states 1 and 2 with a diaphragm at D.

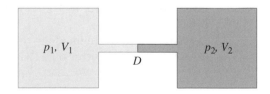

For an adiabatic process things happen quickly so heat remains constant. The first law of thermodynamics yields

$$c_V dT + p dV = 0 \quad \text{and also} \quad p = \frac{RT}{V}$$

so

$$c_V dT + \frac{RT}{V} dV = 0$$

and separating the variables and integrating, using $pV = RT$ yields

$$pv^\gamma = \text{constant.}$$

Exercises

2.1 Given $f(x, y, z) = x^4 y^2 \cos(z)$ find the three first-order partial derivatives.

2.2 A terrain is given by the equation $z = h(x, y) = x^3 + y^3 - 3xy^2$ inside the box $0 \leq x, y \leq 2$. Find the slope of the surface in the x and y directions at the point $(1, 1)$.

2.3 Determine the slope of the surface $z = x^4 - y^2$ at the point $(1, 1)$ in the general direction that makes an angle θ with the positive x-axis. Find a value of θ with $0 \leq \theta \leq \pi/2$ where this slope is zero. [This is a *directional derivative* see Chapter 7 for a general vectorial treatment.]

2.4 Determine the absolute error in the volume of a box with an open top that has dimensions $a = 5$ cm, $b = 6$ cm with height $c = 3$ cm. The errors in these lengths are $\Delta a = \pm 0.2$ cm, $\Delta b = \pm 0.3$ cm and $\Delta c = \pm 0.1$ cm Find also the maximum relative error.

2.5 A rod of length l has a cross section of diameter d. The rod is subjected to a uniform load w. The deflection y at the centre of the rod is given by the formula

$$y = k \frac{w l^3}{d^4}, \quad \text{where } k \text{ is a constant.}$$

a) Using partial differentiation, determine the approximate % error in y given that w is in error by $+2\%$, l is in error by -1% and d is in error by $+3\%$.

b) Determine the maximum approximate % error in y given that w is in error by $\pm 2\%$, l is in error by $\pm 1\%$ and d is in error by $\pm 3\%$.

2.6 In a triangle ABC the angle A is accurately known, but the measurement of the side b is in error by an amount Δb and that of side c by an amount Δc. Find the error in the calculation of side a in terms of $b, c, \Delta b$ and Δc.

2.7 Find $\frac{dy}{dx}$ if $y^x = x^y$.

2.8 Use Leibniz rule to evaluate the following integrals:

a) Given $\displaystyle\int_0^\infty \frac{dx}{1+x^4} = \frac{\pi}{2\sqrt{2}}$ find both

$$\int_0^\infty \frac{dx}{(1+x^4)^2} \quad \text{and} \quad \int_0^\infty \frac{dx}{(1+x^4)^3}.$$

b) Given $\displaystyle\int_{-\pi}^\pi \frac{dx}{a+\sin x} = \frac{2\pi}{\sqrt{a^2-1}}$ find

$$\int_{-\pi}^\pi \frac{dx}{(2+\sin x)^2} \quad \text{and} \quad \int_{-\pi}^\pi \frac{dx}{(3+\sin x)^3}.$$

c) Given $\displaystyle\int_0^1 x^p\,dx = \frac{1}{p+1}$ find

$$\int_0^1 x^p(\ln x)^m\,dx, \quad \text{where } p > -1,$$

where m is a natural number.

2.9 Find the Taylor expansions to second order for the functions below about the points given:
a) $\sin(x+y)$, $(0,0)$;
b) $x\ln y$, $(0,1)$; $\frac{x}{x+y}$, (h,k),
where (h,k) is a general point in the x, y plane.

2.10 By considering $d\phi$ and $d\psi$ show that the curves $\phi(x,y) = a_n$ and $\psi(x,y) = b_n$ where a_n and b_n are constants, $n = 0, \pm1, \pm2, \dots$ intersect at right angles provided ϕ and ψ are conjugate functions.

2.11 Verify that the Cauchy–Riemann equations hold for the following pairs of functions:
a) $\dfrac{1+x}{(1+x)^2+y^2}, \dfrac{-y}{(1+x)^2+y^2}$,
b) $\dfrac{1+x^2+y^2}{(1+x^2-y^2)^2+4x^2y^2}, \dfrac{2xy}{(1+x^2-y^2)^2+4x^2y^2}$.

2.12   Perform the integration of the equation

$$c_V dT + \frac{RT}{V} dV = 0$$

to obtain pv^γ = constant.

2.13   The ideal gas law for m moles is $pV = mRT$ where the other four symbols have their usual meaning (pressure, volume, gas constant and temperature, respectively). Prove that

$$\frac{\partial T}{\partial p} \frac{\partial p}{\partial V} \frac{\partial V}{\partial T} = -1.$$

3

Maxima and Minima

3.1 Introduction

In the last chapter, Taylor's theorem in two variables was proved. In this chapter, we look at two-dimensional surfaces that have extreme values. In fact we shall find points where the local gradients are zero then determine whether these points are extreme. The general name for these are *turning points* as it is quite possible for gradients to be zero yet the points be neither maximum nor minimum. However the name *extreme* has become generic and synonymous with *turning point*. Taylor's theorem in two dimensions is used to derive criteria for classifying whether an extreme point is a minimum, maximum or a saddle point. Consider a function of two variables written $z = z(x, y)$ where in this chapter x, y and z are the standard three-dimensional Cartesian co-ordinates. Such a function therefore represents a surface in space. For each pair of x, y co-ordinates, $(x, y), z(x, y)$ defines a value, and as x and y vary this traces out a surface. It is assumed that the function $z(x, y)$ is well defined so that only one value of z corresponds to a given pair of co-ordinates (x, y). Therefore, for example, a whole sphere could not be represented, but the top hemisphere could. In this chapter, this is not a problem; but it is a bit of a nuisance when multiple integrals are considered later and its resolution can wait until then.

Definition 3.1   A function $z(x, y)$ is said to have a *relative minimum* at the point (x_0, y_0) if there is a disc centred at the point (x_0, y_0) such that $z(x_0, y_0) < z(x, y)$ for all points that lie inside this disc.

Definition 3.2   A function $z(x, y)$ is said to have a *relative maximum* at the point (x_0, y_0) if there is a disc centred at the point (x_0, y_0) such that $z(x_0, y_0) > z(x, y)$ for all points that lie inside this disc.

At a minimum or a maximum the tangent plane is horizontal. Visualise this plane as a small flat square. At a minimum, the surface can (locally) tangentially

Two and Three Dimensional Calculus: With Applications in Science and Engineering, First Edition. Phil Dyke.
© 2018 John Wiley & Sons Ltd. Published 2018 by John Wiley & Sons Ltd.

rest on this square and at a maximum the square can (locally) tangentially rest on the surface; in each case, the square will be level and all lines drawn in the square, in particular, lines parallel to the x and y axes, are horizontal with zero gradient. Therefore, at a minimum or a maximum, it must be the case that

$$\frac{\partial z}{\partial x} = z_x = 0 \quad \text{and} \quad \frac{\partial z}{\partial y} = z_y = 0.$$

These two conditions are enough to ensure that the whole plane of the square is horizontal. Of course, it is possible these days to draw these three-dimensional figures (see Figures 3.1 and 3.2) to inordinate detail, swivel and rotate them, magnify and otherwise interrogate them visually to determine whether they have maxima or minima, but an analytical method will always be preferable. There is also an extremum called the *saddle point* which has yet to be defined, so let's put that right.

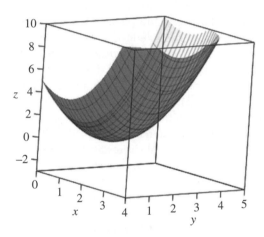

Figure 3.1 The function $z = (x - 1)^2 + (y - 2)^2$ pictured with some transparency has a relative minimum at the point $(1, 2)$.

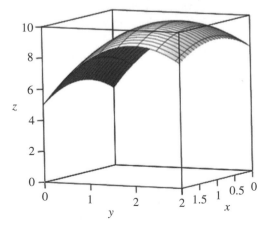

Figure 3.2 The function $z = 5 - x^2 + 2x - y^2 + 4y$ pictured has a relative maximum at the point $(1, 2)$.

3.2   Maxima, Minima and Saddle Points

Here is our definition:

Definition 3.3   If at a point (x_0, y_0) a function $z = z(x, y)$ is such that the two partial derivatives z_x and z_y are both zero, but this point is neither a maximum nor a minimum, then it is termed a *saddle point*.

This definition is an inclusive one and covers points that look like the traditional horse's saddle, but also includes points that resemble the centre of the seat of an upright chair, the centre of a jelly mould or even non-isolated ridges, for example $z = x^3$ along the line $x = 0$ with y arbitrary. The starting point for doing anything analytical and general is Taylor's theorem, this time in two variables. This is taken from the last chapter:

$$z(x + h, y + k) = z(x, y) + \left(h\frac{\partial}{\partial x} + k\frac{\partial}{\partial y} \right) z(x, y)$$

$$+ \frac{1}{2!} \left(h\frac{\partial}{\partial x} + k\frac{\partial}{\partial y} \right)^2 z(x, y) + \cdots + \frac{1}{n!} \left(h\frac{\partial}{\partial x} + k\frac{\partial}{\partial y} \right)^n z(x, y) + \cdots .$$

First of all let us put $(x, y) = (x_0, y_0)$ where (x_0, y_0) is a point where the first partial derivatives z_x and z_y are zero; second, let us only retain the second-order terms. Taylor's theorem applied to a maximum, minimum or saddle point (x_0, y_0) then gives

$$z(x_0 + h, y_0 + k) \approx z(x_0, y_0) + \frac{1}{2!} \left(h\frac{\partial}{\partial x} + k\frac{\partial}{\partial y} \right)^2 z(x, y)|_{x=x_0, y=y_0}.$$

Using the suffix derivative notation, this is

$$z(x_0 + h, y_0 + k) - z(x_0, y_0) \approx \frac{1}{2}[h^2 z_{xx} + 2hk z_{xy} + k^2 z_{yy}],$$

where the derivatives on the right are taken as being evaluated at the point (x_0, y_0). If we can deduce that the right-hand side is negative for all values of h and k, then it must be true that $z(x_0 + h, y_0 + k) - z(x_0, y_0) < 0$ or $z(x_0 + h, y_0 + k) < z(x_0, y_0)$ so (x_0, y_0) has to be a maximum. Similarly, if we can deduce that the right-hand side is always positive, then $z(x_0 + h, y_0 + k) > z(x_0, y_0)$ for all values of h and k and the point (x_0, y_0) has to be a minimum. As it stands, neither of these is true, but some manipulation helps. First of all, assume that $z_{xx} \neq 0$ so that we can divide by it. Then,

$$h^2 z_{xx} + 2hk z_{xy} + k^2 z_{yy} = z_{xx} \left[h^2 + 2hk\frac{z_{xy}}{z_{xx}} + k^2\frac{z_{yy}}{z_{xx}} \right]$$

so completing the square on the right gives

$$h^2 z_{xx} + 2hk z_{xy} + k^2 z_{yy} = z_{xx} \left[\left(h + k\frac{z_{xy}}{z_{xx}} \right)^2 - k^2 \left(\frac{z_{xy}}{z_{xx}} \right)^2 + k^2\frac{z_{yy}}{z_{xx}} \right],$$

which, when rearranged is

$$h^2 z_{xx} + 2hk z_{xy} + k^2 z_{yy} = z_{xx} \left[\left(h + k \frac{z_{xy}}{z_{xx}} \right)^2 + \left(\frac{k^2}{z_{xx}^2} \right) [z_{xx} z_{yy} - z_{xy}^2] \right]$$

so

$$z(x_0 + h, y_0 + k) - z(x_0, y_0)$$

$$\approx z_{xx} \left[\left(h + k \frac{z_{xy}}{z_{xx}} \right)^2 + \left(\frac{k^2}{z_{xx}^2} \right) [z_{xx} z_{yy} - z_{xy}^2] \right].$$

Apologies to those who are impatient through all this algebra, but at last we are now at a point where the following statement can be made. If $z_{xx} z_{yy} - z_{xy}^2 > 0$ then the right-hand side has the same sign as z_{xx} and so if $z_{xx} > 0$ the point (x_0, y_0) must be a minimum and if $z_{xx} < 0$ the point (x_0, y_0) must be a maximum. If $z_{xx} z_{yy} - z_{xy}^2 < 0$, then we have a saddle point. Some may wonder why the sign of the right-hand side *must* change under this circumstance; we will address this in a moment. First, we need to see what happens if $z_{xx} = 0$, but $z_{yy} \neq 0$. Without re-doing all the algebraic manipulation, it is possible to go through all the steps starting with

$$z(x_0 + h, y_0 + k) - z(x_0, y_0) \approx \frac{1}{2} [h^2 z_{xx} + 2hk z_{xy} + k^2 z_{yy}]$$

but dividing by z_{yy} so that

$$h^2 z_{xx} + 2hk z_{xy} + k^2 z_{yy} = z_{yy} \left[k^2 + 2hk \frac{z_{xy}}{z_{yy}} + h^2 \frac{z_{xx}}{z_{yy}} \right]$$

$$= z_{yy} \left[\left(k + h \frac{z_{xy}}{z_{yy}} \right)^2 + \left(\frac{h^2}{z_{yy}^2} \right) [z_{xx} z_{yy} - z_{xy}^2] \right]$$

but with $z_{xx} = 0$ it must be the case that $z_{xx} z_{yy} - z_{xy}^2 < 0$ unless $z_{xy} = 0$ too; so (x_0, y_0) is a saddle unless both $z_{xx} = 0$ and $z_{xy} = 0$ in which case $z_{yy} > 0$ for a minimum and $z_{yy} < 0$ for a maximum. The remaining cases will now be tackled; if both $z_{xx} = 0$ and $z_{yy} = 0$, then the approximation

$$z(x_0 + h, y_0 + k) - z(x_0, y_0) \approx hk z_{xy}$$

holds and provided $z_{xy} \neq 0$ as h and k can be either sign we have a saddle point. Sometimes, it is worth taking a different approach, called tackling the problem from first principles. Let $h = r \cos \theta$ and $k = r \sin \theta$ and allow θ to vary from 0 to 2π while holding r at a constant value. As θ increases, the point $(x_0 + h, y_0 + k)$ describes a circle radius r. In fact, it is as if we are walking around the turning point (x_0, y_0) once in a tight circle. Using this approach often pays dividends for specific cases especially where all three second derivatives z_{xx}, z_{yy} and z_{xy}

are zero. An example will follow soon, but the special case $z_{xx}z_{yy} - z_{xy}^2 = 0$ is worthy of study. Using $h = r\cos\theta$ and $k = r\sin\theta$, we have

$$z(x_0 + h, y_0 + k) - z(x_0, y_0) \approx z_{xx}r\left(\cos\theta + \sin\theta\frac{z_{xy}}{z_{xx}}\right)^2.$$

At first glance, it looks like the squared term on the right-hand side is always positive, but for some value of θ, it will be the case that

$$\tan\theta = -\frac{z_{xx}}{z_{xy}}$$

and at that point the right-hand side vanishes and the approximation whereby third-order terms in the Taylor expansion were ignored breaks down. The letter D is commonly used to denote $z_{xx}z_{yy} - z_{xy}^2$, and it is called the *discriminant*. If this discriminant is negative, then no matter how small its absolute value, there will be, around the value of θ that renders $\tan\theta = -\frac{z_{xx}}{z_{xy}}$, points where the difference $z(x_0 + h, y_0 + k) - z(x_0, y_0)$ changes sign. Therefore, this is why if $D < 0$ we must have a saddle point independent of the value of either z_{xx} or z_{yy}. The following Table 3.1 thus summarises the criteria for classifying turning points of the function $z(x, y)$.

Let us now do a few examples.

Example 3.1   Show that the function

$$f(x, y) = \frac{1}{12}x^3 + xy^2 - 25x - 30y$$

has four extreme points that lie on the ellipse $\frac{1}{4}x^2 + y^2 = 25$. Find them and classify them.

Solution: The first derivatives are

$$f_x = \frac{1}{4}x^2 + y^2 - 25 \quad \text{and} \quad f_y = 2xy - 30.$$

Putting these equal to zero shows that extreme points lie on the ellipse $\frac{1}{4}x^2 + y^2 - 25 = 0$. Solving these by substituting $y = 15/x$ into the equation for the ellipse gives the quartic

Table 3.1   Criteria for turning points.

$z_{xx} > 0$ (or $z_{yy} > 0$)	$D > 0$	Relative minimum
$z_{xx} < 0$ (or $z_{yy} < 0$)	$D > 0$	Relative maximum
	$D < 0$	Saddle point
	$D = 0$	Inconclusive

$$x^4 - 100x^2 + 900 = 0 \quad \text{or} \quad (x^2 - 90)(x^2 - 10) = 0$$

whence $x = \pm\sqrt{10}$ or $x = \pm 3\sqrt{10}$. Solving for y gives the four points:

$$(\sqrt{10}, \frac{3}{2}\sqrt{10}); (-\sqrt{10}, -\frac{3}{2}\sqrt{10}); (3\sqrt{10}, \frac{1}{2}\sqrt{10});$$
$$(-3\sqrt{10}, -\frac{1}{2}\sqrt{10}).$$

The second derivatives are

$$f_{xx} = \frac{1}{2}x, f_{xy} = 2y \quad \text{and} \quad f_{yy} = 2x.$$

Now, consider each turning point in order.

$$(\sqrt{10}, \frac{3}{2}\sqrt{10}) : f_{xx} = \frac{1}{2}\sqrt{10}, f_{yy} = 2\sqrt{10}, f_{xy} = 3\sqrt{10},$$

so $D = f_{xx}f_{yy} - f_{xy}^2 < 0$ therefore it is a saddle point.

$$(-\sqrt{10}, -\frac{3}{2}\sqrt{10}) : f_{xx} = -\frac{1}{2}\sqrt{10}, f_{yy} = -2\sqrt{10}, f_{xy} = -3\sqrt{10},$$

so $D = f_{xx}f_{yy} - f_{xy}^2 < 0$ therefore it is also a saddle point.

$$(3\sqrt{10}, \frac{1}{2}\sqrt{10}) : f_{xx} = \frac{3}{2}\sqrt{10}(> 0), f_{yy} = 6\sqrt{10}, f_{xy} = \sqrt{10},$$

so $D = f_{xx}f_{yy} - f_{xy}^2 > 0$ therefore it is a minimum point.

$$(-3\sqrt{10}, -\frac{1}{2}\sqrt{10}) : f_{xx} = -\frac{3}{2}\sqrt{10}(< 0), f_{yy} = -6\sqrt{10}, f_{xy} = -\sqrt{10},$$

so $D = f_{xx}f_{yy} - f_{xy}^2 > 0$ therefore it is a maximum point. Although in the three-dimensional graph produced using MAPLE, it is very difficult to make out the turning points in Figure 3.3. □

A practical piece of advice: When solving the pair of simultaneous equations $z_x = 0$ and $z_y = 0$, it is worth checking your solutions as it is easy to get spurious ones, especially if square roots are involved. Here is an example of finding a turning point from first principles.

Example 3.2 Determine the nature of the turning points of the surface

$$z(x, y) = x^5 - \frac{1}{32}x^5 y^5 + y^5.$$

Solution: The first derivatives are equated to zero in the pair of equations

$$5x^4 - \frac{5}{32}x^4 y^5 = 0,$$
$$-\frac{5}{32}x^5 y^4 + 5y^4 = 0.$$

Figure 3.3 The surface $z = \frac{1}{12}x^3 + xy^2 - 25x - 30y$.

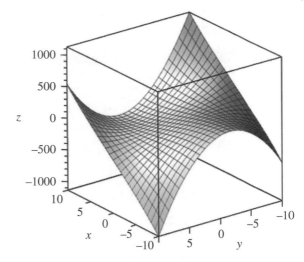

Note that the origin $(0,0)$ is certainly one solution to these equations, and that there is a symmetry in that if x and y are exchanged in $z(x, y)$ the equation for the surface remains the same, so if $x \neq 0$ in the first equation the solution is $y^5 = 32$. Substituting into the second gives $x^5 = 32$ too. We are thus led to the two solutions $(0,0)$ and $(2,2)$ as the only turning points. Back-substitution verifies these assertions. Differentiating again gives

$$z_{xx} = 20x^3 - \frac{20}{32}x^3y^5, \quad z_{yy} = 20y^3 - \frac{20}{32}x^5y^3, \quad \text{and}$$
$$z_{xy} = -\frac{25}{32}x^4y^4.$$

At the origin $(0,0)$, all of these second derivatives are zero, so the tests fail. At the point $(2,2)$, both $z_{xx} = 0$ and $z_{yy} = 0$ but $z_{xy} = -150$ so we conclude that $(2,2)$ is a saddle point. To find out what happens at the origin, investigate from first principles as follows: let $x = 0 + r\cos\theta$ and $y = 0 + r\sin\theta$ with $r \ll 1$ in the formula for $z(x, y)$ so

$$z(r\cos\theta, r\sin\theta) = r^5\cos^5\theta - \frac{1}{32}r^{10}\cos^5\theta\sin^5\theta + r^5\sin^5\theta$$

as r is small this is approximately $r^5(\cos^5\theta + \sin^5\theta)$. This tells us two things. First, as θ increases from 0 to 2π, $z(x, y)$ changes sign so the origin is a saddle. Second, as the lowest order in the small quantity r is the fifth power, all terms of the Taylor's series for $z(x, y)$ about $(0,0)$ up the fifth order are zero. Therefore, a Taylor's series approach to this problem would indeed be very cumbersome. Figure 3.4 is an attempt, not altogether unsuccessful, to display this surface using MAPLE. □

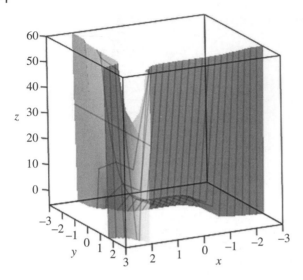

Figure 3.4 The surface $z = x^5 - \frac{1}{32}x^5y^5 + y^5$ shown with some transparency to help with visualising the two saddle points at $(0,0,0)$ and $(2,2,32)$.

Some of you might be wondering whether investigating from first principles might always be worth trying. The answer is usually not. If the criteria of Table 3.1 can be used, then it normally is the easiest way to proceed. The exception is if the second derivatives are particularly lengthy to calculate and expansion of $z(x_0 + r\cos\theta, y_0 + r\sin\theta) - z(x_0, y_0)$ in powers of r is easier. Of course, evaluating the two partial derivatives z_x and z_y still has to be done as solving the pair of equations $z_x = 0$ and $z_y = 0$ is necessary in order to find the turning points (x_0, y_0), which is essential for using either method.

The next example is done purely from first principles, just as an exercise.

Example 3.3   Use first principles to classify the turning points of the function $z = xye^{-xy}$.

Solution: The reason for using first principles for this particular problem will become apparent. It is not a perverse whim, but the best method. Differentiating with respect to x and y and equating both to zero gives the pair of simultaneous equations

$$z_x = (1 - xy)xe^{-xy} = 0,$$
$$z_y = (1 - xy)ye^{-xy} = 0$$

and these are satisfied by the values $x = 0, y = 0$ and by all points on the rectangular hyperbola $xy = 1$. Differentiating the first derivatives again is possible but with non-isolated points the criteria will always give either 'inconclusive' or 'saddle point' neither of which would be particularly informative. At the origin, let $x = r \cos \theta, y = r \sin \theta$ so with $z(0, 0) = 0$

$$z(r \cos \theta, r \sin \theta) - z(0, 0) = r^2 \cos \theta \sin \theta e^{-r^2 \cos \theta \sin \theta}$$

$$= \frac{1}{2} r^2 \sin 2\theta e^{-\frac{1}{2} r^2 \sin 2\theta},$$

and it is obvious that this is a classic saddle point as this difference changes sign as θ takes all values from 0 to 2π. On the curve $xy = 1$, put $x = t$ and $y = \frac{1}{t}$ then the rectangular hyperbola $xy = 1$ is traced as t runs from $-\infty$ to ∞. Negative values of t give one branch, positive values the other. So the technique is to take an arbitrary point on this curve $(t, \frac{1}{t})$ and surround it by a small circle of radius r:

$$x = t + r \cos \theta \quad \text{and} \quad y = \frac{1}{t} + r \sin \theta$$

then substitute into the difference

$$\Delta z = z\left(t + r \cos \theta, \frac{1}{t} + r \sin \theta\right) - z\left(t, \frac{1}{t}\right)$$

and expand in powers of r. The terms in r will cancel due to the conditions $z_x = 0, z_y = 0$ so we retain at least to r^2 and sometimes further. In this problem, retaining terms to r^2 will probably be sufficient as we are analysing non-isolated turning points. If it was the case that all the second-order derivatives were zero, then expanding to third order or even higher would have to be done. Substituting $(t, \frac{1}{t})$ into $z(x, y) = xye^{-xy}$ gives $z = 1/e$ which is constant. This is not a surprise; the rectangular hyperbola $xy = 1$ is either a ridge, trough or a ledge. In each case, a constant height, and that implies a constant z. On the small circle, the difference Δz is given by

$$\Delta z = (t + r \cos \theta)\left(\frac{1}{t} + r \sin \theta\right) e^{-(t + r \cos \theta)\left(\frac{1}{t} + r \sin \theta\right)} - \frac{1}{e}.$$

Now, the product

$$(t + r \cos \theta)\left(\frac{1}{t} + r \sin \theta\right) = 1 + r\left(t \sin \theta + \frac{1}{t} \cos \theta\right) + r^2 \cos \theta \sin \theta \tag{3.1}$$

so

$$e^{-(t + r \cos \theta)\left(\frac{1}{t} + r \sin \theta\right)} = \frac{1}{e} e^{-r\left(t \sin \theta + \frac{1}{t} \cos \theta\right)} e^{-r^2 \cos \theta \sin \theta}$$

or

$$e^{-(t + r \cos \theta)\left(\frac{1}{t} + r \sin \theta\right)}$$

$$= \frac{1}{e} \left[1 - r\left(t \sin \theta + \frac{1}{t} \cos \theta\right) + \frac{1}{2} r^2 \left(t \sin \theta + \frac{1}{t} \cos \theta\right)^2 \right]$$

$$\times [1 - r^2 \cos \theta \sin \theta + \cdots],$$

so

$$e^{-(t+r\cos\theta)\left(\frac{1}{t}+r\sin\theta\right)}$$

$$\approx \frac{1}{e}\left[1 - r\left(t\sin\theta + \frac{1}{t}\cos\theta\right) + \frac{1}{2}r^2\left(t\sin\theta + \frac{1}{t}\cos\theta\right)^2 \atop -r^2\cos\theta\sin\theta\right] \qquad (3.2)$$

and multiplying Equations 3.1 and 3.2 together, one sees that the $O(r)$ terms cancel. Retaining only $O(r^2)$ gives

$$\Delta z = z\left(t + r\cos\theta, \frac{1}{t} + r\sin\theta\right) - z\left(t, \frac{1}{t}\right)$$

$$\approx -\frac{1}{2e}r^2\left(t\sin\theta + \frac{1}{t}\cos\theta\right)^2;$$

the right-hand side is less than zero, apart from where $\tan\theta = -1/t^2$ where it is identically zero. Identically zero here really means *identically* zero, for although an approximation whereby only second-order terms are retained has taken place, we already know that the whole of the hyperbola $xy = 1, z = 1/e$ is exactly flat; no approximation. For a given point on the hyperbola $xy = 1$ that is for every value of t this gives rise to two values of θ in the range $0 \le \theta \le 2\pi$. Therefore, $z\left(t, \frac{1}{t}\right)$ must be a maximum for all t apart from the two points corresponding to the two values of theta where it is exactly zero. What we have here therefore is a ridge, and Figure 3.5 shows this reasonably clearly. The plane $z = 1/e$ is displayed in Figure 3.6. In this figure, the small circle $(x - 0.5)^2 + (y - 2)^2 = 0.1$ the centre of which is $0.5, 2$ corresponding to the parameter value $t = 0.5$ is drawn. Hence, the two particular values of θ are

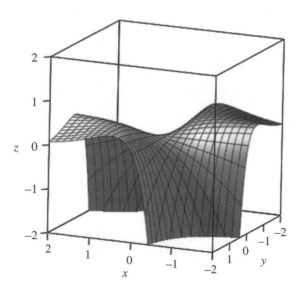

Figure 3.5 The surface $z = xye^{-xy}$ is partially drawn here. The classic saddle at the origin is apparent, and the most elevated parts show sections of the hyperbolic shaped ridge $xy = 1, z = 1/e$.

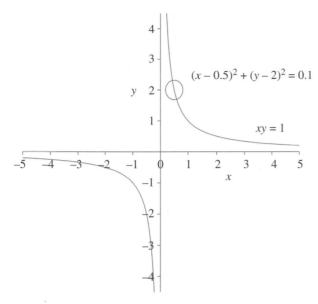

Figure 3.6 The small circle $(x - 0.5)^2 + (y - 2)^2 = 0.1$ is shown and the two points of intersection with $xy = 1$ are where the values of θ (see text) are such that $\tan \theta = -1/t^2$.

$\tan \theta = -4$, that is $\theta = -76°$ and $104°$. These are shown in Figure 3.6. From the equation of the curve $xy = 1$, it is easy to calculate the gradient

$$\frac{dy}{dx} = -\frac{1}{x^2},$$

which at the point $(t, \frac{1}{t})$ is $-1/t^2$ or $\tan \theta$. Therefore, the smaller the radius of the circle around the point $(t, \frac{1}{t})$ is, the closer that part of the curve $xy = 1$ inside this circle is to the straight line $t^2 y + x = 0$. This can be seen visually in Figure 3.6. Try drawing this surface in MAPLE but be careful with the ranges as z quickly gets very large when x and y differ in sign, and very small when they don't. According to our definitions, this does remain a saddle point as it is neither an isolated maximum nor an isolated minimum. By analysing from first principles however much more information on the precise nature of the turning point at $(t, \frac{1}{t})$ comes to light. Finally, in this section, here is a rather different example.

Example 3.4 (Least Squares) Find constants a and b such that the definite integral

$$\phi(a, b) = \int_0^\pi \{\cos x - (ax^2 + b)\}^2 dx$$

is a minimum.

Solution: Although this example looks rather different, the method remains the same, and $\phi(a, b)$ is differentiated with respect to a and b and equated to zero. There is a formula for differentiating integrals with variable limits that was met on page 39, section 2.4.1, but here the limits are constants so, straightforwardly, the following pair:

$$\int_0^\pi \frac{\partial}{\partial a} \{\cos x - (ax^2 + b)\}^2 dx = 0,$$

$$\int_0^\pi \frac{\partial}{\partial b} \{\cos x - (ax^2 + b)\}^2 dx = 0.$$

Multiplying out, differentiating with respect to a for the first and b for the second is not difficult. The resulting pair of equations

$$a \int_0^\pi x^4 dx + b \int_0^\pi x^2 dx = \int_0^\pi x^2 \cos x \, dx,$$

$$a \int_0^\pi x^2 dx + b \int_0^\pi dx = \int_0^\pi \cos x \, dx$$

need to be solved for a and b. Again, the integrals are evaluated by elementary means, and the following pair of equations results

$$a \frac{\pi^5}{5} + b \frac{\pi^3}{3} = 2\pi,$$

$$a \frac{\pi^3}{3} + b\pi = 0$$

giving $a = -\frac{45}{2\pi^4} = -0.231$ and $b = \frac{15}{2\pi^2} = 0.760$. This example finds a parabola $y = ax^2 + b$ that fits the cosine curve $y = \cos x$ in the range $0 \le x \le \pi$ by minimising the 'square area' between them. It is a form of the least-squares algorithm. To check whether we have a minimum and so on, the second derivatives are found as follows

$$\phi_{aa} = 2 \int_0^\pi x^4 dx = \frac{2\pi^5}{5}; \ \phi_{bb} = 2 \int_0^\pi dx = 2\pi; \text{ and}$$

$$\phi_{ab} = 2 \int_0^\pi x^2 dx = \frac{2\pi^3}{3}$$

from which $D = \phi_{aa}\phi_{bb} - \phi_{ab}^2 = (\frac{4}{5} - \frac{4}{9})\pi^6 > 0$, and as $\phi_{aa} > 0$ it is certainly a minimum. In fact, the second derivatives are constant as $\phi(a, b)$ is quadratic in a and b. There is only the one turning point, being the solution of a pair of linear simultaneous equations. $\quad\square$

The final example in this section is a practical one.

Example 3.5 A rectangular strip of metal of width L is bent up at the sides to form a gutter for rainwater, the cross-section of which is shown in Figure 3.7. Find the widths of the side pieces (labelled x in Figure 3.7) in order for the gutter to have maximum cross-sectional area and hence carry maximum capacity.

Solution: This is a reasonably straightforward example, the new part is the interpretation of words into mathematics. It is a modelling problem. The cross-sectional area of the gutter shown in Figure 3.7 is a trapezium hence the area is

$$A(x, \theta) = \frac{1}{2}[L - 2x + L - 2x + 2x \cos \theta][x \sin \theta]$$

sometimes expressed as 'half the sum of the parallel sides times the distance between them'. It is beneficial to simplify this expression for A before going on:

$$A = Lx \sin \theta - 2x^2 \sin \theta + x^2 \sin \theta \cos \theta.$$

The first derivatives of A are found and equated to zero:

$$A_x = L \sin \theta - 4x \sin \theta + x \sin \theta \cos \theta = 0,$$
$$A_\theta = Lx \cos \theta - 2x^2 \cos \theta + x^2(\cos^2\theta - \sin^2\theta) = 0.$$

A factor x can be cancelled from the second of these ($x = 0$ is not a solution of the first). Another factor of $\sin \theta$ can be cancelled from the first as $\sin \theta = 0$ is not a solution to the second. The following pair thus need to be solved

$$L - 4x + 2x \cos \theta = 0,$$
$$L \cos \theta - 2x \cos \theta + x(2\cos^2\theta - 1) = 0$$

putting all in terms of $\cos \theta$. Eliminating $\cos \theta$ from the first via $\cos \theta = (4x - L)/2x$ gives

$$4xL - L^2 - 8x^2 + 2xL + 16x^2 - 8xL + L^2 - 2x^2 = 0$$

or $x = \frac{1}{3}L$. Therefore,

$$\cos \theta = \frac{1}{2} \quad \text{so} \quad \theta = \frac{\pi}{3}$$

This gives the solution shown in Figure 3.8. The actual value of the cross-sectional area of the gutter is

$$\frac{L^2\sqrt{3}}{12}.$$

$\square$

Figure 3.7 The cross-section of the gutter.

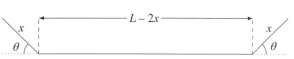

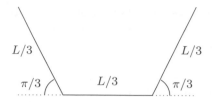

Figure 3.8 The optimal cross-section of the gutter.

3.3 Lagrange Multipliers

In this section, the idea of a *constraint* is introduced. It is reasonably straight-forward to comprehend the idea of a local maximum or a local minimum, but suppose now that whilst finding these, it is also required that another condition holds. For example, imagine walking through a formal park. There is little doubt that the big hillock on your right is the local maximum, but you are constrained to walk along a path due to 'keep-off-the-grass' signs, and your path travels up to its own local maximum, perhaps when closest to the peak. It is a maximum along the path, but it is different from the peak due to the constraint that you have to remain on the path. How do we find such constrained maxima? First of all, let us pose the problem in three variables in the form of an example; use partial differentiation to find a solution, then state the theorem in general.

Example 3.6 A function $f(x, y, z)$ has turning points subject to the constraint $g(x, y, z) = 0$. Find conditions that enable you to find the turning points of f subject to this constraint.

Solution: Theoretically at least, one can eliminate the variable z between $f(x, y, z)$ and $g(x, y, z) = 0$ but let us be systematic. First of all, realising that $z = z(x, y)$ and so z cannot be treated as an independent variable enables the chain rule to be used as follows:

$$\frac{\partial}{\partial x}[f(x, y, z(x, y))] = \frac{\partial f}{\partial x} + \frac{\partial f}{\partial z}\frac{\partial z}{\partial x}, \tag{3.3}$$

$$\frac{\partial}{\partial y}[f(x, y, z(x, y))] = \frac{\partial f}{\partial y} + \frac{\partial f}{\partial z}\frac{\partial z}{\partial y}, \tag{3.4}$$

where it is understood that each first term on the right is differentiating with respect to x and y only considering those xs and ys explicitly in the formula for $f(x, y, z)$, ignoring those inside $z = z(x, y)$. This is practical as we have not, and usually cannot, find $z(x, y)$ explicitly. For this reason, in what follows, the derivatives of z will be eliminated. Now, $g(x, y, z) = 0$ is the constraint, so it is the case that

$$dg = \frac{\partial g}{\partial x}dx + \frac{\partial g}{\partial y}dy + \frac{\partial g}{\partial z}dz = 0$$

but because x, y, z are connected through $z = z(x, y)$, we also have

$$dz = \frac{\partial z}{\partial x}dx + \frac{\partial z}{\partial y}dy$$

and these two equations in differentials give

$$\frac{\partial z}{\partial x} = -\frac{g_x}{g_z} \quad \text{and} \quad \frac{\partial z}{\partial y} = -\frac{g_y}{g_z}$$

reverting for convenience to the suffix derivative notation. For stationary values (that is for turning points) both right-hand sides of Equations (3.3) and (3.4) equate to zero. Elimination of z_x and z_y between these and the equations just derived leads to

$$\frac{f_x}{g_x} = \frac{f_y}{g_y} = \frac{f_z}{g_z}.$$

The denominators are not allowed to be zero of course. If one of the derivatives of g is zero, then because $g(x, y, z) = 0$, the other derivatives are zero too. In the vicinity of such a place, the turning points of $f(x, y, z)$ are the same whether or not $g(x, y, z) = 0$. Later, when the normal to surfaces is calculated, it will be seen that at such points there is no well-defined normal (one such point would be at the point of a cone for example). These singular points are thus excluded from calculation. As written, no coherent criteria have yet been derived. This point can now be addressed. Form the function $F(x, y, z, \lambda) = f(x, y, z) + \lambda g(x, y, z)$, then demand that

$$F_x = f_x + \lambda g_x = 0,$$
$$F_y = f_y + \lambda g_y = 0,$$
$$F_z = f_z + \lambda g_z = 0,$$
$$F_\lambda = g = 0.$$

The last equation is obviously true as it is just the constraint. The first three are true also provided that we write

$$\frac{f_x}{g_x} = \frac{f_y}{g_y} = \frac{f_z}{g_z} = -\lambda. \tag{3.5}$$

Now, let us work back from these equations. Start with the derivative that is Equation (3.3)

$$\frac{\partial}{\partial x}[f(x, y, z(x, y))] = \frac{\partial f}{\partial x} + \frac{\partial f}{\partial z}\frac{\partial z}{\partial x}.$$

As $g = 0$, we still have that

$$\frac{\partial z}{\partial x} = -\frac{g_x}{g_z} \quad \text{and} \quad \frac{\partial z}{\partial y} = -\frac{g_y}{g_z}.$$

These are not dependent upon any assumptions about turning points, they arise only from $g = 0$. Thus,

$$\frac{\partial}{\partial x}[f(x, y, z(x, y))] = \frac{\partial f}{\partial x} + \frac{\partial f}{\partial z}\frac{\partial z}{\partial x} = f_x - f_z\frac{g_x}{g_z} = g_x\left[\frac{f_x}{g_x} - \frac{f_z}{g_z}\right] = 0$$

from Equation 3.5. Similarly, it can be proved that

$$\frac{\partial}{\partial y}[f(x, y, z(x, y))] = 0$$

too. This proves the following theorem.

Theorem 3.1   Given a function $f(x, y, z)$ of the three independent variables x, y and z, write $F(x, y, z, \lambda) = f(x, y, z) + \lambda g(x, y, z)$, then the extrema of F determine the extrema of f subject to the constraint $g = 0$.

This means that we merely form $F = f + \lambda g$ then differentiate with respect to each variable, equate each to zero then solve the equations. This is very similar to the previous procedure for finding turning points of multivariable functions without constraint. The variable λ is termed a *Lagrange multiplier*. The equations to be solved are

$$F_x = f_x + \lambda g_x = 0,$$
$$F_y = f_y + \lambda g_y = 0,$$
$$F_z = f_z + \lambda g_z = 0,$$
$$F_\lambda = g = 0.$$

Now, let us do an example.

Example 3.7   A rectangular tank open at the top is to have a volume of 32 m^3. Determine the dimensions of the tank such that the surface area is a minimum.

Solution: Let us choose the dimensions of the tank as x, y horizontally and z vertically. Therefore, the constraint that the volume must be 32 m^3 leads to $xyz = 32$. The surface area, remembering that the top is open is

$$S = xy + 2xz + 2yz.$$

In this instance, it is possible to eliminate z from the constraint and solve as a standard turning point problem in two variables. However, let us use Lagrange multipliers. The function to be differentiated is $F = xy + 2xz + 2yz + \lambda(xyz - 32)$ so the four equations to be solved are

$$F_x = y + 2z + \lambda yz = 0,$$
$$F_y = x + 2z + \lambda xz = 0,$$
$$F_z = 2x + 2y + \lambda xy = 0,$$
$$F_\lambda = xyz - 32 = 0.$$

By the first two (or by symmetry), $x = y$. The third equation then gives $x = y = -4/\lambda$ and so either of the first two equations then gives $z = -2/\lambda$ and, finally the last equation $xyz = 32$ then becomes $\lambda^3 = -1$ hence $\lambda = -1$, and $x = y = 4$ and $z = 2$. Thus, the dimensions for minimal S are $4 \times 4 \times 2$. Although that this is a minimum has not been formally proved, from practical considerations it has to be a minimum.

□

This is typical of Lagrange multiplier problems; a proof of minimum (or maximum) is seldom necessary. For ease of exposition, three variables and one constraint have been chosen. Generalisation to n variables and m constraints is possible and is now done.

3.3.1 Generalisations

Doing problems in many dimensions with lots of constraints is usually not possible due to the sheer amount of algebra; some real-life problems, for example, the efficient nationwide distribution of broadband internet, are of this type. They are solved using numerical techniques. In addition, some of you would have met operational research problems in the form of finding the best routes through a network or maximising profit (linear programming). This is usually part of the discipline of operational research, or sometimes forms part of a module called optimisation. This is the next step for this kind of multivariable calculus, though operational research usually confines itself to discrete mathematics, lends itself to programming techniques or use of proprietary software, and is free of calculus. Some optimisation is met later in this chapter, but there is no pure operational research.

Here is the generalisation of the method of Lagrange multipliers (shortened from Lagrange's method of undetermined multipliers).

Theorem 3.2 Consider a function of n variables $f(x_1, x_2, \ldots, x_n)$ where the variables $x_1, x_2, \ldots, x_n$ are subject to m constraints, $m < n$:

$$g_1(x_1, x_2, \ldots, x_n) = 0,$$
$$g_2(x_1, x_2, \ldots, x_n) = 0,$$
$$g_3(x_1, x_2, \ldots, x_n) = 0,$$
$$\vdots$$
$$g_m(x_1, x_2, \ldots, x_n) = 0.$$

Assuming that all partial derivatives of f and g_i, $i = 1, 2, \ldots, m$ exist, the turning points (extrema) can be found by forming the function

$$F = f + \sum_{j=1}^{m} \lambda_j g_j = f + \lambda_1 g_1 + \lambda_2 g_2 + \cdots + \lambda_m g_m,$$

where λ_j $(j = 1, 2, \ldots, m)$ are m undetermined constants called Lagrange multipliers. The turning points are then found by solving the equations

$$\frac{\partial F}{\partial x_i} = 0, (i = 1, 2, \ldots, n) \quad \text{together with} \quad \frac{\partial F}{\partial \lambda_j} = 0, \ (j = 1, 2, \ldots, m).$$

in the $m + n$ unknowns x_i, $i = 1, 2, \ldots, n$ and λ_j, $j = 1, 2, \ldots, m$.

The proof of this is not given. It involves knowledge of linear algebra including quadratic forms and details of properties of determinants that may well be outside the knowledge of most of the readers. Those interested can find a proof in the classic text by Gillespie [1] available on line.

For the more mathematical amongst you, here is a geometric example that involves not one but two undetermined (Lagrange) multipliers.

Example 3.8   Find the shortest distance between the origin and the curve of intersection of the surfaces $xyz = a$ and $y = bx$ (z arbitrary) where a and b are given positive constants.

Solution: The surfaces have been drawn in Figure 3.9 with specific values for a and b. It is tempting to do some manipulation of the algebra in the light of seeing this figure. It can be seen that the surface $xyz = a$ is asymptotic to the three co-ordinate axes, resembling one of those plastic chairs. The plane $y = bx$ passes through the origin, and intersects $xyz = a$ in a hyperbola. Is the minimum distance in the plane $y = bx$? We will see later, but let us solve

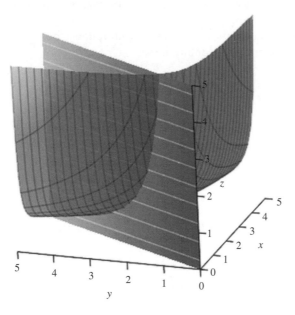

Figure 3.9 The surface $xyz = a$ is partially drawn here with $a = 3$ intersecting with the plane $y = bx$ drawn with $b = 2$.

the problem 'blindly', then examine the solution in the light of Figure 3.9. The distance from the origin to a point (x, y, z) is $\sqrt{x^2 + y^2 + z^2}$ but minimising the square of this $x^2 + y^2 + z^2$ is easier and will give the same correct solution to minimising $\sqrt{x^2 + y^2 + z^2}$. The two surfaces are constraints, so we form the function

$$F(x, y, z, \lambda_1, \lambda_2) = x^2 + y^2 + z^2 + \lambda_1(xyz - a) + \lambda_2(y - bx)$$

and differentiate five times with respect to each of the five variables, equating each derivative to zero, obtaining

$$2x + \lambda_1 yz - \lambda_2 b = 0, \tag{3.6}$$

$$2y + \lambda_1 xz + \lambda_2 = 0, \tag{3.7}$$

$$2z + \lambda_1 xy = 0, \tag{3.8}$$

$$xyz - a = 0, \tag{3.9}$$

$$y - bx = 0. \tag{3.10}$$

From Equation 3.8, $\lambda_1 = -2z/xy$, and substituting this into Equation 3.7 gives

$$\lambda_2 = \frac{2z^2}{y} - 2y,$$

and putting these expressions for λ_1 and λ_2 into Equation 3.6 gives

$$2z^2 = x^2 + y^2. \tag{3.11}$$

From Equation 3.11, the value of $x^2 + y^2 + z^2$ becomes $3z^2$. Since $b = y/x$ from Equation 3.10, Equation 3.9 yields

$$z = \frac{a}{xy} = \frac{a}{bx^2}. \tag{3.12}$$

The last bit of manipulation is to use Equation 3.11 and Equation 3.10 with Equation 3.12 to obtain

$$z^2 = \frac{1}{2}(x^2 + y^2) = \frac{1}{2}x^2(1 + b^2)$$

so

$$z^2 = \frac{a(1 + b^2)}{2bz}$$

giving

$$z^3 = \frac{a(1 + b^2)}{2b}.$$

As the distance required is $z\sqrt{3}$, the required answer is

$$\sqrt{3}\left[\frac{a(1 + b^2)}{2b}\right]^{1/3}.$$

Now, to check this, we can solve the problem more geometrically. The example requires the determination of the shortest distance from the origin to the intersection curve of the surfaces $xyz = a$ and $y = bx$. The origin lies on the plane surface $y = bx$. The other end of the line of shortest distance between the origin and the curve of intersection of the plane $y = bx$ with $xyz = a$ also by construction lies on the plane $y = bx$. As two points on this line are on the plane $y = bx$ so is the whole line. The following is thus a way forward. Let the distance be r, then $r^2 = x^2 + y^2 + z^2$, where (x, y, z) is the intersection specified. As this intersection is on the surface $xyz = a$, we can put $z = a/xy$. Similarly, $y = bx$ too hence $z = a/(bx^2)$ and $y = bx$ so $r^2 = (1 + b^2)x^2 + a^2/(b^2x^4)$. This is only a function of the single variable x so elementary methods can be used.

$$\frac{d}{dx}(r^2) = 2x(1 + b^2) - \frac{4a^2}{b^2x^5} = 0$$

for a minimum. This gives the value

$$x^6 = \frac{2a^2}{b^2(1 + b^2)}.$$

With $r^2 = (1 + b^2)x^2 + a^2/(b^2x^4)$ substitution yields, after algebra, the same answer as obtained using undetermined multipliers. The shortest distance is displayed in Figure 3.10. □

The second method could be said to be easier, though the method of undetermined multipliers needs less thought. This example did have an alternative method of solution; this is seldom the case. The next section could be a

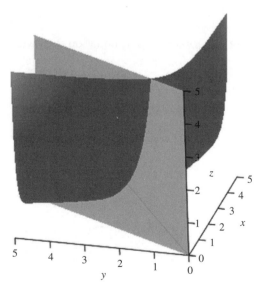

Figure 3.10 This is Figure 3.9 re-drawn with the shortest distance shown.

different chapter as the first paragraphs indicate; however, in a book on two- and three-dimensional calculus, optimisation follows naturally as a different kind of generalisation of finding maxima and minima under constraints.

3.4 Optimisation

The word 'optimisation' indicates finding the best way of doing something. The scope of optimisation is very wide. One can think of the maximisation of profit from a business, or the least number of people to man a fire station consistent with health and safety requirements, or indeed the operation of a large hospital so as to minimise cost, again with health and safety constraints. The solutions to these kind of problems are typically based on linear algebra, the first such technique usually met is called the simplex method and is accessible by those who have studied mathematics to age 16. The problems mentioned above are certainly of this type, but not quite so easy to solve and usually demand the use of numerical methods and software these days. These are however firmly outside the scope of this book which is restricted to the use of two- and three-dimensional calculus. In this chapter, there has indeed already been some optimisation covered, both in finding the maximum and minimum of two-dimensional functions and especially this together with the use of constraints, called the method of Lagrange multipliers. We shall start with the generalisation of the Newton–Raphson method and go on to introduce other methods that use calculus for finding the best way of determining maxima and minima building on the material of Chapters 1 and 2. The kind of practical problem that this prepares the student for is the minimisation of drag on a car or the maximisation of lift on an aeroplane. The use of numerical methods will for the most part be avoided, but there are a few instances where a numerical algorithm is necessary to find a solution. The discretisation of such problems opens up the use of methods based upon linear algebra mentioned above. The emphasis here will always be on the application of calculus. Everywhere else in this book, the methods are focussed on two- or three-dimensional applications. Although this is also true for optimisation, it is more helpful if the techniques are expressed in more general terms. This means that the term 'vector' means matrix with a single column rather than an arrow that points in three dimensions. After this chapter, 'vector' will revert to its three-dimensional definition.

3.4.1 Hill Climbing Techniques

The phrase 'hill-climbing techniques' encompasses methods by which one can reach the top of a hill or bottom of a hollow (but in many dimensions) in an efficient way. There are several ways to do this, but they always start from Taylor's series, but in n dimensions. In two dimensions, Taylor's series is written:

$$f(a + h, b + k) = f(a, b) + hf_x + kf_y + \frac{1}{2}(h^2 f_{xx} + 2hkf_{hk} + k^2 f_{yy}) + \cdots,$$

where all derivatives are evaluated at the values $x = a, y = b$. However, to see that the generalisation to n variables is much easier if matrices are used, write the above $n = 2$ series as

$$f(a + h, b + k) = f(a, b) + (h, k) \begin{pmatrix} f_x \\ f_y \end{pmatrix}$$

$$+ \frac{1}{2}(h, k) \begin{pmatrix} f_{xx} & f_{xy} \\ f_{xy} & f_{yy} \end{pmatrix} \begin{pmatrix} h \\ k \end{pmatrix} + \cdots,$$

where once more all derivatives are evaluated at $x = a, y = b$. Extending this to $n > 3$ is only practical if matrix notation continues to be used. First of all, the following notation is standard and is adopted here. A matrix consisting of a single column is denoted by a boldface lower-case letter:

$$\mathbf{x}^{(r)} = \begin{pmatrix} x_1^{(r)} \\ x_2^{(r)} \\ \vdots \\ x_n^{(r)} \end{pmatrix}$$

often without the superfix (r) which is there for the Newton–Raphson and other optimisation algorithms. The matrix of second-order partial derivatives $\mathbf{G}$ is defined by

$$\mathbf{G} = \begin{vmatrix} \dfrac{\partial^2 f}{\partial x_1^2} & \dfrac{\partial^2 f}{\partial x_1 \partial x_2} & \cdots & \dfrac{\partial^2 f}{\partial x_1 \partial x_n} \\[2mm] \dfrac{\partial^2 f}{\partial x_2 \partial x_1} & \dfrac{\partial^2 f}{\partial x_2^2} & \cdots & \dfrac{\partial^2 f}{\partial x_2 \partial x_n} \\[2mm] \vdots & \vdots & \ddots & \vdots \\[2mm] \dfrac{\partial^2 f}{\partial x_n \partial x_1} & \dfrac{\partial^2 f}{\partial x_n \partial x_2} & \cdots & \dfrac{\partial^2 f}{\partial x_n^2} \end{vmatrix}.$$

Finally, define

$$\nabla f = \left(\frac{\partial f}{\partial x_1}, \frac{\partial f}{\partial x_2}, \dots, \frac{\partial f}{\partial x_n} \right),$$

where $f(x_1, x_2, \dots, x_n)$ is a function of n variables with continuous second-order partial derivatives. The truncated Taylor's series, often called the *Taylor polynomial*, to second order is thus

$$f(\mathbf{x}) \approx f(\boldsymbol{\alpha}) + \delta \mathbf{x}^T \nabla f(\boldsymbol{\alpha}) + \frac{1}{2} \delta \mathbf{x}^T \mathbf{G} \delta \mathbf{x},$$

where

$$\alpha = \begin{pmatrix} \alpha_1 \\ \alpha_2 \\ \vdots \\ \alpha_n \end{pmatrix}, \quad \delta\mathbf{x} = \begin{pmatrix} \delta x_1 \\ \delta x_2 \\ \vdots \\ \delta x_n \end{pmatrix},$$

T denotes transpose, and it is understood that the derivatives on the right are all evaluated at the point α. The question is what the new version of the criteria for maximum, minimum and so on is. Certainly, as all the δx_i are independent; it has to be the case that $\nabla f(\mathbf{x}) = \mathbf{0}$ when $\alpha = \mathbf{x}$ which means that the first-order terms are zero. In n dimensions, it makes sense only to concern ourselves with maxima and minima, but it is still true that there are no simple conditions on the second derivatives that fill the $n \times n$ matrix $\mathbf{G}$ equivalent to the $n = 2$ case. There are however the KKT (Karush–Kuhn–Tucker sometimes just called the Kuhn-Tucker) conditions that generalise the technique of undetermined multipliers but also include inequality-type constraints. The kind of mathematics this involves is outside the scope of this book. However, with a little thought some theoretical progress can be made. We are in n-dimensional space here, but suppose we pick (guess) a point $\mathbf{x}_0$ we suspect has an extremum nearby, how can we improve the guess? It is given that the object is to find minimum or maximum values of the function $f(\mathbf{x}) = f(x_1, x_2, \ldots, x_n)$ where the components are written out explicitly for convenience. Therefore, it is a reasonable strategy to seek a path that is the steepest, after all the quickest way to the top of a hill or the bottom of a depression is along the path that has the greatest slope. The way to find this is to form the function $f(\mathbf{x}_0 + \mathbf{h})$ where $\mathbf{h} = (h_1, h_2, \ldots, h_n)$ and

$$|\mathbf{h}|^2 = h_1^2 + h_2^2 + \cdots + h_n^2$$

is the distance (in n-dimensional space) from the base point, the guess, $\mathbf{x}_0$. The function $f(\mathbf{x}_0 + \mathbf{h})$ is thus constrained by the equation $h^2 = h_1^2 + h_2^2 + \cdots + h_n^2$. Hence, the problem reduces to a standard Lagrange multiplier problem with variables $(h_1, h_2, \ldots, h_n)$ and constraint $h^2 = h_1^2 + h_2^2 + \cdots + h_n^2$. This leads to the n equations:

$$\frac{\partial f}{\partial h_i} - 2\lambda h_i = 0, \quad i = 1, 2, \ldots, N.$$

These equations mean that the vectors $\mathbf{h}$ and $\nabla f(\mathbf{x}_0 + \mathbf{h})$ must be parallel which is exactly saying that the vector $\mathbf{h}$ has to be in the direction of the greatest slope of the function f. This is all very well, but in practice the calculation leading to finding this direction of greatest slope is difficult and very lengthy. A much better strategy is to use approximate methods so let us take a look at them next. It will be just a look as it only peripherally involves calculus so is on the edge of what can be covered in a book on two- and three-dimensional calculus. It is more centrally dependant on numerical techniques as are modern texts on optimisation.

The equation

$$f(\mathbf{x}) \approx f(\boldsymbol{\alpha}) + \delta \mathbf{x}^T \nabla f(\boldsymbol{\alpha}) + \frac{1}{2} \delta \mathbf{x}^T \mathbf{G} \delta \mathbf{x}$$

is satisfied approximately in the vicinity of a maximum or minimum if we set the first guess as $\mathbf{x}^{(r)}$ and use the recurrence

$$\mathbf{x}^{(r+1)} = \mathbf{x}^{(r)} - [\mathbf{G}(\mathbf{x}^{(r)})]^{-1} \nabla f(\mathbf{x}^{(r)}).$$

This is the direct generalisation of the Newton–Raphson method introduced in Chapter 1, and looks a very viable method but in practice inverting the $n \times n$ matrix $\mathbf{G}(\mathbf{x}^{(r)})$ at each step proves too difficult except in a few idealised cases. Instead of abandoning the method, approximations are derived for this inverse that will do and are easier to find. The first such method is the DFP method (from Davidon–Fletcher–Powell, due initially to W.C. Davidon (1927–2013), a physicist and mathematician perhaps more famous for his pacifist activities in connection with the Vietnam war and FBI than for the DFP method), a modified form of which is still useful, but there are many other methods all structured on approximating the Newton–Raphson method and given the general name *quasi-Newton methods*. These are still being updated and modified, so instead of doing a computational example here, this is one instance where the reader is pointed to MAPLE or MATLAB that solves such problems very efficiently. It is not laziness on the part of the author, more that the techniques are not calculus, but numerical algorithms and hence fall outside the remit of this textbook. However, here is one example on the calculus side of optimisation that is purely illustrative.

Example 3.9 Use the method of steepest descent to find an approximation to the minimum of the function $f(x, y, z) = 2(x - 0.2)^4 + (y - 0.1)^4 + 0.5(z - 0.3)^4$.

Solution: A glance at this function tells that the minimum is obviously at the values $x = 0.2, y = 0.1$ and $z = 0.3$ as $f = 0$ there but is otherwise always positive. However, let us run through finding this using steepest descent by first calculating ∇f

$$\nabla f = (8(x - 0.2)^3, 4(y - 0.1)^3, 2(z - 0.3)^3)^T$$

so choosing $(0, 0, 0)$ as the start value pretending we do not know the answer. On the line through the origin, this steepest direction is the vector $\nabla f(0, 0, 0) = (0.064, 0.004, 0.054)^T$ so the next best guess would be

$$\mathbf{x}^{(1)} = (0, 0, 0)^T + \lambda(0.064, 0.004, 0.054)^T = \lambda(0.064, 0.004, 0.054)^T.$$

A brief preview of straight-line vector geometry is to be covered in Chapter 5; reference to the equation of a straight line using vectors shows that the above equation is $\mathbf{x}^{(1)} = \lambda(0.064, 0.004, 0.054)^T$ where λ can take on any value. It represents the 'co-ordinates' in three space of any point on a line through the origin in the direction of the steepest slope. The value of λ is chosen such that at this point the line is at right angles to the direction of steepest descent. This is given (see Chapter 6) by the condition $\mathbf{x}_{(1)}.\nabla f = 0$ or

$$0.512(0.064\lambda - 0.2)^3 + 0.016(0.004\lambda - 0.1)^3$$
$$+ 0.108(0.054\lambda - 0.3)^3 = 0,$$

where the value $\lambda = 0$ has been ignored as it is the origin, our starting point. This is a cubic equation for λ and is solved by iterative methods such as Newton–Raphson or software these days. $\lambda = 4.3$ is the approximate solution. The value of $\mathbf{x}_{(1)} = \lambda(0.064, 0.004, 0.054)^T$ is thus $(0.27, 0.017, 0.23)^T$. The method is then repeated. Find the new value of λ for which

$$\mathbf{x}_{(2)} = (0.27, 0.017, 0.23)^T - \lambda\nabla f(\mathbf{x}_{(1)})$$
$$= (0.27 - 0.0027\lambda, 0.017 + 0.0029\lambda, 0.23 + 0.000686\lambda)^T$$

is the next approximation to the minimum value of f. Once again the value of λ is chosen so that the vector $\mathbf{x}^{(2)}$ is perpendicular to the direction of steepest descent $\nabla f(0.27, 0.017, 0.23)$. Therefore, the new value of λ is found by solving $\mathbf{x}_{(2)}.\nabla f = 0$. The path to a more accurate estimate of the minimum thus follows a zig-zag route and the method is often called an alternating direction method. However, more and more awkward cubics are needed to be solved at each step. As the method itself is finding estimates, a more practical approach is to use an easier-to-find estimate of the new λ. This is acceptable as the whole method is not precise and a less precise estimate of λ will do. These are the quasi-Newton methods mentioned above. For example, if the next value of λ is guessed at 20 the next estimate for the solution is $(0.216, 0.0628, 0.244)^T$ which is quite close to the precise answer $(0.2, 0.1, 0.3)^T$. □

Sometimes, the approximate methods fail to converge or only converge very slowly. For example, the function $f(x, y) = 100(y - x^2)^2 + (1 - x)^2$ called Rosenbrock's function fails to respond very quickly to this technique. What happens with this and similar functions is that they have a minimum inside a long trough and the algorithm to find the minimum gets stuck in the trough and takes forever actually getting close to the minimum. The discussion will be left here as the whys and wherefores of how the algorithms are tweaked and improved is securely numerical optimisation and outside the remit of this text.

Exercises

3.1 Find the extreme values of the following functions and classify them:
a) $f(x, y) = x^2 + 2y^2 - x^2y$;
b) $f(x, y) = x^3 + y^3 - 3x - 3y$;

c) $f(x, y) = 3x^2 - 2xy + y^2 - 8y$;

d) $f(x, y) = 4xy - x^4 - y^4$;

e) $f(x, y) = -xye^{(x^2+y^2)/2}$;

f) $f(x, y) = e^x \cos y$;

g) $f(x, y) = \dfrac{x^2 + 2y^2}{(x + y)^2}$.

3.2 A function is defined by $f(x, y) = \sin x \cos y$. Determine all of the extreme points and classify them as maximum, minimum or saddle point.

3.3 Given that $0 < a < b$ and $x, y > 0$, find the maximum value assumed by the expression:

$$f(x, y) = \frac{xy}{(a + x)(x + y)(b + y)}.$$

3.4 A rectangular box just fits inside an ellipsoid that has equation $2x^2 + 3y^2 + z^2 = 1$ with each edge parallel to one of the co-ordinate axes. Determine its maximum volume.

3.5 Determine the dimensions of a rectangular box, open at the top, having a volume of 100 cubic centimetres having minimum surface area.

3.6 A tea trolley has a rectangular cross section, dimensions $p \times q$ has to be able to get around a right-angled bend in a corridor width b before the bend and width a after the bend. Show that the cross-sectional area of the trolley is a maximum when

$$p = \sqrt{a^2 + b^2} \quad \text{and} \quad q = \frac{ab}{\sqrt{a^2 + b^2}}.$$

3.7 A rectangular box is to be inscribed in the cone $z = 9 - \sqrt{x^2 + y^2}, z \geq 0$. Find the dimensions of the box that maximises its volume.

3.8 Snell's law for the refraction of light states that

$$\frac{\sin \theta_1}{v_1} = \frac{\sin \theta_2}{v_2},$$

where θ_1 and θ_2 are the magnitudes of the angles shown in Figure 3.11. Use Lagrange multipliers to derive this law using the constraint $x + y = a$ where these are defined in Figure 3.11.

Figure 3.11 Snell's law of refraction showing $x + y = a$.

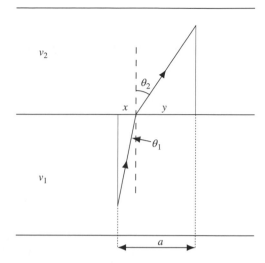

3.9 Use the Newton–Raphson method to maximise the function $f(x, y, z) = x^2 + 2y^2 + 3z^2$ subject to the constraints

$$x + 2y + 4z = 12 \quad \text{and} \quad 2x + y + 3z = 10$$

and comment on its accuracy.

4

Vector Algebra

4.1 Introduction

Obviously, in a book that includes properties of vectors and how they change, one needs to have a knowledge of vectors themselves. The assumption is that most will have some experience of vectors in much the same way as the contents of Chapter 1 assumes that differentiation has been encountered before. In this case, however, the revision is a little more extensive.

To start at the beginning, a scalar is a quantity that has only magnitude; examples include temperature, density and pressure. By contrast, a vector is a quantity that has both magnitude and direction. Examples of vectors include wind, force, velocity and magnetic field. In this chapter, the intention is to familiarise the reader with the algebra of vectors. Some of these will already be familiar, some might well be new. The first topic to sort out is notation. Scalars are universally denoted by standard letters such as T for temperature, ρ for density and p for pressure. For vectors, there is more choice. If the concentration is on the geometry of vectors, then the vector formed by the directed line segment between a point A and B can be denoted by $\overrightarrow{AB}$ where the vector starts at A and ends at B. The reverse vector from B to A would be written $\overrightarrow{BA}$ but never $\overleftarrow{AB}$. Alternatively, the notation $\underline{a}$ is used, but this is usually restricted to the classroom and hand-written mathematics these days. The notation used in books and lecture notes in Word or Adobe Acrobat (.pdf) is to use a bold character, $\mathbf{a}$. Therefore, wind is $\mathbf{W}$, force is $\mathbf{F}$, velocity is $\mathbf{v}$ and magnetic field is $\mathbf{H}$. Figure 4.1 shows all three notations; it is the bold letter that is used in this text. Many will be familiar with the algebra of vectors, but formally a vector is a new quantity that demands a formal definition as well as statements of their properties. For example, all vectors are *commutative* under addition which implies that $\mathbf{a} + \mathbf{b} = \mathbf{b} + \mathbf{a}$ for all pairs of vectors $\mathbf{a}, \mathbf{b}$; they are

Two and Three Dimensional Calculus: With Applications in Science and Engineering, First Edition. Phil Dyke.
© 2018 John Wiley & Sons Ltd. Published 2018 by John Wiley & Sons Ltd.

also *associative* under addition which means that the order that the addition occurs doesn't affect the result, this is expressed as

$$(a + b) + c = a + (b + c).$$

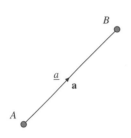

Figure 4.1 The vector having the magnitude equal to the length of the line *AB* and the direction indicated by the arrow over $\overrightarrow{AB}$, together with the alternative $\underline{a}$ or that adopted here **a**.

There is also the *identity* vector **0** that when added to any vector doesn't change its value, and the *inverse* vector $-\mathbf{a}$ that when added to **a** give the identity vector **0**. The generalisation of this is the theory of *linear spaces* sometimes called *vector spaces*. Here, there will be no formal mathematical structures and generalisations, instead the concentration is on practical aspects of calculation and application. This is thus clearly not pure mathematics, but nor is it solely applied mathematics. It is a familiarisation with vectors that is required here so that an understanding is acquired. From this, application will be easier but so will the understanding of the underlying pure mathematical structures. This author prefers the term *mathematical methods* to describe this approach. The next section examines in some detail how vectors are added together.

4.2 Vector Addition

In order to add vectors together, one can either operate geometrically or algebraically. Eventually, algebraic methods will win, but here is a geometric way of adding vectors, and the arrow notation $\overrightarrow{AB}$ is used. Vectors are defined by a magnitude and a direction. If the direction of two vectors $\mathbf{a}, \mathbf{b}$ is the same and their magnitudes are also both equal, then the vectors are equal and we can write $\mathbf{a} = \mathbf{b}$. This means that if vectors are represented by directed line segments, then two parallel lines of the same length can represent the same vector. For example, see Figure 4.2 where the directed line segments $\overrightarrow{AB}$ and $\overrightarrow{DC}$ represent precisely the same vector, so do $\overrightarrow{BC}$ and $\overrightarrow{AD}$. This figure shows how the addition of two vectors works. First of all, represent the two vectors by two directed straight lines $\overrightarrow{AB}$ and $\overrightarrow{AD}$ in Figure 4.2, remembering that this is always possible. One of the lines representing the vectors can always be moved so that they have the same starting point. First of all complete the parallelogram $ABCD$ as shown, then the sum of the two vectors $\overrightarrow{AB} + \overrightarrow{AD}$ is represented by the directed line $\overrightarrow{AC}$ and is in the direction of the main diagonal of the parallelogram $ABCD$. Its magnitude is the length of this diagonal. Vector addition can also be thought of as obeying a triangle law since $\overrightarrow{AB}$ and $\overrightarrow{BC}$ form two sides

Figure 4.2 Adding two vectors geometrically.

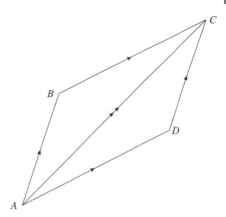

of a triangle, the third side of which is the result of the addition $\overrightarrow{AC}$. If the two vectors are in the same direction, then this diagram still gives the result; it's just that in this case B lies on the line AC and the addition $\overrightarrow{AB} + \overrightarrow{BC} = \overrightarrow{AC}$ happens in a straight line and is just the addition of lengths. The term *resultant* is often used in this context: $\overrightarrow{AC}$ is the resultant $\overrightarrow{AB} + \overrightarrow{BC}$.

There is a way of doing this without geometry, and this is done in the next section. First, some more basics are required. Consider the three Cartesian axes x, y and z that are mutually orthogonal. Suppose there are three vectors $\mathbf{i}$, $\mathbf{j}$ and $\mathbf{k}$ defined to lie along these axes as shown in Figure 4.3. These three vectors are defined as being of unit magnitude or length; they are called *unit vectors*. The magnitude or length of a vector $\mathbf{a}$ is indicated by $|\mathbf{a}|$ and so

$$|\mathbf{i}| = |\mathbf{j}| = |\mathbf{k}| = 1.$$

Figure 4.3 Three mutually orthogonal axes with unit vectors $\mathbf{i}$, $\mathbf{j}$ and $\mathbf{k}$, the vector $\mathbf{a}$ and its components.

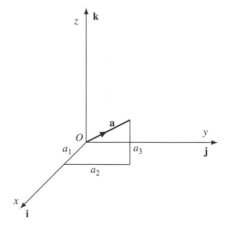

Sometimes, a unit vector is indicated with a carat, $\hat{\mathbf{i}}, \hat{\mathbf{j}}$ and $\hat{\mathbf{k}}$, but this is not done here for these special unit vectors. In general though, if a vector **a** has unit length then it will be written â, and $|\hat{\mathbf{a}}| = 1$.

There is an exception to the rule that parallel lines of equal length represent the same vector. Sometimes, particularly in mechanics where a force is represented by a vector, this force acts along a particular line. A force of the same magnitude acting in the same direction but along a different line is of course different. Such vectors are called *line-bound vectors* and they are certainly necessary for application to mechanics and electromagnetism. The most practical way for dealing with vectors without recourse to geometry is to use *components* and this leads us into using algebra as an alternative.

4.3 Components

Those of you who are familiar with the pure mathematics of linear algebra will know something of linear vector spaces. The use of the term vector is not entirely coincidental, but the definition of a vector space is a lot more general than the vectors defined above. Three-dimensional vectors do form a vector space under the addition rule defined above with scalar multiplication, but the only property taken from vector space theory is that of *linear independence*. For any vector **a**, it is possible to write

$$\mathbf{a} = a_1\mathbf{i} + a_2\mathbf{j} + a_3\mathbf{k},$$

where the scalar quantities a_1, a_2 and a_3 are uniquely defined. These are shown in Figure 4.3 where with some thought, this relation can be seen geometrically too. Using Pythagoras' theorem, the length of the vector **a** can be seen to be given by

$$|\mathbf{a}| = \sqrt{a_1^2 + a_2^2 + a_3^2}.$$

If $\mathbf{a} = \mathbf{0}$, then $a_1 = a_2 = a_3 = 0$ and vice-versa. This property implies $\mathbf{i}, \mathbf{j}$ and $\mathbf{k}$ are linearly independent and that the dimension of the vector space is 3. The three vectors $\mathbf{i}, \mathbf{j}$ and $\mathbf{k}$ form a *basis* for all three vectors (the vector space of ordinary vectors) which means that any vector **a** can be expressed as the sum $a_1\mathbf{i} + a_2\mathbf{j} + a_3\mathbf{k}$ for scalars a_1, a_2 and a_3 and what is more, the expression is unique. If vector spaces are a mystery, then the point to take from this is that **a** and its component form $a_1\mathbf{i} + a_2\mathbf{j} + a_3\mathbf{k}$ can be treated as synonymous. In fact, the triple (a_1, a_2, a_3) is often used as an alternative notation for a vector. Adding two vectors expressed in component form is thankfully straightforward. Consider two vectors $\mathbf{a} = a_1\mathbf{i} + a_2\mathbf{j} + a_3\mathbf{k}$ and $\mathbf{b} = b_1\mathbf{i} + b_2\mathbf{j} + b_3\mathbf{k}$, then

$$\mathbf{a} + \mathbf{b} = (a_1 + b_1)\mathbf{i} + (a_2 + b_2)\mathbf{j} + (a_3 + b_3)\mathbf{k},$$

that is we simply add each corresponding component and form the vector from these. From this, results such as $\mathbf{a} + \mathbf{a} = 2\mathbf{a}$ and $\mathbf{a} - \mathbf{a} = \mathbf{0}$ follow at once; they also follow from the geometric version introduced earlier of course. In general, the addition and subtraction of vectors is quite procedural, so let's just do one example to cement the basic ideas behind addition and subtraction, and finding unit vectors.

Example 4.1   Given the three vectors: $\mathbf{a} = \mathbf{i} + \mathbf{j} + \mathbf{k}, \mathbf{b} = \mathbf{i} - \mathbf{j} + \mathbf{k}$ and $\mathbf{c} = \mathbf{i} + \mathbf{j} - \mathbf{k}$, find (a) the magnitude of $\mathbf{a} + \mathbf{b} + \mathbf{c}$, (b) the unit vector in the direction of $\mathbf{a} - \mathbf{b} + \mathbf{c}$ and (c) a unit vector parallel to $\mathbf{a} + \mathbf{c}$.

Solution:
a) The vector

$$\mathbf{a} + \mathbf{b} + \mathbf{c} = (\mathbf{i} + \mathbf{j} + \mathbf{k}) + (\mathbf{i} - \mathbf{j} + \mathbf{k}) + (\mathbf{i} + \mathbf{j} - \mathbf{k})$$

so

$$\mathbf{a} + \mathbf{b} + \mathbf{c} = 3\mathbf{i} + \mathbf{j} + \mathbf{k}$$

and

$$|\mathbf{a} + \mathbf{b} + \mathbf{c}| = \sqrt{3^2 + 1 + 1} = \sqrt{11}.$$

b) The vector

$$\mathbf{a} - \mathbf{b} + \mathbf{c} = \mathbf{i} + 3\mathbf{j} - \mathbf{k}$$

and since it is not line bound, any vector in this direction will do. Its magnitude is $\sqrt{9 + 1 + 1} = \sqrt{11}$ hence the unit vector in the direction of $\mathbf{a} - \mathbf{b} + \mathbf{c}$ is

$$\frac{1}{\sqrt{11}}(\mathbf{i} + 3\mathbf{j} - \mathbf{k}).$$

c) The vector $\mathbf{a} + \mathbf{c} = 2\mathbf{i} + 2\mathbf{j}$ and again as it is not line bound, similar to part (b) its magnitude is $\sqrt{4 + 4} = 2\sqrt{2}$ so the unit vector in the direction of $2\mathbf{i} + 2\mathbf{j}$ has to be

$$\frac{\mathbf{i} + \mathbf{j}}{\sqrt{2}}.$$

$\square$

Vectors can be a powerful ally in proving theorems of geometry, here is a remarkable example of this.

Example 4.2   Show that the quadrilateral formed by joining the mid-points of an arbitrary quadrilateral must be a parallelogram.

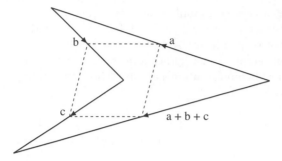

Solution: This is very hard to prove by geometry, especially as an arbitrary quadrilateral need not be planar. Using the properties of vectors, it is easy to prove. Suppose that three of the sides of the quadrilateral are represented by vectors $\mathbf{a}, \mathbf{b}$ and $\mathbf{c}$. Using the property of addition, the fourth can then be $\mathbf{a} + \mathbf{b} + \mathbf{c}$, see Figure 4.4. The four mid-points of the side when joined give a second quadrilateral with sides:

$$\frac{1}{2}(\mathbf{a} + \mathbf{b}), \frac{1}{2}(\mathbf{b} + \mathbf{c}), \frac{1}{2}(\mathbf{a} + \mathbf{b} + \mathbf{c} - \mathbf{c}) \quad \text{and} \quad \frac{1}{2}(\mathbf{a} + \mathbf{b} + \mathbf{c} - \mathbf{a})$$

that is a quadrilateral with pairs of opposite sides parallel and of the same length $\frac{1}{2}(\mathbf{a} + \mathbf{b})$ and $\frac{1}{2}(\mathbf{b} + \mathbf{c})$. The dashed quadrilateral is thus a parallelogram.   □

Many classical theorems found in textbooks devoted to the geometry first done in ancient times by the Greeks can be proved elegantly using vectors, but let us leave the subject here as it is peripheral to the study of two- and three-dimensional calculus. There are some more in the exercises. Vectors can thus be added and subtracted either geometrically, or algebraically via their components, and finding various sums, magnitude and unit vectors are straightforward. The next step is to multiply them.

4.4   Scalar Product

The scalar or dot product of two vectors is most simply defined algebraically as follows:

Definition 4.1   The *scalar product* of two vectors $\mathbf{a} = a_1\mathbf{i} + a_2\mathbf{j} + a_3\mathbf{k}$ and $\mathbf{b} = b_1\mathbf{i} + b_2\mathbf{j} + b_3\mathbf{k}$ is given by

$$\mathbf{a} \cdot \mathbf{b} = a_1 b_1 + a_2 b_2 + a_3 b_3$$

and is a scalar quantity.

An alternative geometrical definition is to define the scalar product as

$$\mathbf{a} \cdot \mathbf{b} = |\mathbf{a}||\mathbf{b}| \cos \theta,$$

Figure 4.5 The triangle *OAB*.

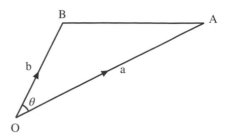

where θ is the angle between the two vectors in the sense of being that angle passed through by turning from the direction of **a** to the direction of **b**. We shall now prove the equivalence of these two definitions with the help of Figure 4.5 where a triangle *OAB* is drawn with the two vectors **a** and **b** clearly shown. First of all, the sides of the triangle are given by

$$OB = |\mathbf{b}| = \sqrt{b_1^2 + b_2^2 + b_3^2},$$

$$OA = |\mathbf{a}| = \sqrt{a_1^2 + a_2^2 + a_3^2},$$

$$AB = |\mathbf{a}-\mathbf{b}| = \sqrt{(a_1 - b_1)^2 + (a_2 - b_2)^2 + (a_3 - b_3)^2},$$

and so using the cosine rule we have

$$(AB)^2 = (OA)^2 + (OB)^2 - 2(OA)(OB)\cos\theta.$$

Expanding the left-hand side:

$$(AB)^2 = (a_1 - b_1)^2 + (a_2 - b_2)^2 + (a_3 - b_3)^2$$
$$= a_1^2 + a_2^2 + a_3^2 + b_1^2 + b_2^2 + b_3^2 - 2(a_1 b_1 + a_2 b_2 + a_3 b_3).$$

Expanding the right-hand side:

$$(OA)^2 + (OB)^2 - 2(OA)(OB)\cos\theta$$
$$= a_1^2 + a_2^2 + a_3^2 + b_1^2 + b_2^2 + b_3^2 - 2|\mathbf{a}||\mathbf{b}|\cos\theta.$$

Equating the two sides, all the squared terms cancel and we are left with

$$-2(a_1 b_1 + a_2 b_2 + a_3 b_3) = -2|\mathbf{a}||\mathbf{b}|\cos\theta,$$

which upon cancelling -2 proves the equivalence of the two definitions of scalar product.

From this more geometric definition of scalar product, one important fact emerges, that is if $\theta = \pi/2$, then $\mathbf{a} \cdot \mathbf{b} = 0$. In other words, if two vectors are at right angles to each other, then their scalar product is zero. The reverse is also true, so that to test whether any two vectors are at right angles, simply calculate their scalar product and see if this is zero. Hence, the following theorem is true.

Theorem 4.1   Two vectors **a** and **b** are at right angles to each other if and only if $\mathbf{a} \cdot \mathbf{b} = 0$.

This leads naturally to determining a procedure for finding the angle between vectors. To do this, it is often useful to define *direction cosines*:

Definition 4.2   If the angles between a vector **a** and the co-ordinate axes **i**, **j** and **k** are respectively α, β and γ, then $\cos \alpha$, $\cos \beta$ and $\cos \gamma$ are called the *direction cosines* of the vector **a**.

Given that the vector **a** is not line bound, it can be thought of as emerging from the origin O so it is not difficult to see that

$$\frac{\mathbf{a}}{|\mathbf{a}|} = \cos(\alpha)\mathbf{i} + \cos(\beta)\mathbf{j} + \cos(\gamma)\mathbf{k}$$

from elementary trigonometry. Hence, the direction cosines $\cos \alpha$, $\cos \beta$ and $\cos \gamma$ must be given by

$$\cos \alpha = \frac{\mathbf{a} \cdot \mathbf{i}}{|\mathbf{a}|}, \cos \beta = \frac{\mathbf{a} \cdot \mathbf{j}}{|\mathbf{a}|}, \quad \text{and} \quad \cos \gamma = \frac{\mathbf{a} \cdot \mathbf{k}}{|\mathbf{a}|}$$

as $\mathbf{i} \cdot \mathbf{j} = \mathbf{k} \cdot \mathbf{i} = \mathbf{j} \cdot \mathbf{k} = 0$.

Before doing an example, it is worth looking more closely at the geometry of vectors. It is tempting to think of a vector such as $\mathbf{i} + 3\mathbf{j} + 5\mathbf{k}$ as the vector equation of a straight line. It is not. Certainly, it can be represented by an arrow in space, but it is a direction and a length. The vector equation to a straight line belongs to a later section. However, it is true that the vector connecting two points (x_1, y_1, z_1) and (x_2, y_2, z_2) has the length $\sqrt{(x_2 - x_1)^2 + (y_2 - y_1)^2 + (z_2 - z_1)^2}$ and direction $(x_2 - x_1)\mathbf{i} + (y_2 - y_1)\mathbf{j} + (z_2 - z_1)\mathbf{k}$ and this does enable us to do an example.

Example 4.3   A cube has vertices $(\pm 1, \pm 1, \pm 1)$. Find the angle of intersection of the two long diagonals $(1, 1, 1)$ to $(-1, -1, -1)$ and $(-1, 1, 1)$ to $(1, -1, -1)$.

Solution: The lengths of the diagonals are both $2\sqrt{3}$ and it is possible to use the definitions of scalar product to calculate the required angle as follows. Set the vectors $\mathbf{a} = (1, 1, 1) - (-1, -1, -1) = 2\mathbf{i} + 2\mathbf{j} + 2\mathbf{k}$ and $\mathbf{b} = (-1, 1, 1) - (1, -1, -1) = -2\mathbf{i} + 2\mathbf{j} + 2\mathbf{k}$. Since

$$\mathbf{a} \cdot \mathbf{b} = a_1 b_1 + a_2 b_2 + a_3 b_3 = |\mathbf{a}||\mathbf{b}| \cos \theta$$

this gives

$$-4 + 4 + 4 = 2\sqrt{3} \cdot 2\sqrt{3} \cos \theta$$

or

$$\cos \theta = \frac{1}{3}.$$

Hence, $\theta = \cos^{-1}(1/3) = 70°30'$. □

Definition 4.3 The *centroid* of a set of points with position vectors $\mathbf{a}_1, \mathbf{a}_2, \mathbf{a}_3, \ldots, \mathbf{a}_n$ is given by the expression $\bar{\mathbf{a}}$, where

$$\bar{\mathbf{a}} = \frac{\mathbf{a}_1 + \mathbf{a}_2 + \mathbf{a}_3 + \cdots + \mathbf{a}_n}{n}.$$

Example 4.4 Show that the position of the centroid is independent of the choice of origin.

Solution: Suppose the point O' is at the position vector $\mathbf{k}$ from the original choice of origin O, therefore any vector with position vector $\mathbf{a}_i$ with respect to O will have position vector $\mathbf{a}_i - \mathbf{k}$ relative to the new origin O'. Hence, the centroid relative to O' will be

$$\bar{\mathbf{a}'} = \frac{\mathbf{a}_1 - \mathbf{k} + \mathbf{a}_2 - \mathbf{k} + \mathbf{a}_3 - \mathbf{k} + \cdots + \mathbf{a}_n - \mathbf{k}}{n} = \bar{\mathbf{a}} - \mathbf{k},$$

which is the original centroid referred to the new origin. Hence, the centroid is independent of choice of origin. □

4.5 Vector Product

The second way to multiply two vectors is via the vector or cross product. The definition equivalent to the one given for the scalar product is in terms of components.

Definition 4.4 The *vector product* of two vectors $\mathbf{a} = a_1\mathbf{i} + a_2\mathbf{j} + a_3\mathbf{k}$ and $\mathbf{b} = b_1\mathbf{i} + b_2\mathbf{j} + b_3\mathbf{k}$ is given by

$$\mathbf{a} \times \mathbf{b} = (a_2 b_3 - a_3 b_2)\mathbf{i} + (a_3 b_1 - a_1 b_3)\mathbf{j} + (a_1 b_2 - a_2 b_1)\mathbf{k}.$$

It is a vector quantity in the direction perpendicular to both $\mathbf{a}$ and $\mathbf{b}$.

Another way of writing this is to use a determinant and to write

$$\mathbf{a} \times \mathbf{b} = \begin{vmatrix} \mathbf{i} & \mathbf{j} & \mathbf{k} \\ a_1 & a_2 & a_3 \\ b_1 & b_2 & b_3 \end{vmatrix}.$$

If **a** is parallel to **b**, then $\mathbf{a} = \lambda\mathbf{b}$ for some scalar λ, in which case $\mathbf{a} \times \mathbf{b} = \mathbf{0}$ as is obvious from either the definition or the determinant form. An alternative definition, equivalent to that above is

$$\mathbf{a} \times \mathbf{b} = \hat{\mathbf{n}}|\mathbf{a}||\mathbf{b}| \sin\theta,$$

where $\hat{\mathbf{n}}$ is the unit vector perpendicular to both **a** and **b**. The magnitude of this vector is the area of the parallelogram that has **a** and **b** as the two parallel sides, $ABCD$ in Figure 4.2. However, in order to show that the two expressions for vector products are the same, algebra alone is used. Start with the expression

$$\begin{aligned}|\mathbf{a} \times \mathbf{b}|^2 &= |\mathbf{a}|^2|\mathbf{b}|^2\sin^2\theta, \\ &= |\mathbf{a}|^2|\mathbf{b}|^2(1 - \cos^2\theta), \\ &= |\mathbf{a}|^2|\mathbf{b}|^2 - |\mathbf{a}|^2|\mathbf{b}|^2\cos^2\theta, \\ &= [\mathbf{a}|^2|\mathbf{b}|^2 - (\mathbf{a} \cdot \mathbf{b})^2.\end{aligned}$$

Then, insert the co-ordinate expression for each vector into the right-hand side that becomes

$$(a_1^2 + a_2^2 + a_3^2)(b_1^2 + b_2^2 + b_3^2) - (a_1b_1 + a_2b_2 + a_3b_3)^2.$$

Without completely multiplying this expression out, one can spot that the terms $a_1^2b_1^2, a_2^2b_2^2$ and $a_3^2b_3^2$ generated by the first and second groups of terms cancel. What remains can be regrouped as

$$(a_1b_2 - a_2b_1)^2 + (a_1b_3 - a_3b_1)^2 + (a_2b_3 - a_3b_2)^2,$$

which is $|\mathbf{a} \times \mathbf{b}|^2$ straight from the definition of cross product, so the magnitudes of the two expressions for cross products are equal. Of course, for vectors to be the same, the direction needs to be the same too. The determinant version of the definition comes to our aid here. Take the scalar product of **a** with the determinant

$$\mathbf{a} \cdot (\mathbf{a} \times \mathbf{b}) = \mathbf{a} \cdot \begin{vmatrix} \mathbf{i} & \mathbf{j} & \mathbf{k} \\ a_1 & a_2 & a_3 \\ b_1 & b_2 & b_3 \end{vmatrix} = \begin{vmatrix} a_1 & a_2 & a_3 \\ a_1 & a_2 & a_3 \\ b_1 & b_2 & b_3 \end{vmatrix} = 0$$

as determinants with two rows the same are zero. Similarly, $\mathbf{b} \cdot (\mathbf{a} \times \mathbf{b}) = 0$ thus $\mathbf{a} \times \mathbf{b}$ is perpendicular to both **a** and **b**. As this is the same direction as $\hat{\mathbf{n}}$, the proof is complete and so as for scalar products the two definitions of cross product can be used interchangeably. The normal direction $\hat{\mathbf{n}}$ is such that $\mathbf{a}, \mathbf{b}$ and $\hat{\mathbf{n}}$ form a right-handed system in that order.

Example 4.5 Show that for all vectors **a**, **b** and **c**

$$\mathbf{a} \times (\mathbf{b} + \mathbf{c}) = \mathbf{a} \times \mathbf{b} + \mathbf{a} \times \mathbf{c}.$$

Solution: Define the vector $\mathbf{p} = \mathbf{b} + \mathbf{c}$ so that the components are related by $p_i = b_i + c_i$ where $i = 1, 2, 3$. By definition

$$\mathbf{a} \times \mathbf{p} = \begin{vmatrix} \mathbf{i} & \mathbf{j} & \mathbf{k} \\ a_1 & a_2 & a_3 \\ p_1 & p_2 & p_3 \end{vmatrix}.$$

The first component of the right-hand side is

$$\mathbf{i}(a_2 p_3 - a_3 p_2) = \mathbf{i}(a_2(b_3 + c_3) - a_3(b_2 + c_2))$$
$$= \mathbf{i}(a_2 b_3 - a_3 b_2) + \mathbf{i}(a_2 c_3 - a_3 c_2)$$

and similarly with the other two components. Grouping the three right-hand sides give the vector $\mathbf{a} \times \mathbf{b} + \mathbf{a} \times \mathbf{c}$ and this proves the result.   □

One of the most useful properties of vector products is their ability to model rotation. Figure 4.6 shows a disc spinning with angular speed ω. The quantity ω is a vector with magnitude equal to the rate at which the angle made by a radius and a fixed line changes with time, and by convention its direction is perpendicular to the disc in the direction of a right-handed screw thread. This is shown in Figure 4.6. The vector product $\omega \times \mathbf{r}$ is a quantity perpendicular to both ω and $\mathbf{r}$, so it is in the plane of the grey spinning disc pointing into the paper/screen as indicated in the figure. Its magnitude is $\omega r \sin \theta$, so it is the velocity $\mathbf{v}$ of any point on the rim of the rotating disc. The vector product $\omega \times \mathbf{r}$ also describes the velocity *anywhere* on the disc as both $\mathbf{r}$ and θ change. At the centre of the disc $\theta = 0$, and the disc's velocity is zero. If in Figure 4.6 $OA = a$, then simple trigonometry gives $r = a \sec \theta$ so $|\omega \times \mathbf{r}| = \omega r \sin \theta = \omega a \tan \theta$, and

Figure 4.6 A rotating disc shown as grey with an arbitrary point on its rim P at position vector $\mathbf{r}$ rotating with angular velocity ω.

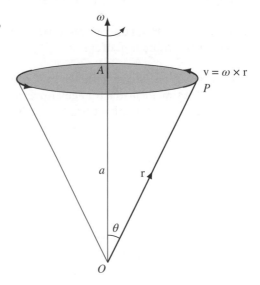

this formula shows that as θ approaches $\pi/2$ the speed approaches ∞. This is consistent with the disc getting very large and the rim speed getting larger and larger as the angle θ gets close to $\pi/2$ although of course it never reached.

Example 4.6   A disc is spinning with speed 6 rad s^{-1} about an axis in the direction of unit vector $\frac{1}{3}(\mathbf{i} + 2\mathbf{j} - 2\mathbf{k})$. A point on its rim has position vector $\mathbf{r}$. Determine the velocity there.

Solution: The vector expressing the angular velocity of the disc is

$$\omega = 6 \times \frac{1}{3}(\mathbf{i} + 2\mathbf{j} - 2\mathbf{k}) = 2\mathbf{i} + 4\mathbf{j} - 4\mathbf{k}$$

and the velocity of the point on the rim with position vector $\mathbf{r}$ is $\omega \times \mathbf{r}$ and so is given by the determinant

$$\begin{vmatrix} \mathbf{i} & \mathbf{j} & \mathbf{k} \\ 2 & 4 & -4 \\ x & y & z \end{vmatrix} = (4z + 4y)\mathbf{i} - (4x + 2z)\mathbf{j} + (2y - 4x)\mathbf{k}.$$

$\square$

Just to continue with the last example, using the formula generated, it is easy to find the velocity of any point on the disc. For example, if the point $(1, 1, 1)$ was on its rim, then the velocity there is $8\mathbf{i} - 6\mathbf{j} - 2\mathbf{k}$. A slightly trickier vector calculation would be to calculate the speed of the rim given the radius of the disc. At this point, it would be easier not to use vectors and to spot that with a rotation rate of 6 rad s^{-1} the rim has to be travelling at $6a$ ms^{-1}. To do this using vectors and magnitudes would be long; spotting the alternative is the essence of applied mathematics, which helps solving real problems in the simplest way.

Example 4.7   Find the shortest distance between the non-intersecting edges of a regular tetrahedron of side a.

Solution: A tetrahedron, shown in Figure 4.7, is the smallest of the Platonic solids, sometimes called a triangular pyramid. The opposite edges are what are called skew lines; lines that do not meet. This in fact is another example where the power of vectors is used to solve a geometry problem. The sides can be labelled as vectors, and the edges OC labelled vector $\mathbf{c}$ and BA labelled vector $\mathbf{f}$ are the sides chosen for the problem, they are of course line bound in this instance. The cross product $\mathbf{c} \times \mathbf{f}$ is at right angles to both $\mathbf{c}$ and $\mathbf{f}$ so the shortest distance labelled PQ in Figure 4.7 is in this direction. The method now is to find a line that joins the two skew lines OC and BA, OB will do, as would OA. Then, by taking the scalar product of either $\mathbf{a}$ or $\mathbf{b}$ with $\mathbf{c} \times \mathbf{f}$ one gets the projection of the vector $\mathbf{a}$ on to the vector $\mathbf{c} \times \mathbf{f}$ which is PQ the required length. Here are the details. Choose the origin of co-ordinates as the point in the base that has the apex C above it. The z-axis thus goes through C; choose the y-axis

Figure 4.7 The tetrahedron
$O'ABC$ with PQ the shortest
distance between the sides BA
represented by vector **f**, and
$O'C$ represented by vector **c**.
The point Q is behind and
below P.

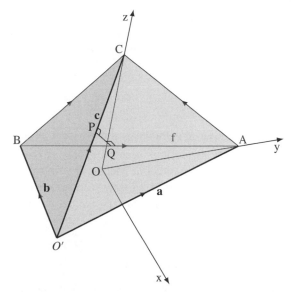

to pass through the vertex A and then the x-axis will be parallel to the edge
AB as shown in Figure 4.7. Some trigonometry reveals that the points O', A, B
and C are respectively $\left(\frac{1}{2}a, -\frac{1}{6}a\sqrt{3}, 0\right)$, $\left(0, \frac{1}{3}a\sqrt{3}, 0\right)$, $\left(-\frac{1}{2}a, -\frac{1}{6}a\sqrt{3}, 0\right)$ and
$\left(0, 0, \frac{1}{3}a\sqrt{6}\right)$. Now, use vector algebra to find the vector representations of the
sides.

$$\mathbf{a} = \overrightarrow{OA} - \overrightarrow{OO'} = \left(0, \frac{1}{3}a\sqrt{3}, 0\right) - \left(\frac{1}{2}a, -\frac{1}{6}a\sqrt{3}, 0\right) = \left(-\frac{1}{2}a, \frac{1}{2}a\sqrt{3}, 0\right),$$

$$\mathbf{b} = \overrightarrow{OB} - \overrightarrow{OO'} = \left(-\frac{1}{2}a, -\frac{1}{6}a\sqrt{3}, 0\right) - \left(\frac{1}{2}a, -\frac{1}{6}a\sqrt{3}, 0\right) = (-a, 0, 0),$$

$$\mathbf{c} = \overrightarrow{OC} - \overrightarrow{OO'} = \left(0, 0, \frac{1}{6}a\sqrt{3}\right) - \left(\frac{1}{2}a, -\frac{1}{6}a\sqrt{3}, 0\right)$$
$$= \left(-\frac{1}{2}a, \frac{1}{6}a\sqrt{3}, \frac{1}{3}a\sqrt{6}\right),$$

$$\mathbf{f} = \mathbf{a} - \mathbf{b} = \left(\frac{1}{2}a, \frac{1}{2}a\sqrt{3}, 0\right).$$

Therefore, the cross product

$$\mathbf{c} \times \mathbf{f} = \begin{vmatrix} \mathbf{i} & \mathbf{j} & \mathbf{k} \\ -\frac{1}{2}a & \frac{1}{6}a\sqrt{3} & \frac{1}{3}a\sqrt{6} \\ \frac{1}{2}a & \frac{1}{2}a\sqrt{3} & 0 \end{vmatrix} = \left(-\frac{1}{2}a^2\sqrt{2}, \frac{1}{6}a^2\sqrt{6}, -\frac{1}{3}a^2\sqrt{3}\right)$$

gives the vector in the direction of the required perpendicular. To get the length,
simply project any vector that joins **c** and **f** on to this direction, choose **b**.
The scalar product $(\mathbf{c} \times \mathbf{f}) \cdot \mathbf{b} = |\mathbf{c} \times \mathbf{f}||\mathbf{b}| \cos\theta$ where θ is the angle between

these two vectors. However, $|PQ| = |\mathbf{b}| \cos\theta$ using the simple trigonometry of a right-angled triangle, therefore

$$PQ = \frac{(\mathbf{c} \times \mathbf{f}) \cdot \mathbf{b}}{|\mathbf{c} \times \mathbf{f}|} = \frac{\left(-\frac{1}{2}a^2\sqrt{2}, \frac{1}{6}a^2\sqrt{6}, -\frac{1}{3}a^2\sqrt{3}\right) \cdot (-a, 0, 0)}{a^2}$$

$$= \frac{1}{2}a\sqrt{2}$$

gives the required distance. Just to be clear about what has happened here, the vector $\mathbf{b}$ as representing the side $\overrightarrow{O'B}$ is line bound, but $\mathbf{b}$ itself is not. Therefore, mentally move it parallel to itself so that the end at B is now at P. The triangle calculation $|PQ| = |\mathbf{b}| \cos\theta$ is now done as is the elimination of $\mathbf{b}\cos\theta$ to give the formula that enables the calculation of PQ. Now, move $\mathbf{b}$ back to $\overrightarrow{O'B}$ and all is well.

□

4.5.1 Scalar Triple Product

In this subsection, let us consider first the quantities $\mathbf{a} \cdot (\mathbf{b} \times \mathbf{c}), \mathbf{c} \cdot (\mathbf{a} \times \mathbf{b})$ and $\mathbf{b} \cdot (\mathbf{c} \times \mathbf{a})$ where $\mathbf{a} = a_1\mathbf{i} + a_2\mathbf{j} + a_3\mathbf{k}, \mathbf{b} = b_1\mathbf{i} + b_2\mathbf{j} + b_3\mathbf{k}$ and $\mathbf{c} = c_1\mathbf{i} + c_2\mathbf{j} + c_3\mathbf{k}$. From the component form of these three vectors, these three *scalar triple products* as they are called are given by

$$\begin{vmatrix} a_1 & a_2 & a_3 \\ b_1 & b_2 & b_3 \\ c_1 & c_2 & c_3 \end{vmatrix}, \quad \begin{vmatrix} c_1 & c_2 & c_3 \\ a_1 & a_2 & a_3 \\ b_1 & b_2 & b_3 \end{vmatrix}, \quad \text{and} \quad \begin{vmatrix} b_1 & b_2 & b_3 \\ c_1 & c_2 & c_3 \\ a_1 & a_2 & a_3 \end{vmatrix},$$

respectively. From the property of a determinant that one can cyclically permute rows without changing its value, these three expressions must be equal. Hence,

$$\mathbf{a} \cdot (\mathbf{b} \times \mathbf{c}) = \mathbf{c} \cdot (\mathbf{a} \times \mathbf{b}) = \mathbf{b} \cdot (\mathbf{c} \times \mathbf{a}),$$

which on the face of it looks a remarkable result, but only remarkable until the geometrical interpretation that follows. In fact, each scalar triple product equals the same volume called a parallelepiped, see Figure 4.8. In this figure, the base shaded has area $|\mathbf{a} \times \mathbf{b}|$. The vector product $\mathbf{a} \times \mathbf{b}$ has a direction perpendicular to the grey parallelogram therefore the scalar triple product $(\mathbf{a} \times \mathbf{b}) \cdot \mathbf{c}$ which is

$$|\mathbf{a} \times \mathbf{b}||\mathbf{c}|\hat{\mathbf{n}} \sin\theta$$

is the product of the area of the base (shaded) times the perpendicular height of the solid. Hence, it is equal to the volume of the parallelepiped. As can be confirmed by looking at Figure 4.8, this is the product of the area shaded grey, the sine of the angle θ between the normal to this parallelogram and the magnitude of the vector $\mathbf{c}$. This is the area of the parallelogram times the vertical distance

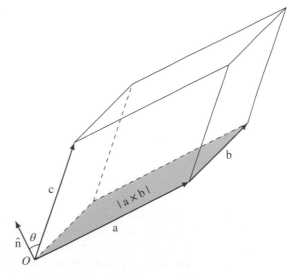

Figure 4.8 The parallelepiped formed from the three vectors **a**, **b** and **c**.

between this parallelogram and the opposite parallelogram face, and this is precisely the formula for the volume of the parallelepiped shown. Obviously, as scalar products are commutative, this volume is also given by $\mathbf{c} \cdot (\mathbf{a} \times \mathbf{b})$. There is a symmetry here, as the areas of the other faces given by $|(\mathbf{b} \times \mathbf{c})|$ and $|(\mathbf{c} \times \mathbf{a})|$ multiplied by the sine of the angle between the normal and the third vector will also give this same volume. This makes the equality of the scalar products more obvious. Thus:

$$\mathbf{a} \cdot (\mathbf{b} \times \mathbf{c}) = \mathbf{c} \cdot (\mathbf{a} \times \mathbf{b}) = \mathbf{b} \cdot (\mathbf{c} \times \mathbf{a}) = (\mathbf{b} \times \mathbf{c}) \cdot \mathbf{a} = (\mathbf{a} \times \mathbf{b}) \cdot \mathbf{c}$$
$$= (\mathbf{c} \times \mathbf{a}) \cdot \mathbf{b}.$$

In addition, since $(\mathbf{a} \times \mathbf{b}) = -(\mathbf{b} \times \mathbf{a})$ this volume is also equal to

$$-\mathbf{a} \cdot (\mathbf{c} \times \mathbf{b}) = -\mathbf{c} \cdot (\mathbf{b} \times \mathbf{a}) = -\mathbf{b} \cdot (\mathbf{a} \times \mathbf{c}) = -(\mathbf{c} \times \mathbf{b}) \cdot \mathbf{a} = -(\mathbf{b} \times \mathbf{a}) \cdot \mathbf{c}$$
$$= -(\mathbf{a} \times \mathbf{c}) \cdot \mathbf{b}.$$

One fact that emerges from all these equalities is $\mathbf{a} \cdot (\mathbf{b} \times \mathbf{c}) = (\mathbf{a} \times \mathbf{b}) \cdot \mathbf{c}$ that is the dot and the cross operations can be interchanged, a useful result in some computations.

Example 4.8   Determine the volume of a tetrahedron in terms of a scalar triple product given on vertex is the origin and the other three have position vectors **a**, **b** and **c**.

Solution: The formula for the volume of a tetrahedron, or of any pyramid in fact is one-third base area times the height. The Figure 4.9 shows the tetrahedron, the base OAB is a triangle of area $\frac{1}{2}OA \cdot OB \sin \theta$ where θ is the angle

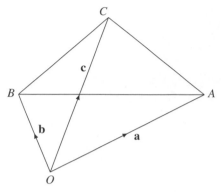

Figure 4.9 The tetrahedron *OABC*

$A\hat{O}B$. As $\mathbf{a} \times \mathbf{b} = |\mathbf{a}||\mathbf{b}|\sin\theta\,\hat{\mathbf{n}}$ this area can be written in vector form $\frac{1}{2}\mathbf{a} \times \mathbf{b}$. The direction $\hat{\mathbf{n}}$ is perpendicular to the triangle OAB. The height of the tetrahedron assuming OAB as the base is the projection of $\mathbf{c}$ on to this perpendicular direction $\hat{\mathbf{n}}$, that is $\mathbf{c} \cdot \hat{\mathbf{n}}$. Hence, the volume of the tetrahedron is

$$\frac{1}{3}\mathbf{c} \cdot \hat{\mathbf{n}}\left(\frac{1}{2}OA \cdot OB\sin\theta\right) = \frac{1}{3}\mathbf{c} \cdot \left(\frac{1}{2}\mathbf{a} \times \mathbf{b}\right)$$

or $\frac{1}{6}\mathbf{c} \cdot (\mathbf{a} \times \mathbf{b})$, a sixth of the scalar triple product of the vectors $\mathbf{a}, \mathbf{b}$ and $\mathbf{c}$. To see how to fit six of these into the parallelepiped of Figure 4.8 is challenging. $\square$

Starting with three non-coplanar, non-zero vectors $\mathbf{a}, \mathbf{b}$ and $\mathbf{c}$, define three more $\mathbf{a}', \mathbf{b}'$ and $\mathbf{c}'$ through the relationships:

$$\mathbf{a}' = \frac{\mathbf{b} \times \mathbf{c}}{\mathbf{a} \cdot \mathbf{b} \times \mathbf{c}}, \mathbf{b}' = \frac{\mathbf{c} \times \mathbf{a}}{\mathbf{a} \cdot \mathbf{b} \times \mathbf{c}}, \text{ and } \mathbf{c}' = \frac{\mathbf{a} \times \mathbf{b}}{\mathbf{a} \cdot \mathbf{b} \times \mathbf{c}}$$

then the set $\mathbf{a}', \mathbf{b}'$ and $\mathbf{c}'$ are the *reciprocal vectors* to the set $\mathbf{a}, \mathbf{b}$ and $\mathbf{c}$. Various properties can be proved easily such as

$$\mathbf{a}' \cdot \mathbf{a} = \mathbf{a} \cdot \mathbf{a}' = 1,$$
$$\mathbf{b}' \cdot \mathbf{b} = \mathbf{b} \cdot \mathbf{b}' = 1,$$
$$\mathbf{c}' \cdot \mathbf{c} = \mathbf{c} \cdot \mathbf{c}' = 1,$$
$$\mathbf{a}' \cdot \mathbf{b} = \mathbf{a} \cdot \mathbf{b}' = 0,$$
$$\mathbf{b}' \cdot \mathbf{c} = \mathbf{b} \cdot \mathbf{c}' = 0,$$
$$\mathbf{c}' \cdot \mathbf{a} = \mathbf{c} \cdot \mathbf{a}' = 0.$$

In addition,

$$\mathbf{a} \cdot \mathbf{b} \times \mathbf{c} = \frac{1}{\mathbf{a}' \cdot \mathbf{b}' \times \mathbf{c}'}$$

the origin of the term reciprocal for these sets of vectors. It is also true that

$$\mathbf{r} = (\mathbf{r} \cdot \mathbf{a}')\mathbf{a} + (\mathbf{r} \cdot \mathbf{b}')\mathbf{b} + (\mathbf{r} \cdot \mathbf{c}')\mathbf{c}.$$

4.5.2 Vector Triple Product

The second triple product is the *vector triple product* $\mathbf{a} \times (\mathbf{b} \times \mathbf{c})$. The result is a vector, and the parentheses are essential as the order of execution of the cross products certainly matters since $\mathbf{a} \times (\mathbf{b} \times \mathbf{c}) \neq (\mathbf{a} \times \mathbf{b}) \times \mathbf{c}$. This can be seen at once as the vectors $\mathbf{a}$ and $\mathbf{b} \times \mathbf{c}$ are perpendicular to the vector $\mathbf{a} \times (\mathbf{b} \times \mathbf{c})$ and this is not the same direction as the vector $(\mathbf{a} \times \mathbf{b}) \times \mathbf{c}$ which is perpendicular to the two different vectors $\mathbf{a} \times \mathbf{b}$ and $\mathbf{c}$. A very useful result for vector triple products can be stated as the following theorem.

Theorem 4.2   For any three vectors, $\mathbf{a}, \mathbf{b}$ and $\mathbf{c}$, it is true that

$$\mathbf{a} \times (\mathbf{b} \times \mathbf{c}) = \mathbf{b}(\mathbf{a} \cdot \mathbf{c}) - \mathbf{c}(\mathbf{a} \cdot \mathbf{b}).$$

Solution: The vector $\mathbf{a} \times (\mathbf{b} \times \mathbf{c})$ has to be perpendicular to both $\mathbf{a}$ and $\mathbf{b} \times \mathbf{c}$. Now, there are only three perpendicular directions in our physical world (excluding quantum mechanics and relativity) and the vector $\mathbf{b} \times \mathbf{c}$ is also perpendicular to both $\mathbf{b}$ and $\mathbf{c}$. It has to be true therefore that the triple vector product $\mathbf{a} \times (\mathbf{b} \times \mathbf{c})$ is in the same plane as $\mathbf{b}$ and $\mathbf{c}$. Hence, there exist two scalars λ and μ such that

$$\mathbf{a} \times (\mathbf{b} \times \mathbf{c}) = \lambda \mathbf{b} + \mu \mathbf{c}.$$

Take the scalar product of this equation with the vector $\mathbf{a}$. The left-hand side is immediately zero as $\mathbf{a} \times (\mathbf{b} \times \mathbf{c})$ is perpendicular to $\mathbf{a}$ and remember that the scalar product of two perpendicular vectors is zero. Therefore,

$$0 = \lambda(\mathbf{a} \cdot \mathbf{b}) + \mu(\mathbf{a} \cdot \mathbf{c}). \tag{4.1}$$

This last equation is of the form $ax + by = 0$ for all x, y and so it must be possible to write $a = \alpha y, b = -\alpha x$ and satisfy the equation for all α. Thus, to solve Equation 4.1 write $\lambda = \alpha(\mathbf{a} \cdot \mathbf{c}), \mu = -\alpha(\mathbf{a} \cdot \mathbf{b})$ so that

$$\mathbf{a} \times (\mathbf{b} \times \mathbf{c}) = \alpha[\mathbf{b}(\mathbf{a} \cdot \mathbf{c}) - \mathbf{c}(\mathbf{a} \cdot \mathbf{b})].$$

This is true for all vectors $\mathbf{a}, \mathbf{b}$ and $\mathbf{c}$ called an identity in mathematics, so to find the constant α special values can be chosen such as $\mathbf{a} = \mathbf{b} = \mathbf{i}, \mathbf{c} = \mathbf{j}$ where $\mathbf{i}, \mathbf{j}$ and $\mathbf{k}$ are the unit vectors along the x, y and z-axes respectively. This leads to Equation 4.1 yielding $-\mathbf{j} = -\alpha \mathbf{j}$, so $\alpha = 1$ and the theorem is proved.   □

Here is an example that might be surprising:

Example 4.9   Show that any vector $\mathbf{p}$ can be expressed in the form:

$$\frac{((\mathbf{p} \times \mathbf{b}) \cdot \mathbf{c})\mathbf{a} + ((\mathbf{p} \times \mathbf{c}) \cdot \mathbf{a})\mathbf{b} + ((\mathbf{p} \times \mathbf{a}) \cdot \mathbf{b})\mathbf{c}}{(\mathbf{a} \times \mathbf{b}) \cdot \mathbf{c}}.$$

Solution: This is solved by first considering the triple vector product $(\mathbf{a} \times \mathbf{b}) \times \mathbf{q}$ where $\mathbf{q} = \mathbf{c} \times \mathbf{p}$. Using the expansion of vector triple product derived earlier,

$$(\mathbf{a} \times \mathbf{b}) \times \mathbf{q} = (\mathbf{a} \cdot \mathbf{q})\mathbf{b} - (\mathbf{b} \cdot \mathbf{q})\mathbf{a}.$$

Writing $\mathbf{s} = \mathbf{a} \times \mathbf{b}$ the product of the four vectors can also be written

$$(\mathbf{a} \times \mathbf{b}) \times \mathbf{q} = (\mathbf{a} \times \mathbf{b}) \times (\mathbf{c} \times \mathbf{p}) = \mathbf{s} \times (\mathbf{c} \times \mathbf{p})$$

and the right-hand side is $(\mathbf{s} \cdot \mathbf{p})\mathbf{c} - (\mathbf{s} \cdot \mathbf{c})\mathbf{p}$. Therefore, now equate the two different expansions

$$(\mathbf{a} \cdot \mathbf{q})\mathbf{b} - (\mathbf{b} \cdot \mathbf{q})\mathbf{a} = (\mathbf{s} \cdot \mathbf{p})\mathbf{c} - (\mathbf{s} \cdot \mathbf{c})\mathbf{p}$$

or

$$\mathbf{a} \cdot (\mathbf{c} \times \mathbf{p})\mathbf{b} - \mathbf{b} \cdot (\mathbf{c} \times \mathbf{p})\mathbf{a} = ((\mathbf{a} \times \mathbf{b}) \cdot \mathbf{p})\mathbf{c} - ((\mathbf{a} \times \mathbf{b}) \cdot \mathbf{c})\mathbf{p}.$$

The last term in this equation is now made the subject:

$$(\mathbf{a} \times \mathbf{b} \cdot \mathbf{c})\mathbf{p} = (\mathbf{b} \cdot \mathbf{c} \times \mathbf{p})\mathbf{a} - (\mathbf{a} \cdot \mathbf{c} \times \mathbf{p})\mathbf{b} + (\mathbf{a} \times \mathbf{b} \cdot \mathbf{p})\mathbf{c};$$

unnecessary parentheses have been removed, that makes it easier to see the next and last step. Exchange dot and cross products, and swap the order of the second vector product on the right, changing the sign to plus to give

$$(\mathbf{a} \times \mathbf{b} \cdot \mathbf{c})\mathbf{p} = (\mathbf{p} \times \mathbf{b} \cdot \mathbf{c})\mathbf{a} + (\mathbf{p} \times \mathbf{c} \cdot \mathbf{a})\mathbf{b} + (\mathbf{p} \times \mathbf{a} \cdot \mathbf{b})\mathbf{c},$$

which upon division by the scalar triple product $\mathbf{a} \times \mathbf{b} \cdot \mathbf{c}$ gives the required result:

$$\mathbf{p} = \frac{(\mathbf{p} \times \mathbf{b} \cdot \mathbf{c})\mathbf{a} + (\mathbf{p} \times \mathbf{c} \cdot \mathbf{a})\mathbf{b} + (\mathbf{p} \times \mathbf{a} \cdot \mathbf{b})\mathbf{c}}{\mathbf{a} \times \mathbf{b} \cdot \mathbf{c}}.$$

It does look strange without the parentheses, but it is perfectly well defined. □

Put $\mathbf{a} = \mathbf{i}, \mathbf{b} = \mathbf{j}$ and $\mathbf{c} = \mathbf{k}$ in the last example and see what you get.

Exercises

4.1 Determine the vector that joins the tip of $\mathbf{a} = \mathbf{i} + 2\mathbf{j} + 3\mathbf{k}$ to the tip of $\mathbf{b} = 3\mathbf{i} + 2\mathbf{j} + \mathbf{k}$.

4.2 With $\mathbf{a}$ and $\mathbf{b}$ as in the last question, show that the vector $\frac{1}{2}(\mathbf{a} + \mathbf{b})$ is from the origin to the mid-point of the third side of the triangle.

4.3 Find the lengths of the sides of the triangle of Exercise 4.1.

4.4 If $\cos \alpha$, $\cos \beta$ and $\cos \gamma$ are the direction cosines, show that

$$\cos^2\alpha + \cos^2\beta + \cos^2\gamma = 1.$$

4.5 Determine the angles of the triangle of Exercise 4.1.

4.6 A triangle has vertices $A(1, 2, 4)$, $B(-2, 2, 1)$ and $C(2, 4, -3)$. Show that $\cos A = \sqrt{3}/3$, $\cos C = \sqrt{6}/3$ and that B is a right angle.

4.7 Define the *inner product* $\langle \mathbf{a}, \mathbf{b} \rangle$ as an operation between vectors that satisfy the following four laws:
a) $\langle \mathbf{a}, \mathbf{a} \rangle \geq 0$ for all $\mathbf{a}, \mathbf{b}$,
b) $\langle \mathbf{a}, \mathbf{a} \rangle = 0$ only if $\mathbf{a} = \mathbf{0}$,
c) $\langle \mathbf{a}, \mathbf{b} \rangle = \langle \mathbf{b}, \mathbf{a} \rangle$ for all vectors $\mathbf{a}, \mathbf{b}$,
d) $\langle (a\mathbf{a} + b\mathbf{b}), \mathbf{c} \rangle = a\langle \mathbf{a}, \mathbf{c} \rangle + b\langle \mathbf{b}, \mathbf{c} \rangle$ for real scalars a, b for all vectors $\mathbf{a}, \mathbf{b}$ and $\mathbf{c}$.
Show that the scalar product of vectors $\mathbf{a} \cdot \mathbf{b}$ with a, b as real numbers satisfy these axioms, hence show that the scalar product is an inner product.

4.8 A triangle has vertices $(1, 1, 1)$, $(1, 2, 3)$ and $(3, 2, 1)$. Use scalar products to determine the three angles of the triangle.

4.9 Use the scalar product to prove the cosine rule: $c^2 = a^2 + b^2 - 2ab \cos C$ using the usual convention for labelling the vertices and sides of a triangle.

4.10 Show that $\mathbf{a} \cdot \mathbf{b} = \frac{1}{4}(|\mathbf{a} + \mathbf{b}|^2 - |\mathbf{a} - \mathbf{b}|^2)$.

4.11 Show that for any two vectors $\mathbf{a} \times \mathbf{b} = -\mathbf{b} \times \mathbf{a}$.

4.12 Show that $\mathbf{a} \times (\mathbf{b} \times \mathbf{c}) + \mathbf{b} \times (\mathbf{c} \times \mathbf{a}) + \mathbf{c} \times (\mathbf{a} \times \mathbf{b}) = \mathbf{0}$.

4.13 Show that the vector $\mathbf{n} = \mathbf{i} + 2\mathbf{j} - 6\mathbf{k}$ is a vector normal to the plane $x + 2y - 6z = 0$.
[Hint: Determine two vectors that lie in the plane first.]

4.14 Generalise this last result by finding a normal to any plane through the origin $ax + by + cz = 0$.

4.15 The paraboloid reflector of shape $4z = x^2 + y^2$ has a normal $\mathbf{n} = -x\mathbf{i} - y\mathbf{j} + 2\mathbf{k}$. Show that any ray parallel to the z-axis that reflects from the concave surface always passes through the same point on the z-axis and find this point. It is the focus of the paraboloid.

4.16 Use the vector cross product to prove the sine rule: that for a triangle ABC with sides a, b and c

$$\frac{\sin A}{a} = \frac{\sin B}{b} = \frac{\sin C}{c}.$$

4.17 Prove the orthogonality conditions for reciprocal vectors $\mathbf{a}, \mathbf{b}$ and $\mathbf{c}$ and $\mathbf{a}', \mathbf{b}'$ and $\mathbf{c}'$. Prove also the two results that if $\mathbf{a} \cdot \mathbf{b} \times \mathbf{c} = V$ then $\mathbf{a}' \cdot \mathbf{b}' \times \mathbf{c}' = 1/V$, and for the position vector $\mathbf{r}$,

$$\mathbf{r} = (\mathbf{r} \cdot \mathbf{a}')\mathbf{a} + (\mathbf{r} \cdot \mathbf{b}')\mathbf{b} + (\mathbf{r} \cdot \mathbf{c}')\mathbf{c}.$$

4.18 Prove that for vectors $\mathbf{a}, \mathbf{b}, \mathbf{c}$ and $\mathbf{d}$[(a) is called Lagrange's identity:]

a)

$$(\mathbf{a} \times \mathbf{b}) \cdot (\mathbf{c} \times \mathbf{d}) = (\mathbf{a} \cdot \mathbf{c})(\mathbf{b} \cdot \mathbf{d}) - (\mathbf{a} \cdot \mathbf{d})(\mathbf{b} \cdot \mathbf{c})$$

and

b)

$$(\mathbf{a} \times \mathbf{b}) \cdot (\mathbf{c} \times \mathbf{d}) + (\mathbf{b} \times \mathbf{c}) \cdot (\mathbf{a} \times \mathbf{d}) + (\mathbf{c} \times \mathbf{a}) \cdot (\mathbf{b} \times \mathbf{d}) = 0.$$

5

Vector Differentiation

5.1 Introduction

In the last chapter, the algebra of vectors was introduced. Their power to solve geometrical problems was demonstrated through several examples. The algebra of vectors can be very useful for analysing other areas on the more applied side of mathematics, but there is so much more scope if the vectors are allowed to vary. This means allowing both direction and magnitude to vary continuously with position and time. For example, a force or velocity is often time dependent in a mechanics problem, and in electromagnetism it is essential to allow magnetic and electric fields to vary in order to calculate important forces. That being so, this chapter introduces the way that calculus is used to analyse how vectors change. Suppose a vector $\mathbf{F}(t)$ depends upon time, then the definition of the rate of change of $\mathbf{F}$ would be

$$\frac{d\mathbf{F}}{dt} = \lim_{\delta t \to 0} \left\{ \frac{\mathbf{F}(t + \delta t) - \mathbf{F}(t)}{\delta t} \right\}$$

and this limit looks straightforward enough. Except that, since $\mathbf{F}$ is a vector, its value at the time $t + \delta t$ will have a different direction as well as magnitude to its value at t so the limiting process hides a subtlety. If a variable vector is expressed in terms of its components, $\mathbf{F}(t) = F_1(t)\mathbf{i} + F_2(t)\mathbf{j} + F_3(t)\mathbf{k}$, then as the unit vectors are constant, it is the case that

$$\frac{d\mathbf{F}}{dt} = \mathbf{i}\frac{dF_1}{dt} + \mathbf{j}\frac{dF_2}{dt} + \mathbf{k}\frac{dF_3}{dt}.$$

On the other hand, it is not always convenient or even possible to express a varying vector in terms of its components. In this case, one treats the vector as the product of a variable scalar and its varying direction, $\mathbf{F}(t) = |\mathbf{F}(t)|\hat{\mathbf{F}}(t)$, and the product rule is used:

$$\frac{d\mathbf{F}(t)}{dt} = \frac{d|\mathbf{F}(t)|}{dt}\hat{\mathbf{F}}(t) + |\mathbf{F}(t)|\frac{d\hat{\mathbf{F}}(t)}{dt}.$$

Two and Three Dimensional Calculus: With Applications in Science and Engineering, First Edition. Phil Dyke.
© 2018 John Wiley & Sons Ltd. Published 2018 by John Wiley & Sons Ltd.

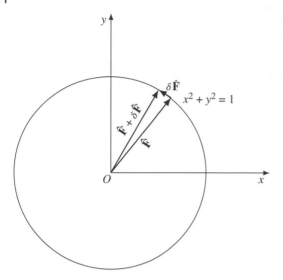

Figure 5.1 The unit vector $\hat{\mathbf{F}}$ with the unit circle $|\mathbf{r}| = 1$ or $x^2 + y^2 = 1$ and the direction of its derivative.

Books on the pure mathematics behind this will contain proofs required to justify using the product rule. This text is more concerned with the practicalities behind differentiating vectors. For example, consider the scalar product of a unit vector with itself, which has to equal 1, namely $\hat{\mathbf{F}} \cdot \hat{\mathbf{F}} = 1$ as a function of t and differentiate it with respect to t. The result, using the product rule is

$$\hat{\mathbf{F}} \cdot \frac{d\hat{\mathbf{F}}}{dt} = 0;$$

a result that tells us that the vector $\hat{\mathbf{F}}$ and its derivative are perpendicular vectors. Figure 5.1 displays the situation: Any point on the unit circle can represent the end point of the vector $\hat{\mathbf{F}}$; the triangle of forces is shown and as the angle between $\hat{\mathbf{F}}$ and $\hat{\mathbf{F}} + \delta\hat{\mathbf{F}}$ tends to zero, the direction of the vector $\delta\hat{\mathbf{F}}$ tends to a tangent to the unit circle, and so it is verified geometrically that $\hat{\mathbf{F}}$ and its derivative are indeed perpendicular. For a general curve, if a point P on the curve has position vector $\mathbf{r}$, then the derivative $d\mathbf{r}/dt$ will be in the direction of the tangent to the curve at P, see Figure 5.2. The angle between $\mathbf{r}$ and its derivative is only a right angle if the curve $\mathbf{r} = \mathbf{r}(t)$ is a circle with its centre at the origin. There are several natural vehicles for introducing vector calculus, one uses geometry and is called *differential geometry* where vectors are used to describe curves and surfaces and the above introductory material is already carrying us along this route. The others are *mechanics* where vectors describe the motion of particles and rigid bodies under the influence of forces; *electromagnetism* where the fields are continuously varying in all three dimensions as well as with time; *fluid mechanics* where vectors help in the understanding of how fluids move, with many applications including

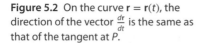

Figure 5.2 On the curve $\mathbf{r} = \mathbf{r}(t)$, the direction of the vector $\frac{d\mathbf{r}}{dt}$ is the same as that of the tangent at P.

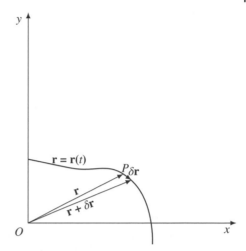

flight and Formula 1 racing; *elasticity* where vectors describe the deformation of solids as well as waves within them. This has applications to materials science, mechanical and civil engineering as well as the geology of the earth, in particular earthquakes and aftershocks. All except differential geometry and mechanics has to wait a chapter. Here, some introductory mechanics examples follow a brief practical foray into differential geometry.

5.2 Differential Geometry

Differential geometry is a vast subject, but it includes the description of curves and surfaces through the use of calculus. It is possible to do this without vectors of course, and indeed all the books on differential geometry published before the early twentieth century are vector free, see for example Eisenhart's 1909 [2] classic text 'A Treatise on the Differential Geometry of Curves and Surfaces', nearly 500 pages in length with no mention of vectors. The Dover edition is still available. However, the aspects of differential geometry included in the book by Eisenhart do lend themselves to the application of vectors, their absence owes more to the era in which it was written; the present-day vector notation was in its infancy then. What cannot be attempted here is a full blown account using all the power of mathematical structures such as simplectic manifolds and Lie groups. To do that would be totally out of place; hence, the modest claim is to provide the student with an application of vectors to reasonably familiar geometry of the type met at school or just after, and geometry covers both curves and surfaces. Let us begin with using vectors to study curves.

5.2.1 Space Curves

The starting point here is the *position vector* **r**. This is the vector joining the origin $(0, 0, 0)$ to a point with co-ordinates (x, y, z). It is the vector

$$\mathbf{r} = x\mathbf{i} + y\mathbf{j} + z\mathbf{k}.$$

For geometry, the point (x, y, z) is fixed, but now suppose that x, y and z are functions of a variable t. This means that

$$\mathbf{r}(t) = x(t)\mathbf{i} + y(t)\mathbf{j} + z(t)\mathbf{k},$$

where t is a scalar that can run through a range of values continuously. It is convenient to think of t as time, and the position vector $\mathbf{r}(t)$ as pointing to a point that moves around with time; a fly perhaps. Flies are not being studied here, so to trace out a particular curve, $\mathbf{r}(t)$ is given a correct functional form called a *parametric* form by specifying the functions $x(t), y(t)$ and $z(t)$. These functions need to be smooth, so normally they are differentiable functions of the variable t. The derivative vector $\mathbf{r}'(t) = (x'(t), y'(t), z'(t))$ is formed by differentiating each component of the position vector $\mathbf{r}(t)$. The vector $\mathbf{r}'(t)$ is called the *velocity* vector. In mechanics, that is precisely the definition of velocity; here, this definition is generalised. The condition $\mathbf{r}'(t) \neq \mathbf{0}$ has to be imposed. $\mathbf{r}'(t) = \mathbf{0}$ means that the tangent vector has no length and, importantly, no direction and is therefore a singular point. The plane curve $\mathbf{r}(t) = (t^2, t\sin t, 0)$ has such a point at $t = 0$ as it starts there; in addition, this curve is also symmetric as $\mathbf{r}(t) = (t^2, t\sin t, 0) = \mathbf{r}(-t)$. A useful quantity to define and find is the *arc length s* of a curve and the definition follows:

Definition 5.1 The *arc length s* of a curve is the distance between two fixed points along the arc of the curve itself. Using Pythagoras' theorem for an infinitesimal length of the curve ds so that $ds^2 = dx^2 + dy^2 + dz^2$, the arc length is defined by the integral

$$s = \int_a^b ds = \int_{t_0}^{t_1} \frac{ds}{dt} dt = \int_{t_0}^{t_1} \sqrt{\left(\frac{dx}{dt}\right)^2 + \left(\frac{dy}{dt}\right)^2 + \left(\frac{dz}{dt}\right)^2} \, dt,$$

where $a = s(t_0)$ and $b = s(t_1)$.

Let us look at a few examples now. All the curves here will be defined parametrically. For example, the position vector $\mathbf{r}(t) = (a\cos t, a\sin t, c)$ would trace a circle radius a centred at $(0, 0, c)$ on the plane $z = c$, see Figure 5.3. The velocity vector in this case is in the plane of the curve and is given by $\mathbf{r}' = (-a\sin t, a\cos t, 0)$, and the arc length of the curve is $2\pi a$, the circumference of the circle. Verify this using the equation for s above.

The equation of a straight line has already been mentioned, and it is now shown that the general vector expression $\mathbf{r} = \mathbf{a} + t\mathbf{b}$, where $\mathbf{a}$ and $\mathbf{b}$ are arbitrary

Figure 5.3 The circle centre $(0, 0, c)$ radius a on the plane $z = c$.

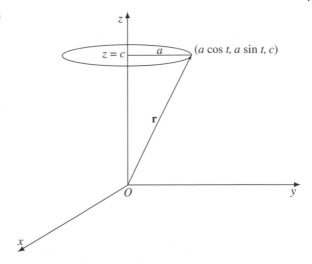

Figure 5.4 The straight line $\mathbf{r} = \mathbf{a} + t\mathbf{b}$ labelled l.

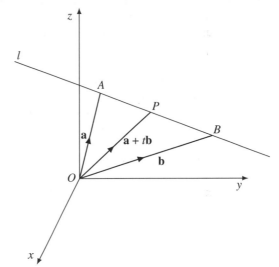

constant vectors, is a straight line. Exercise 4.15 in Chapter 4 has already used the concept of a vector representation of a line. This line, labelled l, is shown in Figure 5.4. The vector $\overrightarrow{OP}$ is $\mathbf{r} = \mathbf{a} + t\mathbf{b}$. If $t = 0$, then this is the vector $\overrightarrow{OA}$; if $t = \infty$, then this is the vector $\overrightarrow{OB}$. Negative values of t will give position vectors on the line l outside the points A and B; hence, the expression $\mathbf{a} + t\mathbf{b}$ represents the position vector of any point on l, a particular point corresponding to a particular value of t where $-\infty < t < \infty$. Of course, other parameterisations are possible; for example, $u\mathbf{a} + \mathbf{b}$ where this time B corresponds to $u = 0$ and A to $u = \infty$, or $(1 + \lambda)\mathbf{a} + (1 - \lambda)\mathbf{b}$ with $\lambda = 1$ corresponding to A and $\lambda = -1$

to B. The important point is that curves are described by a single parameter. In Cartesians, two equations in x, y and z are required to specify a curve: The straight line is given by the three equations

$$x = a_1 + tb_1, \tag{5.1}$$

$$y = a_2 + tb_2, \tag{5.2}$$

$$z = a_3 + tb_3. \tag{5.3}$$

The parameter t can be eliminated using the first two equations, and again using the last two equations. This gives two linear equations connecting x, y and z, and these represent the cartesian equations of two planes. Two planes intersect in a straight line; hence, the three equations do indeed represent a straight line. However, this is a clumsy procedure and eliminating from a Cartesian representation is poor practice, and is only possible on rare occasions. It is best to utilise the single-parameter description, and the rest of this section does exactly this. Sometimes, it is possible to choose the distance along the curve as this single parameter, and this distance is denoted by s. There will be much more about s later when general results are derived. Here is another example of a curve described parametrically, this time the curve does not lie in a plane:

$$\mathbf{r}(t) = a \cos t\mathbf{i} + a \sin t\mathbf{j} + ct\mathbf{k}.$$

It is called a circular helix and has the shape of a coiled spring with axis along the z-axis, see Figure 5.5. The distance between coils, called the *pitch* is c, and the radius of the helix as viewed from the end is a. The velocity or tangent vector is $\mathbf{r}'(t) = -a \sin t\mathbf{i} + a \cos t\mathbf{j} + c\mathbf{k}$ and has a constant upward slope equal to the pitch c at all points of the helix. This is shown in Figure 5.5. In reality of course the helix is infinitely long. For this curve, the arc length between two

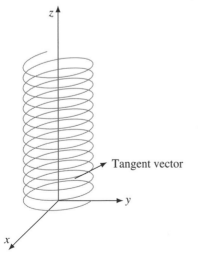

Figure 5.5 The circular helix $\mathbf{r}(t) = a \cos t\mathbf{i} + a \sin t\mathbf{j} + ct\mathbf{k}$.

Tangent vector

Figure 5.6 The twisted cubic $\mathbf{r} = (3t, 3t^2, 2t^3)$.

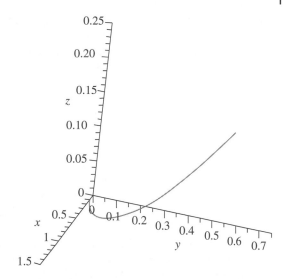

points directly above each other will be $2\pi\sqrt{a^2 + c^2}$; the integration is very straightforward. Another three-dimensional curve, with equation $\mathbf{r}(t) = 3t\mathbf{i} + 3t^2\mathbf{j} + 2t^3\mathbf{k}$ is called a *twisted cubic*. Here, the general twisted cubic has been avoided in favour of this special one that leads to the velocity $\mathbf{r}' = (3, 6t, 6t^2)$ and an arc length $3t_0 + 2t_0^3$ between points when $t = 0$ and $t = t_0$. Most choices of twisted cubic give rise to arc lengths only expressible in terms of elliptic functions. The curve is drawn in Figure 5.6.

Rather than pursue examples, let us instead leave this totally co-ordinate-based approach and examine a local intrinsic view of curves. There are five quantities associated with curves, three are vectors and two are scalars. Even though the curve may be three dimensional, take an arbitrary point on the curve, and provided it is smooth, just either side of this point the curve will lie in a plane, see Figure 5.7. This plane is sometimes called the *osculating plane*,

Figure 5.7 The curve $\mathbf{r} = \mathbf{r}(t)$ with the mutually orthogonal triad of unit vectors $\hat{\mathbf{T}}, \hat{\mathbf{N}}$ and $\hat{\mathbf{B}}; \hat{\mathbf{B}}$ is pointing towards you.

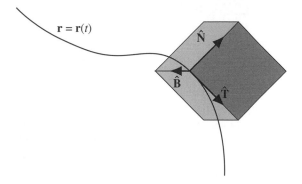

a name that means 'kissing plane' which is an apt description. The vectors can be described reasonably easily and they should be no surprise; they are the *unit tangent* $\hat{\mathbf{T}}$ which is in this local plane, tangent to the curve at the chosen point, the *unit normal* $\hat{\mathbf{N}}$ also in this plane but at right angles to the tangent, and the *unit binormal* $\hat{\mathbf{B}}$ at right angles to the local plane and therefore at right angles to both the tangent and normal. The three vectors $\hat{\mathbf{T}}$, $\hat{\mathbf{N}}$ and $\hat{\mathbf{B}}$ form a right-handed triad of unit vectors, analogous to $\mathbf{i}, \mathbf{j}$ and $\mathbf{k}$; but of course they twist and turn as the arbitrary point is allowed to travel along the curve. The scalars are the *curvature* κ and the *torsion* τ. If $\tau = 0$, then the curve lies entirely in a plane; the straight line and the conics would be examples of such curves. A curve with $\tau = 0$ and constant κ is the circle. The radius ρ of the circle is the reciprocal of the curvature $\rho = 1/\kappa$. It is worth spending a little time on what the these scalars represent geometrically, then the relationships that are derived using vector calculus will have more meaning. The circle is a curve with zero torsion and constant curvature. The larger the radius is, the smaller the curvature will be. This gives a direct meaning to curvature, and it coincides with the layman's view that it is a measure of how severe a curve bends in its own plane. The circular helix $\mathbf{r}(t) = a \cos t\mathbf{i} + a \sin t\mathbf{j} + ct\mathbf{k}$; Figure 5.5 displays the only curve with both a constant curvature and a constant torsion. Mathematical expressions are derived later, but as torsion is the only distinction between circle and circular helix, the meaning can be deduced. In the circular helix, the osculating plane changes as the point moves. By contrast, for the circle, the osculating plane is unique and defines the plane of the curve itself. Torsion therefore describes how the curve twists out of its osculating plane, the locally defined plane of the curve. As with the name curvature, the name torsion describes this well. An earlier result shows that $\dfrac{d\mathbf{r}(t)}{dt}$ is in the direction of the tangent to the curve $\mathbf{r} = \mathbf{r}(t)$. It is also therefore true that

$$\frac{d\mathbf{r}(t)}{dt} = \left| \frac{d\mathbf{r}(t)}{dt} \right| \hat{\mathbf{T}} = \frac{ds}{dt}\hat{\mathbf{T}}$$

so in fact

$$\frac{d\mathbf{r}(t)}{ds} = \hat{\mathbf{T}},$$

where s is the arc length. This follows because as has already been seen, $|d\mathbf{r}| = \sqrt{(dx)^2 + (dy)^2 + (dz)^2} = ds$ (Pythagoras' theorem). As $\mathbf{r}(t)$ is given as a function of t not s, although this is a neat expression, it is not a particularly useful formula for calculation and the previous 'function-of-a-function' rule is used instead. Specialist texts on differential geometry [3] refer to parameterisation in terms of arc length s as curves of unit speed which is a way of expressing that $|\mathbf{v}| = |d\mathbf{r}/ds| = 1$. The practical consequences of this will be clearer when an example is done later. A unit vector that is variable can only vary as in Figure 5.3, a line of unit length, the end of which describes a sphere. Infinitesimally, it describes the arc of a circle. Therefore, it is no surprise that

$$\hat{\mathbf{T}}(t) \cdot \frac{d\hat{\mathbf{T}}(t)}{dt} = 0 \quad \text{and} \quad \hat{\mathbf{N}}(t) \cdot \frac{d\hat{\mathbf{N}}(t)}{dt} = 0,$$

results obtained by differentiating $|\hat{\mathbf{T}}|^2 = 1$ and $|\hat{\mathbf{N}}|^2 = 1$. However, as $\hat{\mathbf{T}}, \hat{\mathbf{N}}$ and $\hat{\mathbf{B}}$ form a right-handed triad of unit vectors, it is certainly true that

$$\hat{\mathbf{T}} \cdot \hat{\mathbf{N}} = 0, \quad \hat{\mathbf{N}} \cdot \hat{\mathbf{B}} = 0 \quad \text{and} \quad \hat{\mathbf{B}} \cdot \hat{\mathbf{T}} = 0.$$

As $\frac{d\hat{\mathbf{T}}(t)}{dt}$ lies in the osculating plane defined by unit vectors $\hat{\mathbf{T}}$ and $\hat{\mathbf{N}}$ but was shown to be perpendicular to $\hat{\mathbf{T}}$, it must be in the same direction as $\hat{\mathbf{N}}$ so

$$\frac{d\hat{\mathbf{T}}(t)}{dt} = \frac{ds}{dt}\kappa(t)\hat{\mathbf{N}}(t);$$

the scalar of proportionality is the curvature κ times ds/dt.

Now, consider the relationship $\hat{\mathbf{B}} = \hat{\mathbf{T}} \times \hat{\mathbf{N}}$ and differentiate this with respect to t:

$$\frac{d\hat{\mathbf{B}}}{dt} = \frac{d}{dt}\{\hat{\mathbf{T}} \times \hat{\mathbf{N}}\} = \frac{d\hat{\mathbf{T}}}{dt} \times \hat{\mathbf{N}} + \hat{\mathbf{T}} \times \frac{d\hat{\mathbf{N}}}{dt}$$

using the product rule. The first term is zero because $\frac{d\hat{\mathbf{T}}}{dt}$ is parallel to $\hat{\mathbf{N}}$. Since $\hat{\mathbf{N}} \cdot \frac{d\hat{\mathbf{N}}}{dt} = 0$, $\frac{d\hat{\mathbf{N}}}{dt}$ is perpendicular to $\hat{\mathbf{N}}$. However, $\hat{\mathbf{T}}$ is also perpendicular to $\hat{\mathbf{N}}$. The cross product of the two vectors, both perpendicular to $\hat{\mathbf{N}}$ must therefore be parallel to $\hat{\mathbf{N}}$ and

$$\frac{d\hat{\mathbf{B}}}{dt} = -\frac{ds}{dt}\tau\hat{\mathbf{N}},$$

where τ is the torsion. The negative sign is there by convention, but it is also explained in that a positive value of τ denotes the direction of the binormal $\hat{\mathbf{B}}$ that twists clockwise which is against the right-handed nature of the triad $\hat{\mathbf{T}}, \hat{\mathbf{N}}$ and $\hat{\mathbf{B}}$. Finally,

$$\frac{d\hat{\mathbf{N}}}{dt} = \frac{d}{dt}\{\hat{\mathbf{B}} \times \hat{\mathbf{T}}\} = \frac{d\hat{\mathbf{B}}}{dt} \times \hat{\mathbf{T}} + \hat{\mathbf{B}} \times \frac{d\hat{\mathbf{T}}}{dt}$$

$$= -\tau\frac{ds}{dt}\hat{\mathbf{N}} \times \hat{\mathbf{T}} + \kappa\frac{ds}{dt}\hat{\mathbf{B}} \times \hat{\mathbf{N}} = -\frac{ds}{dt}\kappa\hat{\mathbf{T}} + \frac{ds}{dt}\tau\hat{\mathbf{B}}.$$

Tidying the equations up by division by ds/dt leads to the three equations:

$$\frac{d\hat{\mathbf{T}}}{ds} = \kappa\hat{\mathbf{N}}, \tag{5.4}$$

$$\frac{d\hat{\mathbf{N}}}{ds} = -\kappa\hat{\mathbf{T}} + \tau\hat{\mathbf{B}}, \tag{5.5}$$

$$\frac{d\hat{\mathbf{B}}}{ds} = -\tau\hat{\mathbf{N}}, \tag{5.6}$$

and these are called the *Serret–Frenet formulae* or sometimes the *Frenet Apparatus* of a curve. They can be written in matrix form:

$$\frac{d}{ds}\begin{pmatrix} \hat{\mathbf{T}} \\ \hat{\mathbf{N}} \\ \hat{\mathbf{B}} \end{pmatrix} = \begin{pmatrix} 0 & \kappa & 0 \\ -\kappa & 0 & \tau \\ 0 & -\tau & 0 \end{pmatrix}\begin{pmatrix} \hat{\mathbf{T}} \\ \hat{\mathbf{N}} \\ \hat{\mathbf{B}} \end{pmatrix};$$

perhaps, this is slightly more memorable. It is true that in the way that these have been derived, the scalar quantities κ and τ have a mathematically precise definition. The descriptions given earlier are still right; κ is the magnitude of $\frac{d\hat{\mathbf{T}}}{ds}$ so it is indeed a measure of the change in the direction of the tangent which is curvature. Similarly, τ is the magnitude of $\frac{d\hat{\mathbf{B}}}{ds}$ so it is a measure of how the curve twists, and this can be correctly called torsion. The emphasis here is on calculation rather than placing differential geometry in terms of pure mathematical structures, so here is an example where the Frenet apparatus is found.

Example 5.1   Find the five quantities $\hat{\mathbf{T}}, \hat{\mathbf{N}}, \hat{\mathbf{B}}, \kappa$ and τ for the curve defined parametrically by

$$x = ae^t \cos t, \quad y = ae^t \sin t \quad \text{and} \quad z = \sqrt{2}a(e^t - 1).$$

Solution: To begin the process, let us find the derivative of $\mathbf{r}$ with respect to t:

$$\frac{d\mathbf{r}}{dt} = ae^t(\cos t - \sin t)\mathbf{i} + ae^t(\cos t + \sin t)\mathbf{j} + \sqrt{2}ae^t\mathbf{k}.$$

The tangent $\hat{\mathbf{T}}$ is the unit vector in this direction,

$$\left|\frac{d\mathbf{r}}{dt}\right| = \frac{ds}{dt}$$

$$= \sqrt{a^2e^{2t}(\cos t - \sin t)^2 + a^2e^{2t}(\cos t + \sin t)^2 + 2a^2e^{2t}} = 2ae^t,$$

and so

$$\hat{\mathbf{T}} = \frac{d\mathbf{r}}{dt}\bigg/\left|\frac{d\mathbf{r}}{dt}\right| = \frac{1}{2}[(\cos t - \sin t)\mathbf{i} + (\cos t + \sin t)\mathbf{j} + \sqrt{2}\mathbf{k}].$$

Differentiating this with respect to t gives

$$\frac{d\hat{\mathbf{T}}}{dt} = \frac{1}{2}[(-\sin t - \cos t)\mathbf{i} + (-\sin t + \cos t)\mathbf{j}]$$

and so

$$\frac{d\hat{\mathbf{T}}}{ds} = \frac{d\hat{\mathbf{T}}}{dt}\bigg/\frac{ds}{dt} = \frac{e^{-t}}{4a}[(-\sin t - \cos t)\mathbf{i} + (-\sin t + \cos t)\mathbf{j}]$$

and the magnitude of this is the curvature κ. The key point here is to remember that t the parameter is not unique and that it is important to find the derivatives with respect to the arc length s which is unique. Therefore,

$$\kappa = \frac{e^{-t}}{4a}\sqrt{(\sin t + \cos t)^2 + (-\sin t + \cos t)^2} = \frac{\sqrt{2}e^{-t}}{4a}.$$

Since $\frac{d\hat{\mathbf{T}}}{dt} = \kappa\hat{\mathbf{N}}$, this enables us to write

$$\hat{\mathbf{N}} = \frac{1}{\sqrt{2}}[(-\sin t - \cos t)\mathbf{i} + (-\sin t + \cos t)\mathbf{j}].$$

Next, we find $\hat{\mathbf{B}}$ through $\hat{\mathbf{B}} = \hat{\mathbf{T}} \times \hat{\mathbf{N}}$:

$$\hat{\mathbf{B}} = \frac{1}{2\sqrt{2}}\begin{vmatrix} \mathbf{i} & \mathbf{j} & \mathbf{k} \\ \cos t - \sin t & \cos t + \sin t & \sqrt{2} \\ -\cos t - \sin t & \cos t - \sin t & 0 \end{vmatrix}$$

so

$$\hat{\mathbf{B}} = \frac{1}{2}[(\sin t - \cos t)\mathbf{i} - (\cos t + \sin t)\mathbf{j} + \sqrt{2}\mathbf{k}].$$

Differentiating with respect to t gives

$$\frac{d\hat{\mathbf{B}}}{dt} = \frac{1}{2}[(\cos t + \sin t)\mathbf{i} + (\sin t - \cos t)\mathbf{j}].$$

Hence,

$$\frac{d\hat{\mathbf{B}}}{ds} = \frac{e^{-t}}{4a}[(\cos t + \sin t)\mathbf{i} + (\sin t - \cos t)\mathbf{j}].$$

Therefore, using the third Serret–Frenet formula,

$$\frac{d\hat{\mathbf{B}}}{ds} = -\tau\hat{\mathbf{N}} = -\frac{\tau}{\sqrt{2}}[(-\sin t - \cos t)\mathbf{i} + (-\sin t + \cos t)\mathbf{j}].$$

This gives

$$\tau = \frac{\sqrt{2}e^{-t}}{4a}.$$

Taking the modulus $|d\hat{\mathbf{B}}/ds|$ also works as τ is always positive. Therefore, here are the three unit vectors:

$$\hat{\mathbf{T}} = \frac{1}{2}[(\cos t - \sin t)\mathbf{i} + (\cos t + \sin t)\mathbf{j} + \sqrt{2}\mathbf{k}],$$

$$\hat{\mathbf{N}} = \frac{1}{\sqrt{2}}[(-\sin t - \cos t)\mathbf{i} + (-\sin t + \cos t)\mathbf{j}],$$

$$\hat{\mathbf{B}} = \frac{1}{2}[(\sin t - \cos t)\mathbf{i} - (\cos t + \sin t)\mathbf{j} + \sqrt{2}\mathbf{k}].$$

together with the two scalars

$$\kappa = \tau = \frac{\sqrt{2}e^{-t}}{4a}.$$

This is a spiral-shaped curve that expands exponentially along the z-axis; the pitch (spacing between coils) is also growing exponentially. It is hard to display this graphically, even with modern software. □

In the next subsection, the use of vectors to describe surfaces is explored.

5.2.2 Surfaces

The overall objective in the mathematical treatment of surfaces is to be able to treat the surface as if it was a plane and perform all the usual calculus operations on it. Thus, it is usual that mathematicians and students of mathematics concern themselves with being able to transfer planar calculus to the computation of lines and normals on a surface. Thus, the term *patch* is defined as a regular mapping from an open set of points on the plane to that small part of three-dimensional space that is a surface. Once this is done, it is possible to work locally within a surface, and in what follows local co-ordinate-free computation is the ultimate goal. To put this into some kind of context, let us start with a general description of surfaces. Whereas differential geometry is used to describe the geometry of surfaces, these are also usually defined parametrically. In order to do this, two independent parameters are required; these two parameters describe the image of the transformation. Here is an introductory example that will be useful later: the spherical surface. The Cartesian equation of the sphere is easily derived from its geometry, it is the locus of a point that is always the same distance a from the origin so the position vector has to have magnitude a that is $r = a$. This is $r^2 = a^2$ or in Cartesian co-ordinates $x^2 + y^2 + z^2 = a^2$. A parameterisation could be $x = a \cos \theta \sin \phi, y = a \sin \theta \sin \phi, z = a \cos \phi$. The geographically inclined could use latitude and longitude to pinpoint particular locations on the surface of this sphere, here $\pi/2 - \phi$ is equivalent to latitude and θ to longitude.

In general, therefore, a parameterisation $x = x(u, v), y = y(u, v)$ and $z = z(u, v)$ with x, y and z treated as differentiable functions of the variables u, v will define a surface. The opening sentences above define a patch, and this parameterisation is a manifestation of this. Surface co-ordinates u and v are parameters, a local x, y defined on a patch. In one dimension, only one parameter t was required; here, we need two. For the sphere, $u = \theta$ and $v = \phi$ is the usual notation. It is not always possible to eliminate u and v between the three equations $x = x(u, v), y = y(u, v)$ and $z = z(u, v)$; but when you can, an equation $f(x, y, z) = 0$ results and this is called the equation of the surface. The sphere radius a and the centre at the origin with an equation $x^2 + y^2 + z^2 = a^2$ provides a good example. Sometimes, it is possible to make z the subject of this formula $z = F(x, y)$, and this is another way of expressing the equation of a surface. This was useful in Chapter 3 where maxima and minima were found. It is the preferred way to consider contour maps in geography or pressure distribution maps in meteorology, but again it is rarely possible. Where it is possible, one can set $x = u, y = v$ and $z = F(u, v)$ with parameterisation $(u, v, F(u, v))$. It is often possible to do this for part of a surface, for example, the hemisphere $z = \sqrt{a^2 - x^2 - y^2}$ can be expressed as $(u, v, \sqrt{a^2 - u^2 - v^2})$ but it is only valid for $z \geq 0$ so it is the surface of the upper hemisphere. These kind of representations are called Monge patches after the French mathematician Gaspard Monge (1746–1818), one of the fathers of differential geometry who lived through the French revolution initially to some advantage as he backed the winning side, but later when Napoleon was deposed, he lost everything. The Monge patch enables some specific kind of calculus to take place. For example, the normal takes a simple form. Consider a Monge patch $z = F(x, y)$, then define the position vector $\mathbf{r}$ of any point on this patch as $\mathbf{r} = (u, v, F(u, v))$. Lines $(u, v_0, F(u, v_0))$ and $(u_0, v, F(u_0, v))$ can be drawn on this patch. It is usually the case that as u and v are allowed to vary, these lines are orthogonal to each other, though for what follows this need not be true. Later, when this geometry on a surface is used for integration in Chapter 10, the curves $u = $ const. and $v = $ const. on a surface will be at right angles. The chapter on curvilinear co-ordinates, Chapter 7, introduces this. The lines $(u, v_0, F(u, v_0))$ and $(u_0, v, F(u_0, v))$ drawn on the patch will have tangent lines $\mathbf{r}_u = (1, 0, F_u)$ and $\mathbf{r}_v = (0, 1, F_v)$ in the osculating plane at the point $(u_0, v_0, F(u_0, v_0))$ on the patch. Hence, the normal to the patch at this point will be

$$\mathbf{r}_u \times \mathbf{r}_v = \begin{vmatrix} \mathbf{i} & \mathbf{j} & \mathbf{k} \\ 1 & 0 & F_u \\ 0 & 1 & F_v \end{vmatrix} = (-F_u, -F_v, 1),$$

where all derivatives are evaluated at $u = u_0, v = v_0$. This is uniquely defined for differentiable functions $F(x, y)$; hence, Monge patches can be called *smooth* or *regular*. Here is an example.

Example 5.2 Calculate the unit normal to the paraboloid $4z = x^2 + y^2$ at an arbitrary point.

Solution: Using the above formula, the normal will be $(-F_u, -F_v, 1)$ where $F(u, v) = \frac{1}{4}(u^2 + v^2)$. Straightforwardly, $F_u = \frac{1}{2}u$ and $F_v = \frac{1}{2}v$; hence, the unit normal is

$$\hat{\mathbf{n}} = \left(-\frac{u}{\sqrt{u^2 + v^2 + 4}}, -\frac{v}{\sqrt{u^2 + v^2 + 4}}, \frac{2}{\sqrt{u^2 + v^2 + 4}} \right).$$

Therefore, at the point (x, y, z) on the paraboloid $4z = x^2 + y^2$, the unit normal is the vector

$$\frac{1}{2\sqrt{z + 1}}(-x, -y, 2).$$

It can be shown that any ray parallel to **k** that strikes a mirrored concave surface of this shape reflects and passes through a fixed point, called the focus, on the z-axis, see Exercise 4.15. If you didn't try this exercise before, it is a good idea to try it now. □

It might have escaped your notice that the two vectors $\mathbf{r}_u = (1, 0, F_u)$ and $\mathbf{r}_v = (0, 1, F_v)$ although both tangential to the surface $z = F(x, y)$ and therefore at right angles to the normal vector $(-F_u, -F_v, 1)$ are not at right angles to each other. $\mathbf{r}_u \cdot \mathbf{r}_v = F_u F_v \neq 0$ in general. This is not surprising as there are many choices for parameterising a given surface. A sphere parameterised using longitude and co-latitude as parameters, see above, will have these tangent vectors orthogonal to each other. That was done by design. So one has to be aware that this situation is different to the differential geometry of curves where for the curve $\mathbf{r} = \mathbf{r}(t)$ the tangent in the osculating plane **T** is definitely parallel to $\mathbf{r}_t$.

It is therefore reasonable to ask how to define the curvature of a surface. Unlike a curve in space that has one well-defined scalar curvature, a surface can have many. Consider for simplicity a cylinder: the circle that generates it has a curvature equal to the reciprocal of the radius, but also on the cylinder, at right angles to this is a line parallel to the axis of the cylinder that is straight, with zero curvature. Lines on the surface of the cylinder will have curvatures somewhere between these two values, see Figure 5.8. Similar lines drawn on a sphere will all have the same curvature, of course, the reciprocal of the radius of the sphere.

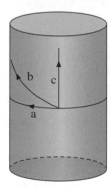

Figure 5.8 A cylinder with lines of minimum (**c**), maximum (**a**) and intermediate (**b**) curvature drawn.

The co-ordinate $(f(u), g(u), v)$ gives the general form for a cylinder the axis of which is in the z direction. In fact, this is a two-dimensional curve with z arbitrary. The surface tangent curves are $\left(\dfrac{df}{du}, \dfrac{dg}{du}, 0 \right)$ and $(0, 0, 1)$. These tangent curves are at right angles to each other and are an example of a *ruled surface*. For the circular cylinder, these lines are $(a \sin u, -a \cos u, 0)$ and $(0, 0, 1)$. Another kind of surface can be obtained by rotating a curve about a fixed axis, this is called a surface of revolution. These will have the form $(f(u) \cos v, f(u) \sin v, h(u))$ provided the axis of revolution is the z-axis. Elimination gives an equation connecting x, y and z of the general type $x^2 + y^2 = \alpha(z)$ where α is the function $f^2 h^{-1}$. A constructive example follows:

Example 5.3 Investigate the surface of revolution:

$$((R + r \cos u) \cos v, (R + r \cos u) \sin v, r \sin u) \quad R > r.$$

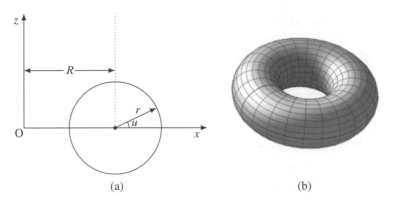

Figure 5.9 A circle radius r rotated about an axis distance $R > r$ from its centre parallel to the z-axis, shown in (a) on the left generates the torus on the right (b) shown in isometric view.

A torus is a doughnut-shaped solid of revolution generated by revolving a circle about an axis in the plane of the circle but not intersecting it, see Figure 5.9. From the equations straightforwardly, $x^2 + y^2 = (R + r \cos u)^2$ so elimination of both u and v can be done and results in

$$(x^2 + y^2 + z^2 - R^2 - r^2)^2 = 4R^2(r^2 - z^2)$$

but this is of little practical use. Far better is to calculate each tangent to the lines that have generated this surface. Start from the parametric equations to the torus:

$$\mathbf{r}(u, v) = ((R + r \cos u) \cos v, (R + r \cos u) \sin v, r \sin u)$$

so differentiating to get the tangents gives

$$\mathbf{r}_u = (-r \sin u \cos v, -r \sin u \sin v, r \cos u)$$

and

$$\mathbf{r}_v = (-(R + r \cos u) \sin v, (R + r \cos u) \cos v, 0)$$

that immediately tells us that $\mathbf{r}_u \cdot \mathbf{r}_v = 0$ so these lines are at right angles to each other. Moreover, treating $u = u_0 = $ constant as curves in space, these are circles $x^2 + y^2 = (R + r \cos u_0)^2$ in the planes $z = r \sin u_0$ and so have arc length $(R + r \cos u_0)v$. Hence, using the methods associated with the Serret–Frenet formulas, the curvature of this curve is say κ_v. To find this takes a bit of reasonably routine vector calculus and algebra, here are the details. The unit tangent vector along the $u = u_0$ curves is

$$\hat{\mathbf{T}}_v = \frac{\mathbf{r}_v}{|\mathbf{r}_v|} = (-(R + r \cos u_0) \sin v, (R + r \cos u_0) \cos v, 0)$$

$$\times \frac{1}{(R + r \cos u_0)},$$

which is

$$= (-\sin v, \cos v, 0)$$

so differentiating this unit tangent to the curve which is in fact no longer a partial derivative as $u = u_0$ everywhere on the curve,

$$\frac{\partial \hat{\mathbf{T}}_v}{\partial v} = \frac{d\hat{\mathbf{T}}_v}{dv} = \frac{d\hat{\mathbf{T}}_v}{dv} \bigg/ \frac{dv}{ds} = \frac{1}{(R + r\cos u_0)}(-\cos v, -\sin v, 0)$$

since $s = (R + r\cos u_0)v$,

$$\left|\frac{d\hat{\mathbf{T}}_v}{ds}\right| = \kappa_v = \frac{1}{(R + r\cos u_0)}$$

and this is the curvature of the generating curves $u = u_0$. Similarly, the curvature of the lines $v = v_0$ is given by $\kappa_u = 1/r$ and is independent of the choice of v_0. This can be seen geometrically in this case, as the lines $v = v_0$ correspond to the planes $y = x \tan v_0$ that contain the z-axis. No matter the value of v, the plane $y = x \tan v$ will intersect the torus in two vertical (in a plane through the z-axis) circles of radius r with centre distance $\sqrt{x^2 + y^2} = R$ either side of the origin. Similarly, the geometry of the torus reveals that the lines $u = u_0 = $ constant are planes $z = r \sin u_0$ and intersect the torus in circles with radial distances $R \pm \sqrt{r^2 - z^2}$ from the z-axis. When $u_0 = 0, z = 0$ and these two radii are $R \pm r$; when $u_0 = \pi/2$, the annulus closes to become the single circle of radius R and the plane $z = r$ is then tangent to the top of the torus. These two families are shown as lines on the torus in Figure 5.9 and are, as proved above, orthogonal. They can be used to generate the surface and typically do so in software such as MAPLE's ParametricPlot 3D command. The torus certainly has many curvatures and this subject, the curvature of surfaces, is explored next. □

For a sphere radius r, there is obviously a single curvature $1/r$. The curvature for a plane is zero, which should come as no surprise. However, the curvature for the cylinder $x^2 + y^2 = R^2$ with z arbitrary is not well defined. The curvature of a curve in space is zero for a straight line or positive for a curved line. Curves on a surface are different. A reasonably trivial distinction might be made between curves on a convex surface and curves on a concave surface, even curves that are otherwise identical and this can be done with reference to the normal to the surface, which *is* well defined. In the case mentioned, they will differ only by sign. In mathematical terms, to do a proper job to describe how a surface curves, a *shape operator* that in essence classifies how the normal to the surface behaves around a given location on the surface has to be introduced. It is defined as a directional derivative, but in terms independent of co-ordinates. This kind of derivative is called a *covariant derivative*, a notion that more easily generalises and so is attractive to those more interested in mathematical structures. Although it would be out of place to venture into the pure mathematics behind the differential geometry of surfaces as in the classic text by O'Neill [3] *Elementary Differential Geometry* by Barrett O'Neill, Academic Press 1966, it is constructive to see a shape operator 'in action' as it were. First, there are a few bits of the standard notation to get through. Denoted by **r** is the position vector of any point on the surface, so $\mathbf{r} = \mathbf{r}(u, v)$. The derivatives $\mathbf{r}_u$ and $\mathbf{r}_v$ have already been met, now define the following functions of u and v through scalar products:

$$E = \mathbf{r}_u \cdot \mathbf{r}_u \quad F = \mathbf{r}_u \cdot \mathbf{r}_v \quad \text{and} \quad G = \mathbf{r}_v \cdot \mathbf{r}_v$$

Using the result from Chapter 4: $|\mathbf{a} \times \mathbf{b}|^2 = |\mathbf{a}|^2 |\mathbf{b}|^2 - (\mathbf{a} \cdot \mathbf{b})^2$, it can be seen that

$$||\mathbf{r}_u \times \mathbf{r}_v||^2 = EG - F^2,$$

where the double vertical line is used for a modulus, the standard notation for a norm or length in abstract algebra, and commonly used in differential geometry. The unit normal to any point on the surface is denoted by **U** and is given by

$$\mathbf{U} = \frac{\mathbf{r}_u \times \mathbf{r}_v}{||\mathbf{r}_u \times \mathbf{r}_v||} = \frac{\mathbf{r}_u \times \mathbf{r}_v}{EG - F^2}.$$

The *shape function* $S_p(\mathbf{v})$ is a function defined on a vector **v** as follows:

Definition 5.2 If **p** is the position vector of a point P and **v** is a tangent vector to a surface, then for each **v**, the function

$$S_p(\mathbf{v}) = -\nabla_\mathbf{v} \mathbf{U},$$

where **U** a unit normal field in the neighbourhood of the vector **p** and the notation $-\nabla_\mathbf{v}$ is a covariant derivative (a generalisation of a directional derivative), is called the *shape function* of the surface at the point **p**.

The covariant derivative $-\nabla_\mathbf{v}$ behaves like a directional derivative but applied not to a scalar field but to a vector field. Here is a simple example that should help you understand this better:

Example 5.4 Define a vector field **W** not in terms of Cartesian co-ordinates but as $\mathbf{W}(t) = W_1 \hat{\mathbf{T}} + W_2 \hat{\mathbf{N}} + W_3 \hat{\mathbf{B}}$ where all six factors on the right can depend on the parameter t. Select $W_1 = x^2$, $W_2 = 0$ and $W_3 = yz$, and set $\mathbf{p} = (2, 1, 0)$ with $\mathbf{v} = (-1, 0, 2)$. Calculate $-\nabla_\mathbf{v} W$.

Solution: With the data given, the tangent vector is $\mathbf{p} + t\mathbf{v} = (2 - t, 1, 2t)$ so that $\mathbf{W}(t) = (2 - t)^2 \hat{\mathbf{T}} + 2t\hat{\mathbf{B}}$ is the value of the given vector field on this tangent vector. The covariant derivative is simply the derivative of this $-2(2 - t)\hat{\mathbf{T}} + 2\hat{\mathbf{B}}$ evaluated at **p** itself, that is $-4\hat{\mathbf{T}} + 2\hat{\mathbf{B}}$, where of course both $\hat{\mathbf{T}}$ and $\hat{\mathbf{B}}$ take their magnitudes and directions from those at **p**. As t varies, so do both $\hat{\mathbf{T}}$ and $\hat{\mathbf{B}}$ and the magnitude and direction of **W**. This 'double' dependence on vectors is what is captured by the covariant derivative, but is not catered for by the notation of a directional derivative of a scalar field. □

Let us apply these results to a specific surface and to do this, one can do a lot worse than returning to the previous example of the torus and calculate the shape operator which has enough structure to show what is going on. First of all, it is essential to calculate a general expression for the normal to the surface, and to do this take the cross product of the two unit tangents $\hat{\mathbf{T}}_u$ and $\hat{\mathbf{T}}_v$:

$$\mathbf{U} = \hat{\mathbf{T}}_u \times \hat{\mathbf{T}}_v = (\cos u \cos v, \cos u \sin v, -\sin u)$$

assuming enough cross products have been done without having to make a song and dance about the process. The notation $\mathbf{U}$ is retained for the normal as this is a vector field on the patch in the vicinity of the point on the surface, and not the kind of normal ($\hat{\mathbf{N}}$ or $\hat{\mathbf{B}}$) found on curves. Now, take the two partial derivatives of $\mathbf{U}$ and these will be the shape operators acting in the direction of $\mathbf{r}_v$ and $\mathbf{r}_u$ in the direction of these normals. The negative sign involved has not yet been mentioned. There is always an ambiguity when taking covariant derivatives as to sign, and it turns out that the choice is really cosmetic and doesn't matter. The negative sign avoids accumulations of them in calculations. Therefore, differentiating gives

$$\mathbf{U}_u = (-\sin u \cos v, -\sin u \sin v, -\cos u) = -S(\mathbf{r}_u)$$
$$\mathbf{U}_v = (-\cos u \sin v, \cos u \cos v, 0) = -S(\mathbf{r}_v),$$

note also that

$$\mathbf{r}_u = (-r \sin u \cos v, -r \sin u \sin v, -r \cos u)$$
$$\mathbf{r}_v = (-(R + r \cos u) \sin v, (R + r \cos u) \cos v, 0).$$

Of course, from the definition of shape function as derivatives in the directions of the tangents, $\mathbf{r}_u$ is in the same direction as $S(\mathbf{r}_u)$ and $\mathbf{r}_v$ is in the same direction as $S(\mathbf{r}_v)$. By insisting that $\mathbf{U}$ is a unit vector ensures that it is S that captures the shape of the surface itself in the vicinity of a particular location. The letters l, m and n are defined by

$$l = S(\mathbf{r}_u) \cdot \mathbf{r}_u,$$
$$m = S(\mathbf{r}_u) \cdot \mathbf{r}_v = 0, \quad \text{for the torus,}$$
$$n = S(\mathbf{r}_v) \cdot \mathbf{r}_v$$

that are scalar functions that completely define the shape operator. Note that m could just as well be defined by $S(\mathbf{r}_v) \cdot \mathbf{r}_u$ as the vectors comprising the scalar product are parallel. The whole point of the shape operator is not to use remote co-ordinates but to express it in terms of local co-ordinate free terms. This is always possible using the covariant derivatives. For this example, we have $l = \dfrac{1}{r}$ and $n = \dfrac{\cos u}{R + r \cos u}$ so

$$S(\mathbf{r}_u) = -l\mathbf{r}_u = -\frac{1}{r}\mathbf{r}_u,$$

$$S(\mathbf{r}_v) = -n\mathbf{r}_v = \frac{\cos u}{R + r \cos u}\mathbf{r}_v$$

and this expresses each shape operator acting on a tangent in terms of properties local to the torus. This can always be done, so that it works here is not just luck. For those interested, a *tangent bundle* is the usual term for the collection of all such tangents, and all tangents to a surface are in fact an example of a differentiable manifold. It is not the definition but a result that follows directly from it that is now used. The equivalent to curvature is this shape operator acting on the covariant derivatives of the vectors in the local tangent space of the surface, and the result needed is that this is the same as the scalar product $S(\mathbf{r}_p) \cdot \mathbf{r}_p$ where p denotes a specific member of the tangent bundle. As said previously, it is a generalisation of a directional derivative. Explicitly, for the torus, the following are obtained:

$$S(\mathbf{r}_u) \cdot \mathbf{r}_u = -\frac{1}{r} = l,$$

$$S(\mathbf{r}_v) \cdot \mathbf{r}_v = -\frac{\cos u}{R + r \cos u} = n.$$

with $m = 0$ as $\mathbf{r}_u$ and $\mathbf{r}_v$ are orthogonal. Having come this far, expressing this in matrix form follows.

$$S(\mathbf{r}) = \begin{pmatrix} -\dfrac{\cos u}{R + r \cos u} & 0 \\ 0 & -\dfrac{1}{r} \end{pmatrix}$$

and this matrix is diagonal because $\mathbf{r}_u$ and $\mathbf{r}_v$ are orthogonal. A scalar called the *Gaussian curvature K* is the determinant of this matrix and is given by

$$K = \frac{\cos u}{r(R + r \cos u)}.$$

The columns contain the two formulas $\dfrac{\cos u}{R + r \cos u}$ and $\dfrac{1}{r}$. These are both curvatures in the single space curve sense already derived in Example 5.3, the latter expresses the curvature $1/r$ of the (vertical) slice through the torus, and since $v = $ constant, the normal is always radially outward, the leftmost fan in Figure 5.10. The former curvature $\dfrac{\cos u}{R + r \cos u}$ is more interesting. When $u = 0$, the normal *is* perpendicular to the (horizontal) slice through the torus; these are shown in Figure 5.10 in two clusters, a divergent group on the lower right

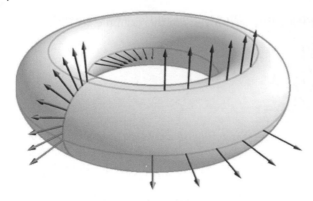

Figure 5.10 An assortment of normals on a torus. When they fan out, the curvature is positive; parallel means zero curvature, fanning inwards, negative curvature.

and a convergent group emerging from the inner ring of the torus. This is the case of the curvature in the traditional sense being the same but the sign of the normal dictating whether it is positive (on the outside) or negative (on the inside). While u increases and the slice moves vertically through the torus, the local normal has an increasing component in the z direction; this is the reason for the factor $\cos u$ in the numerator. At $u = \pi/2$, the plane is tangent to the top of the torus, the normals are all pointing in the z direction; they are all parallel so there is no local curvature associated with it, hence this curvature is zero. This is also shown in Figure 5.10. Note that the Gaussian curvature, actually defined as the product of the two principal curvatures, has a dimension of $1/(\text{length})^2$; so for a sphere of radius r, for example, it is $1/r^2$. Another quantity called the *mean curvature* can also be defined. It is half the trace of the diagonal matrix S. For the torus, it is not a useful measure, but for the sphere it is simply $1/r$. If the mean curvature is zero, then the surface is called a *minimal surface*.

In general, the Gaussian and mean curvatures, K and H respectively, on a patch $\mathbf{p}$, are given by

$$K(\mathbf{p}) = \frac{ln - m^2}{EG - F^2} \quad \text{and} \quad H(\mathbf{p}) = \frac{Gl + En - 2Fm}{2(EG - F^2)};$$

these are left as an exercise to prove (see Exercise 5.9 at the end of this chapter).

Some of the differential geometry of surfaces will re-emerge both in the treatment of curvilinear co-ordinates and in the evaluation of surface integrals. Before leaving surfaces completely, however, consider a surface of the form $F(x, y, z) = \text{constant}$. On this surface, the value of F does not change, so $dF = 0$. Expanding this using the chain rule:

$$dF = \frac{\partial F}{\partial x}dx + \frac{\partial F}{\partial y}dy + \frac{\partial F}{\partial z}dz = 0.$$

The small increment of a line in the surface $F = \text{constant}$ can be written $d\mathbf{r}$ and this can be expanded as

$$d\mathbf{r} = \mathbf{i}dx + \mathbf{j}dy + \mathbf{k}dz.$$

Looking at these two expressions, it can be seen that the scalar product

$$\left(\frac{\partial F}{\partial x}\mathbf{i} + \frac{\partial F}{\partial y}\mathbf{j} + \frac{\partial F}{\partial z}\mathbf{k} \right) \cdot (\mathbf{i}dx + \mathbf{j}dy + \mathbf{k}dz)$$

is simply

$$\frac{\partial F}{\partial x}dx + \frac{\partial F}{\partial y}dy + \frac{\partial F}{\partial z}dz = dF = 0.$$

Writing

$$\frac{\partial F}{\partial x}\mathbf{i} + \frac{\partial F}{\partial y}\mathbf{j} + \frac{\partial F}{\partial z}\mathbf{k} = \mathbf{\nabla} F,$$

this implies that on the surface $F(x, y, z) = $ constant

$$\mathbf{\nabla} F \cdot d\mathbf{r} = 0,$$

where $\mathbf{\nabla} F$ is read as 'grad' F or sometimes 'del' F. Recalling that a zero scalar product means the two vectors are at right angles, together with $d\mathbf{r}$ always being tangential to the surface $F(x, y, z) = $ constant, this means that $\mathbf{\nabla} F$ has to be normal to the surface $F(x, y, z) = $ constant. This fact will be very useful later. However, rather than pursue differential geometry any further, let us look at a different application, mechanics.

5.3   Mechanics

Mechanics is the use of mathematics to describe the motion of particles and rigid bodies under the influence of forces see the author's *Guide to Mechanics* [10]. The layman has a tendency to think of mechanics as being a branch of physics, but Newton's laws of motion that underpin classical mechanics are a set of rules that Newton turned into mathematical axioms; it was over two hundred years before the relativistic modifications of Einstein, so mechanics certainly belongs in mathematics, and at this level, applied mathematics. Relativity corrects the axioms for speeds that approach that of light. There is also quantum mechanics needed to describe the physics of motion on a sub-atomic scale. Everyday scales are still well described by Newtonian mechanics. Probably the finest book placing classical mechanics in a mathematical framework is *The Mathematical Methods of Classical Mechanics* by Arnold (1937–2010) [4] one of the best twentieth-century Russian mathematicians; so those inspired by the rest of this chapter can move on to Arnold's excellent book, but it is advanced mathematics using the mathematical structures of manifolds and the language of advanced differential geometry. To start at a more elementary level, mechanics begins with a fixed origin and a set of axes. Particle mechanics is concerned with the movement of masses considered to be so small as to occupy zero volume, whereas in rigid body mechanics, bodies are considered homogeneous and fixed in shape. As with differential geometry, but more so, there are plenty of textbooks devoted to mechanics so only a brief account can be given here. In this subsection, t definitely means time, and derivatives with respect to time are denoted by a dot, so $dx/dt = \dot{x}; dr/dt = \dot{r}$. Newton's second law of motion involves acceleration, the rate of change of velocity with respect to time, and velocity is in turn the rate of change of position with respect to time $\dot{\mathbf{r}}$. Differentiating with respect to t but utilising the function-of-a-function rule

$$\mathbf{v} = \dot{\mathbf{r}} = \frac{d\mathbf{r}}{ds}\frac{ds}{dt} = \dot{s}\hat{\mathbf{T}}$$

upon using the result from the previous subsection. Differentiating this again with respect to t using the product rule and function of a function gives

$$\mathbf{a} = \ddot{\mathbf{r}} = \dot{\mathbf{v}} = \ddot{s}\hat{\mathbf{T}} + \dot{s}\frac{d\hat{\mathbf{T}}}{ds}\frac{ds}{dt} = \ddot{s}\hat{\mathbf{T}} + \kappa\dot{s}^2\hat{\mathbf{N}} \tag{5.7}$$

using the first of the Serret–Frenet formulae. Two special cases are worth looking at: motion in a straight line means the path $\mathbf{r} = \mathbf{r}(t)$ has zero curvature and the second term is therefore zero so

$$\mathbf{a} = \ddot{s}\hat{\mathbf{T}} = \ddot{x}\mathbf{i}$$

if motion is along the x axis. The second case is motion in a circle radius r_0. Here, $r = r_0 = $ constant, so in this case the first term is now zero. $\kappa = 1/r_0$ and $\dot{s} = v = \omega r_0$ and so

$$\mathbf{a} = \kappa\dot{s}^2\hat{\mathbf{N}} = -\frac{v^2}{r_0}\hat{\mathbf{r}} = -\omega^2 r_0\hat{\mathbf{r}}$$

These may be familiar, dependent on background, but Equation 5.8 is a useful general expression for the acceleration of a particle travelling along any curve

$$\mathbf{a} = \ddot{s}\hat{\mathbf{T}} + \kappa\dot{s}^2\hat{\mathbf{N}}. \tag{5.8}$$

Example 5.5 Derive the version of Equation 5.8 valid for a general plane curve but using plane polar co-ordinates.

Solution: The most natural way to do this is to start from $\mathbf{r}$ expressed in polar co-ordinates, then differentiate twice: $\mathbf{r} = r\hat{\mathbf{r}}$ so differentiating using the product rule gives

$$\frac{d}{dt}[r\hat{\mathbf{r}}] = \dot{r}\hat{\mathbf{r}} + r\frac{d\hat{\mathbf{r}}}{dt}.$$

Now $\hat{\mathbf{r}}$ is a unit vector constrained to lie on the unit circle, hence its rate of change must be in the $\hat{\theta}$ direction and its magnitude is $\dot{\theta}$ as will now be shown. Figure 5.11 shows the situation, and from this figure, the directed arc $\delta\hat{\mathbf{r}} = |\hat{\mathbf{r}}|\delta\theta\hat{\theta}$. With $|\hat{\mathbf{r}}| = 1$, dividing by the increment in time, δt, and taking the limit

Figure 5.11 The plane polar co-ordinates r, θ with their directions; the angle between $\hat{\theta}$ and $\hat{\theta} + \delta\,\hat{\theta}$ is also $\delta\theta$.

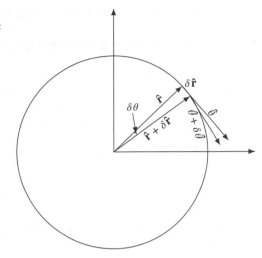

as $\delta t \to 0$, this gives

$$\frac{d\hat{\mathbf{r}}}{dt} = \frac{d\theta}{dt}\hat{\theta} \quad \text{or} \quad \dot{\hat{\mathbf{r}}} = \dot{\theta}\hat{\theta}.$$

Hence,

$$\frac{d\mathbf{r}}{dt} = \dot{\mathbf{r}} = \dot{r}\hat{\mathbf{r}} + r\dot{\theta}\hat{\theta}.$$

Differentiating this expression with respect to t then gives after much use of the product rule,

$$\frac{d^2\mathbf{r}}{dt^2} = \ddot{\mathbf{r}} = \ddot{r}\hat{\mathbf{r}} + \dot{r}\dot{\hat{\mathbf{r}}} + \dot{r}\dot{\theta}\hat{\theta} + r\ddot{\theta}\hat{\theta} + r\dot{\theta}\dot{\hat{\theta}}.$$

The derivative $\dot{\hat{\theta}}$ is given by

$$\dot{\hat{\theta}} = -\dot{\theta}\hat{\mathbf{r}}$$

as $\hat{\theta}$ like $\hat{\mathbf{r}}$ is a unit vector constrained to sweep out an arc of a circle in time δt, see Figure 5.11, so using this result and the previous one for $\dot{\hat{\mathbf{r}}}$ gives

$$\ddot{\mathbf{r}} = \ddot{r}\hat{\mathbf{r}} + 2\dot{r}\dot{\theta}\hat{\theta} + r\ddot{\theta}\hat{\theta} - r\dot{\theta}^2\hat{\mathbf{r}}$$

and this can be rewritten after combining the two $\hat{\theta}$ terms as

$$\ddot{\mathbf{r}} = (\ddot{r} - r\dot{\theta}^2)\hat{\mathbf{r}} + \frac{1}{r}\frac{d}{dt}(r^2\dot{\theta})\hat{\theta}. \tag{5.9}$$

□

The second term on the right-hand side of Equation 5.9 is worth a closer look, and what comes next does exactly that. What now follows is a short foray into mechanics before displaying the usefulness of vector calculus to the subject.

Newton's second law states that 'force=mass times acceleration' written $\mathbf{F} = m\ddot{\mathbf{r}}$ in terms of mathematics. Many forces are directed towards the origin, that is $\mathbf{F} = k\mathbf{r}$; these are given the name *central forces*. Mechanics has a few key quantities, but perhaps the most important one is *angular momentum* defined as the moment of momentum, $\mathbf{L} = \mathbf{r} \times m\dot{\mathbf{r}}$. A related quantity is $\mathbf{r} \times \mathbf{F}$ called the *torque*. These quantities will now be utilised to derive interesting formulae. A central force has zero torque as the cross product is identically zero. Let us calculate the rate of change of angular momentum:

$$\frac{d}{dt}[\mathbf{r} \times m\dot{\mathbf{r}}] = \dot{\mathbf{r}} \times m\dot{\mathbf{r}} + \mathbf{r} \times m\ddot{\mathbf{r}}$$

provided the mass m is constant. The first term is zero as the cross product of a vector with itself vanishes. Newton's second law thus tells us that 'Torque = rate of change of angular momentum' and moreover, under a central force the right-hand side of the last equation is identically zero, so angular momentum does not change:

$$\frac{d}{dt}[\mathbf{r} \times m\dot{\mathbf{r}}] = 0 \quad \text{or} \quad \mathbf{r} \times m\dot{\mathbf{r}} = \text{constant.}$$

Here is an application from astrophysics. Planets travel around the Sun under the influence of gravitational force, the Newtonian view superseded by Einstein's theory, but still a very good approximation. This force is central; it points towards the Sun taken as the origin. Therefore, the angular momentum of planets must be constant. Using polar co-ordinates and dipping into the last example, the angular momentum can be calculated as follows:

$$\mathbf{L} = \mathbf{r} \times m\dot{\mathbf{r}} = r\hat{\mathbf{r}} \times (m\dot{r}\hat{\mathbf{r}} + r\dot{\theta}\hat{\boldsymbol{\theta}})$$

So

$$\mathbf{L} = mr^2\dot{\theta}\hat{\mathbf{r}} \times \hat{\boldsymbol{\theta}} = mr^2\dot{\theta}\hat{\mathbf{k}},$$

where $\hat{\mathbf{k}}$ is the direction perpendicular to the plane of the solar system. The second term of Equation 5.9 recognises this as under a central force the acceleration in the $\hat{\boldsymbol{\theta}}$ direction must be zero by Newton's second law. This implies $r^2\dot{\theta} = h = $ constant for a given planet, and h is the magnitude of the angular momentum. Moreover, as the direction as well as the magnitude of angular momentum is constant, this explains why the solar system lies in a plane and the (constant) direction of angular momentum is perpendicular to this plane. Variations from this ideal state is the business of astronomy, but is due to a variety of factors including the Sun being so close to Mercury and hence so large that the force it exerts cannot be reasonably thought central. Pluto is now no longer a planet and may be a captured object, all other planets have close to planar orbits consistent with the above theory.

For the second application, an expression for the angular momentum of a rigid body is derived. A rigid body is assumed to be a collection of particles and to analyse a rigid body one looks at the motion of the centre of mass and motion relative to the centre of mass. As this is not a mechanics textbook, being comprehensive is not the point. This is an example of using vectors, mainly vector algebra rather than calculus. Consider motion relative to the centre of mass which is a rotation, captured by the angular velocity vector $\boldsymbol{\omega}$. The velocity of any point P in the body referred to a stationary centre of mass O is $\mathbf{v}$ where

$$\mathbf{v} = \boldsymbol{\omega} \times \mathbf{r}.$$

The angular momentum of a rigid body is the angular momentum of a single particle of the rigid body, but summed over the entire rigid body. This is by convention rather imprecisely written using a summation sign, integral signs would be better, but they come later:

$$\mathbf{L} = \sum (\mathbf{r} \times m\mathbf{v}) = \sum (\mathbf{r} \times m(\boldsymbol{\omega} \times \mathbf{r}))$$

Using the expansion of the vector triple product, this is

$$\mathbf{L} = \sum m\mathbf{r}^2 \boldsymbol{\omega} - \sum (m\mathbf{r} \cdot \boldsymbol{\omega})\mathbf{r}$$

Given that it is usual to choose an axis such that the second term on the right is zero, and that $I = \sum m\mathbf{r}^2$ is a quantity whose value is dictated by the shape of the rigid body called the *moment of inertia*, this expression becomes the simple

$$\mathbf{L} = I\boldsymbol{\omega}$$

that can be written in words: angular momentum is the product of the moment of inertia about that axis times the angular velocity. Taking the scalar product of this with the angular velocity $\boldsymbol{\omega}$ then

$$\mathbf{L} \cdot \boldsymbol{\omega} = \sum (\mathbf{r} \times m\mathbf{v}) \cdot \boldsymbol{\omega}$$

so

$$\mathbf{L} \cdot \boldsymbol{\omega} = \sum \boldsymbol{\omega} \cdot (\mathbf{r} \times m\mathbf{v})$$

and using the property of the scalar triple product that the dot and cross can be interchanged without changing the value, this is

$$\mathbf{L} \cdot \boldsymbol{\omega} = \sum (\boldsymbol{\omega} \times \mathbf{r}) \cdot m\mathbf{v}.$$

However, $\boldsymbol{\omega} \times \mathbf{r} = \mathbf{v}$ hence

$$\mathbf{L} \cdot \boldsymbol{\omega} = \sum \mathbf{v} \cdot m\mathbf{v} = \sum mv^2 = 2T,$$

where T is the total kinetic energy of the rigid body.

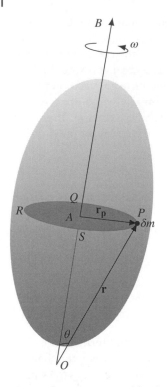

Figure 5.12 A rigid body rotating about its axis of symmetry with angular velocity ω.

Example 5.6 A rigid body rotates about a fixed axis with angular velocity ω prove that the kinetic energy of rotation is

$$T = \frac{1}{2}I\omega^2.$$

Solution: Figure 5.12 shows the rigid body, and the vector $\mathbf{r}$ connecting an arbitrary point P to the fixed origin O on the axis of rotation OB. As the body rotates this vector will describe the circle $PQRS$ with velocity $\mathbf{v} = \boldsymbol{\omega} \times \mathbf{r} = \omega |\mathbf{r}| \sin \theta = \omega r_p$. Therefore, $v = \omega r_p$. The kinetic energy δT of the infinitesimal mass δm at the point P is thus

$$\delta T = \frac{1}{2}\delta m \omega^2 r_p^2.$$

Summing this over the entire solid, using the summation sign rather than the more appropriate integral thus gives

$$T = \sum \delta T = \frac{1}{2} \sum \delta m \omega^2 r_p^2 = \frac{1}{2} \left(\sum \delta m r_p^2 \right) \omega^2 = \frac{1}{2}I\omega^2.$$

using the definition of moment of inertia I given earlier. □

In general, if a rigid body is both rotating as in Figure 5.12 and moving, then the total kinetic energy is that due a mass equal to the mass of the body but travelling at the speed of travel of the centre of mass plus the kinetic energy due to the rotation as derived in the above example:

$$T_{total} = T_{\text{of centre of mass}} + T_{\text{relative to the centre of mass}} = \frac{1}{2}mv_g^2 + \frac{1}{2}I\omega^2.$$

Exercises

5.1 Establish the following rules of differentiation for vectors:

a) $\dfrac{d}{dt}(\mathbf{A} + \mathbf{B}) = \dfrac{d\mathbf{A}}{dt} + \dfrac{d\mathbf{B}}{dt}$,

b) $\dfrac{d}{dt}(\mathbf{A} \cdot \mathbf{B}) = \dfrac{d\mathbf{A}}{dt} \cdot \mathbf{B} + \mathbf{A} \cdot \dfrac{d\mathbf{B}}{dt}$,

c) $\dfrac{d}{dt}(\mathbf{A} \times \mathbf{B}) = \dfrac{d\mathbf{A}}{dt} \times \mathbf{B} + \mathbf{A} \times \dfrac{d\mathbf{B}}{dt}$,

d) $\dfrac{d}{dt}(\phi\mathbf{A}) = \dfrac{d\phi}{dt}\mathbf{A} + \phi\dfrac{d\mathbf{A}}{dt}$.

5.2 Show from first principles that

$$\frac{d}{dt}(\mathbf{A} \cdot \mathbf{B} \times \mathbf{C}) = \frac{d\mathbf{A}}{dt} \cdot \mathbf{B} \times \mathbf{C} + \mathbf{A} \cdot \frac{d\mathbf{B}}{dt} \times \mathbf{C} + \mathbf{A} \cdot \mathbf{B} \times \frac{d\mathbf{C}}{dt}.$$

5.3 If $a = |\mathbf{a}|$, show that

$$\mathbf{a} \cdot \frac{d\mathbf{a}}{dt} = a\frac{da}{dt}.$$

5.4 Find $\mathbf{r} = \mathbf{r}(t)$ if $\mathbf{a}$ and $\mathbf{b}$ are given constant vectors and

a)

$$\frac{d^2\mathbf{r}}{dt^2} = \mathbf{a},$$

b)

$$\mathbf{a} \times \frac{d^2\mathbf{r}}{dt^2} = \mathbf{b}, \quad \text{with} \quad \mathbf{a} \cdot \mathbf{b} = 0.$$

5.5 Determine $\hat{\mathbf{T}}, \hat{\mathbf{N}}, \hat{\mathbf{B}}, \kappa$ and τ for the parabola $y^2 = 4ax$ where a is a constant.

5.6 By first showing that

$$\frac{ds}{d\theta} = 3(2\theta^2 + 1),$$

where s is the arc length, find the quantities $\hat{\mathbf{T}}, \hat{\mathbf{N}}, \hat{\mathbf{B}}, \kappa$ and τ for the twisted cubic given parametrically by $\mathbf{r} = (2\theta^3, 3\theta^2, 3\theta)$.

5.7 For the curve $\mathbf{r} = (2a \cos t, 2a \sin t, bt^2)$, show that $\dot{\mathbf{r}} \cdot \ddot{\mathbf{r}} = 4b^2 t$ and

$$|\dot{\mathbf{r}} \times \ddot{\mathbf{r}}| = 4a\sqrt{a^2 + b^2 + b^2 t^2},$$

where the dot denotes differentiation with respect to t, and determine $\hat{\mathbf{T}}$.

5.8 Using the method in the text following Example 5.3, determine the shape operators and Gaussian curvature for
a) the cylinder $x^2 + y^2 = R^2$,
b) the sphere $x^2 + y^2 + z^2 = r^2$.

5.9 Establish the two formulas for the Gaussian curvature K and mean curvature H:

$$K(\mathbf{p}) = \frac{ln - m^2}{EG - F^2} \quad \text{and} \quad H(\mathbf{p}) = \frac{Gl + En - 2Fm}{2(EG - F^2)},$$

where l, m, n, E, F and G are defined in the text.

5.10 From the definitions in the text, show that

$$l = \mathbf{U} \cdot \mathbf{r}_{uu},$$
$$m = \mathbf{U} \cdot \mathbf{r}_{uv},$$
$$n = \mathbf{U} \cdot \mathbf{r}_{vv}.$$

5.11 Use the previous question to calculate the Gaussian and mean curvatures for the helicoid $\mathbf{r} = (u \cos v, u \sin v, bv), \ b \neq 0$.

5.12 From the general formulas for velocity and acceleration derived in the text, show that

$$\frac{d\mathbf{r}}{dt} \times \frac{d^2\mathbf{r}}{dt^2} = \kappa \left(\frac{ds}{dt}\right)^3 \hat{\mathbf{B}}$$

and deduce that

$$\hat{\mathbf{B}} = \frac{\frac{d\mathbf{r}}{dt} \times \frac{d^2\mathbf{r}}{dt^2}}{\left|\frac{d\mathbf{r}}{dt} \times \frac{d^2\mathbf{r}}{dt^2}\right|} \quad \text{and} \quad \kappa = \frac{\left|\frac{d\mathbf{r}}{dt} \times \frac{d^2\mathbf{r}}{dt^2}\right|}{\left|\frac{d\mathbf{r}}{dt}\right|^3}.$$

5.13 A particle is constrained to move with a constant speed v along a wire the shape of the cardioid $r = a(1 + \cos \theta)$. Use Equation 5.8 to find v and show that the radial component of the acceleration is constant.

5.14 A particle moves in a curve given by $r = a(1 - \cos\theta)$ with $d\theta/dt = 3$. Find the components of velocity and acceleration, and show that the velocity vanishes where $\theta = 0$ but find the acceleration there.

5.15 (*A knowledge of mechanics is required for this exercise.) A top is spinning about its axis of symmetry such that the angle between its axis and the vertical is a constant α. Use rigid body mechanics, in particular the conservation of angular momentum, to show that the top precesses with a constant angular velocity.

6

Gradient, Divergence, and Curl

6.1 Introduction

In this section, there are operators introduced that further our knowledge of vector differentiation. The notions of scalar and vector fields have already been covered; temperature and density are examples of scalar functions of position, and force and velocity examples of vector fields. As far as calculus is concerned of course, Cartesian co-ordinates both scalar and vector functions of position are simply functions of the three variables x, y and z, and partial derivatives have already been covered. However, it turns out that there are three operations that can be defined independent of choice of co-ordinates. These operations help with the overall understanding of the behaviour of scalar and vector fields, properties that are obscured by simply dealing in one-dimensional partial derivatives. The first of these operations is called the gradient operator.

6.2 Gradient

This is perhaps the most straightforward of these operators. Simply take the three partial derivatives with respect to x, y and z, and form a vector sum. Here is the definition in terms of Cartesian co-ordinates:

Definition 6.1 (Grad.) Given a scalar field $\phi = \phi(x, y, z)$, the *gradient* operator ∇ is defined by

$$\nabla \phi(x, y, z) = \mathbf{i}\frac{\partial \phi}{\partial x} + \mathbf{j}\frac{\partial \phi}{\partial y} + \mathbf{k}\frac{\partial \phi}{\partial z}.$$

This is a vector operator on a scalar field, the result of which is a vector, in some books it is written 'grad ϕ'. A co-ordinate free version is much more satisfactory, but this has to wait until after integrals have been met in Chapter 10. There are a plethora of applications for this operator, so let us run through few now. To

Two and Three Dimensional Calculus: With Applications in Science and Engineering, First Edition. Phil Dyke.
© 2018 John Wiley & Sons Ltd. Published 2018 by John Wiley & Sons Ltd.

recap from earlier, if a function $f(x, y, z)$ is a constant, then this defines a surface. On this surface, $df = 0$. Using the chain rule, this is

$$df = \frac{\partial f}{\partial x}dx + \frac{\partial f}{\partial y}dy + \frac{\partial f}{\partial z}dz = \nabla f \cdot d\mathbf{r} = 0.$$

This implies that ∇f is perpendicular to $d\mathbf{r}$. However, $d\mathbf{r}$ is tangential to the surface $f(x, y, z) = 0$, hence ∇f must be normal to the surface. Calculating its value therefore is a convenient way of calculating the normal to a surface. Here is an example:

Example 6.1   Calculate the unit normal to the surface of the cone $x^2 + y^2 = z^2$ at the point $(3, 4, 5)$.

Solution: Writing the surface as $f(x, y, z) = x^2 + y^2 - z^2 = 0$; the three derivatives are

$$\frac{\partial f}{\partial x} = 2x, \quad \frac{\partial f}{\partial y} = 2y, \quad \frac{\partial f}{\partial z} = -2z$$

so $\nabla(x^2 + y^2 - z^2) = 2x\mathbf{i} + 2y\mathbf{j} - 2z\mathbf{k}$. The unit normal is thus

$$\hat{\mathbf{n}} = \frac{\nabla(x^2 + y^2 - z^2)}{|\nabla(x^2 + y^2 - z^2)|} = \frac{1}{\sqrt{x^2 + y^2 + z^2}}[x\mathbf{i} + y\mathbf{j} - z\mathbf{k}].$$

Thus, inserting $x = 3, y = 4$ and $z = 5$ gives the unit normal as

$$\hat{\mathbf{n}} = \frac{1}{\sqrt{50}}[3\mathbf{i} + 4\mathbf{j} - 5\mathbf{k}].$$

Note that this is completely mechanistic, drawing diagrams, or even knowing that the surface is a cone is unnecessary. However, at the point $(0, 0, 0)$ the normal $\hat{\mathbf{n}}$ does not exist, this is the vertex of the cone. Never be completely mechanistic.   □

Another important use of grad. is in expressing the rate of change of a quantity in a given direction, called finding the *directional derivative*. Geometrically if the line $\overrightarrow{AA'}$ denotes the direction of the vector $\mathbf{n}$, then the change in value of a scalar ϕ in this direction is

$$\lim_{A' \to A} \left\{ \frac{\phi_{A'} - \phi_A}{AA'} \right\}$$

or written algebraically

$$\frac{\partial \phi}{\partial n} \quad \text{or even} \quad \mathbf{n} \cdot \frac{\partial \phi}{\partial \mathbf{n}}$$

though a notation whereby division by a vector is even just implied can give a green light to poor students inclined to go in that wrong direction. A more calculus based definition would be to define directional derivative as the following limit:

Definition 6.2 The directional derivative for a function $\phi(\mathbf{r})$ in the direction $\mathbf{n}$ is the limit:

$$\lim_{h \to 0} \left\{ \frac{\phi(\mathbf{r} + h\mathbf{n}) - \phi(\mathbf{r})}{h} \right\}.$$

Taylor's theorem to first order is

$$\phi(\mathbf{r} + h\mathbf{n}) = \phi(\mathbf{r}) + h\mathbf{n} \cdot \nabla\phi + \cdots$$

so the above definition can be written

$$\lim_{h \to 0} \left\{ \frac{\phi(\mathbf{r} + h\mathbf{n}) - \phi(\mathbf{r})}{h} \right\} = \mathbf{n} \cdot \nabla\phi.$$

In Cartesian co-ordinates if $\mathbf{n} = n_1\mathbf{i} + n_2\mathbf{j} + n_3\mathbf{k}$, the directional derivative is

$$\mathbf{n} \cdot \nabla\phi = \frac{\partial\phi}{\partial n} = n_1\frac{\partial\phi}{\partial x} + n_2\frac{\partial\phi}{\partial y} + n_3\frac{\partial\phi}{\partial z}.$$

This can be written

$$\left(n_1\frac{\partial}{\partial x} + n_2\frac{\partial}{\partial y} + n_3\frac{\partial}{\partial z} \right)\phi = (\mathbf{n} \cdot \nabla)\phi$$

as the directional derivative of ϕ in the direction of $\mathbf{n}$. This last expression is better for vector functions, so the directional derivative of $\mathbf{F}$ in the direction of $\mathbf{n}$ is $(\mathbf{n} \cdot \nabla)\mathbf{F}$. This avoids us having to define the grad. of a vector, but directional derivatives of vectors can wait until the scalar version has been explored further. The vector $\mathbf{n}$ is assumed to be a unit vector, the effect of scaling does not change the direction of the vector but scales the directional derivative, so replacing $\mathbf{n}$ by $\alpha\mathbf{n}$ does the same to the directional derivative that changes from $\nabla\phi \cdot \mathbf{n}$ to $\alpha\nabla\phi \cdot \mathbf{n}$. It is not good practice for definitions to be scale dependent, hence the insistence that $\mathbf{n}$ is a unit vector.

Example 6.2 Find the directional derivative of the function $\phi(x, y, z) = x^2y - 3yz + 2xz$ in the direction $\mathbf{n} = 4\mathbf{i} - 7\mathbf{j} + 2\mathbf{k}$ at the point $(3, -2, 1)$.

Solution: First of all the direction needs to be a unit vector, so

$$\hat{\mathbf{n}} = \frac{4\mathbf{i} - 7\mathbf{j} + 2\mathbf{k}}{\sqrt{16 + 49 + 4}} = \frac{1}{\sqrt{69}}(4\mathbf{i} - 7\mathbf{j} + 2\mathbf{k}).$$

Calculate

$$\nabla\phi = (2xy + 2z)\mathbf{i} + (x^2 - 3z)\mathbf{j} + (2x - 3y)\mathbf{k} = -10\mathbf{i} + 6\mathbf{j} + 12\mathbf{k}$$

at the point $(3, -2, 1)$. Finally, the scalar product is

$$\hat{\mathbf{n}} \cdot \boldsymbol{\nabla}\phi = \frac{1}{\sqrt{69}}(-40 - 42 + 24) = -\frac{58}{\sqrt{69}},$$

which is the required directional derivative. Note it is just a numerical value as ϕ is scalar. □

Therefore, the directional derivative tells how the value of the scalar field $\phi(x, y, z)$ changes in a certain direction. In Chapter 5, the covariant derivative was mentioned as the mathematical generalisation of directional derivative, defined there as a two-dimensional field tangential to a surface. The temptation to introduce differentiable manifolds was resisted there, and the resistance will be continued here. Instead recall that a curve can be parameterised by $\mathbf{r}(t) = (x(t), y(t), z(t))$ as was seen in the last chapter. However, it is also true that

$$\frac{d}{dt}\phi(\mathbf{r}(t)) = \boldsymbol{\nabla}\phi \cdot \mathbf{r}'(t),$$

where the prime denotes differentiation with respect to t. The direction of $\mathbf{r}'(t)$ is locally the tangent vector to the curve $\mathbf{r}(t)$ and so the above expression for the directional derivative along this particular path would usually not be straight. Another useful fact is that the directional derivative can be written

$$\mathbf{n} \cdot \boldsymbol{\nabla}\phi = |\boldsymbol{\nabla}\phi| \cos\theta,$$

where $\mathbf{n}$ is a unit vector, and θ is the angle between the two vectors. When $\theta = 0$ the cosine is unity and the directional derivative assumes its maximum value. When $\theta = \pi$ the directional derivative assumes its minimum value. Hence these maximum and minimum values are attained when $\boldsymbol{\nabla}\phi$ is parallel and antiparallel to $\mathbf{n}$, respectively. This may seem a little odd given the calculation earlier that confirmed that $\boldsymbol{\nabla}\phi$ is normal to the surface $\phi = $ constant, but think about this. *On* the surface, ϕ itself has the value zero, so if $\phi(x, y, z)$ is thought of as a scalar field taking values at points throughout three-dimensional space with the surface $\phi = $ constant cutting through on which the value of the field happens to be zero, it is then less surprising that the maximum values of the changes in the values of ϕ occur along directions that are normal to this surface, in fact anything else would be surprising.

The gradient of a scalar field, ϕ is a vector field $\boldsymbol{\nabla}\phi$. If this vector field is a practical quantity like a force, then the associated scalar is called a potential field. The easiest example of one would be the gravitational field due to the earth. Any mass m close to the earth is attracted to its centre and the mass experiences a weight mg directed toward the centre of the earth. Using local Cartesians for convenience, the force on the mass m is $\mathbf{F} = -mg\mathbf{k}$ where $\mathbf{k}$ points directly up. Put this force equal to $-\boldsymbol{\nabla}\phi$ where the minus sign is there for practical convenience, then as there is only z dependence, we have

$$\mathbf{F} = -\boldsymbol{\nabla}\phi = -\frac{d\phi}{dz}\mathbf{k} = -mg\mathbf{k}$$

and integration gives $\phi = mgz + $ constant. This is recognised as the potential energy of the mass m and the constant merely indicates an arbitrary level of potential energy, the surface of the earth perhaps where the potential energy is set to zero. The application to the spherical earth will have to wait until curvilinear co-ordinates have been introduced in the next chapter. In electromagnetic theory and fluid mechanics, the scalar potential function plays a very important part. So far nothing has been mentioned about the co-ordinate free aspect of $\nabla\phi$. It is only after curvilinear co-ordinates have been covered that serious calculations can take place, so here a notational development that has consequences is introduced. Taking advantage of the Cartesian definitions using the suffix derivative notation

$$\nabla\phi = \mathbf{i}\phi_x + \mathbf{j}\phi_y + \mathbf{k}\phi_z,$$

but even this is rather clumsy. The notation that leads to tensors and is easy to generalise is to substitute i for one of x, y or z and a comma for derivative, so $\phi_{,i}$ becomes an alternative notation for $\nabla\phi$. The presence of one index indicates a vector, and the comma a derivative, this is pursued in Section 7.4.1. This notation is called Cartesian tensors, a precursor to full blown tensor notation that will be a requirement for those wanting to go on to study general relativity.

6.3   Divergence

The next operation to acquaint ourselves with is *divergence*, abbreviated to 'div.' and written mathematically as ∇. the dot being very important. This operator is scalar but operates on a vector, and the result is a scalar. Here is the definition in terms of Cartesian co-ordinates:

Definition 6.3   (Div.)   Given a vector field $\mathbf{F} = \mathbf{F}(x, y, z)$, the *divergence* operator ∇. is defined by

$$\nabla.\mathbf{F}(x, y, z) = \frac{\partial F_1}{\partial x} + \frac{\partial F_2}{\partial y} + \frac{\partial F_3}{\partial z},$$

where $\mathbf{F}(x, y, z) = \mathbf{i}F_1 + \mathbf{j}F_2 + \mathbf{k}F_3$.

This is a scalar quantity derived from a vector field. In layman's terms, picture a vector field as a collection of arrows of varying length (magnitude) and pointing in a variety of directions. Put a closed surface, say a sphere for simplicity and sum the amount of the vector emanating from its surface. This will be the scalar product of a surface element $r^2 \sin\phi d\phi d\theta$ (see Chapter 7) with normal pointing radially out and the local vector representing the vector field. Sum this over the entire surface of the sphere and you get a measure of the flux of the vector field emanating from the sphere. Now reduce the radius of the sphere to an infinitesimal value. This represents the divergence of the vector field at the point that is the centre of the shrinking sphere. Put crudely, the divergence of a vector field represents the flux, or amount of stuff, crossing a surface and it plays a crucial role in fluid mechanics and electrodynamics. The name 'divergence' betrays this. Consider a simple one-dimensional vector field $x\mathbf{i}$, the divergence of this is easily calculated to be 1, similarly the divergence of the field $x\mathbf{i} + y\mathbf{j}$ is 2 and

$$\nabla \cdot \mathbf{r} = \nabla \cdot (x\mathbf{i} + y\mathbf{j} + z\mathbf{k}) = \frac{\partial x}{\partial x} + \frac{\partial y}{\partial y} + \frac{\partial z}{\partial z} = 1 + 1 + 1 = 3.$$

The vector field $\mathbf{r}$ looks like a spiky ball at the origin, the ultimate koosh ball, and the divergence is 3 everywhere. Usually, the value of $\nabla \cdot \mathbf{F}$ will depend on position; a negative value of divergence is certainly possible, for example in fluid mechanics $\nabla \cdot \mathbf{u} < 0$ where $\mathbf{u}$ is the velocity of the fluid in a region denotes the presence of a sink, rather like a drain in the road. A more colourful example from astrophysics would be a black hole, perhaps the ultimate sink. Here is a computational example

Example 6.3   Determine the divergence of the vector fields:

a) $\mathbf{F}_1(x, y, z) = x^2 yz\mathbf{i} + xy^2 z\mathbf{j} + xyz^2\mathbf{k}$;
b) $\mathbf{F}_2(x, y, z) = x^2 yz\mathbf{i} + xy^2 z\mathbf{j} - 2xyz^2\mathbf{k}$.

Solution: Differentiation yields for (a)

$$\nabla \cdot \mathbf{F}_1 = 2xyz + 2xyz + 2xyz = 6xyz;$$

and for (b)

$$\nabla \cdot \mathbf{F}_2 = 2xyz + 2xyz - 4xyz = 0.$$

$\square$

Part (b) of the last example yields a field $\mathbf{F}_2$ whose divergence is identically zero. Such fields are called *solenoidal* and they will feature later after the third and final vector operator has been defined. Before this is done, the divergence can also be contracted using Cartesian tensor notation. First of all, a vector $\mathbf{F}$ is denoted by F_i where the index runs through 1, 2 and 3 to denote components in the x, y and z directions, respectively. In fact, the three co-ordinates are relabelled x_1, x_2 and x_3 to facilitate this and generalisations into higher dimensions. The divergence $\nabla \cdot \mathbf{F}$ becomes $F_{i,i}$, where the repeated index indicates a summation, this is called the *summation convention*. Remember the comma denotes differentiation. Therefore, $F_{i,i} = F_{1,1} + F_{2,2} + F_{3,3} = F_{1x} + F_{2y} + F_{3z}$ in standard suffix derivative notation. There will be more on this in Section 7.4.1.

6.4 Curl

The third operator operates on a vector and produces another vector, it is called the 'curl' and it is not short for anything:

Definition 6.4 (Curl) Given a vector field $\mathbf{F} = \mathbf{F}(x, y, z)$, the *curl* operator $\nabla \times$ is defined by

$$\nabla \times \mathbf{F}(x, y, z) = \begin{vmatrix} \mathbf{i} & \mathbf{j} & \mathbf{k} \\ \dfrac{\partial}{\partial x} & \dfrac{\partial}{\partial y} & \dfrac{\partial}{\partial z} \\ F_1 & F_2 & F_3 \end{vmatrix},$$

where $\mathbf{F}(x, y, z) = \mathbf{i}F_1 + \mathbf{j}F_2 + \mathbf{k}F_3$.

Expanded out, this definition is:

$$\nabla \times \mathbf{F}(x, y, z) = \left(\frac{\partial F_3}{\partial y} - \frac{\partial F_2}{\partial z} \right) \mathbf{i} + \left(\frac{\partial F_1}{\partial z} - \frac{\partial F_3}{\partial x} \right) \mathbf{j} + \left(\frac{\partial F_2}{\partial x} - \frac{\partial F_1}{\partial y} \right) \mathbf{k}.$$
$$(6.1)$$

The physical meaning of curl is not straightforward but for a vector field it is equivalent to angular momentum in mechanics defined in the last chapter. It sometimes, but not always, is a twisting of the fields exemplified in the name 'curl'. The curl of both $\mathbf{r}$ and $\mathbf{a}$ can be calculated as zero. However, suppose we calculate the curl of the velocity of a rotating rigid body $\boldsymbol{\omega} \times \mathbf{r}$. To keep calculation simple, suppose the rotation vector is in the z direction: $\boldsymbol{\omega} = \omega \mathbf{k}$ and $\mathbf{v} = \boldsymbol{\omega} \times \mathbf{r} = -\omega y \mathbf{i} + \omega x \mathbf{j}$. Hence

$$\nabla \times (\boldsymbol{\omega} \times \mathbf{r}) = \omega \begin{vmatrix} \mathbf{i} & \mathbf{j} & \mathbf{k} \\ \dfrac{\partial}{\partial x} & \dfrac{\partial}{\partial y} & \dfrac{\partial}{\partial z} \\ -y & x & 0 \end{vmatrix} = 2\omega \mathbf{k}.$$

Therefore, the curl of the velocity of a rigid body is twice the rotation rate of the rigid body. This does give some insight into the meaning of curl, but it is only a cartoon caricature. In the study of both electromagnetic fields and fluids, curl plays a central role and embedding the calculation of curl into these practical applications leads to a greater understanding of its meaning. It is unfortunately not the place for an excursion into these fields here, but an example of its calculations goes some way to help understand its meaning.

Example 6.4 Do the following two exercises:

a) Find the value of $\nabla \times \mathbf{F}$ if $\mathbf{F} = xz^2 \mathbf{i} - 2yz \mathbf{j} + (2x + 4z) \mathbf{k}$ at the point $(1, 1, 1)$.
b) Show that the curl of the vector $(2x + yz, 2y + xz, 2z + xy)$ is always zero.

Solution: For part (a) simply calculate the determinate:

$$\begin{vmatrix} \mathbf{i} & \mathbf{j} & \mathbf{k} \\ \dfrac{\partial}{\partial x} & \dfrac{\partial}{\partial y} & \dfrac{\partial}{\partial z} \\ xz^2 & -2yz & (2x+4z) \end{vmatrix} = 2y\mathbf{i} + 2(xz-1)\mathbf{j} + 0\mathbf{k}.$$

At the point $(1,1,1)$, the value of the curl is $2\mathbf{i}$ or $(2,0,0)$ with no components in the y or z directions. For part (b), evaluating the determinant

$$\begin{vmatrix} \mathbf{i} & \mathbf{j} & \mathbf{k} \\ \dfrac{\partial}{\partial x} & \dfrac{\partial}{\partial y} & \dfrac{\partial}{\partial z} \\ 2x+yz & 2y+xz & 2z+xy \end{vmatrix} = (x-x)\mathbf{i} + (y-y)\mathbf{j} + (z-z)\mathbf{k},$$

which is the zero vector. □

If the curl of a vector field is always zero, then the field is called *irrotational* and this crops up in many applications. In mechanics, an 'irrotational' force is called conservative and ensures the conservation of mechanical energy. In fluid mechanics if the velocity is irrotational, then there exists a velocity potential and this helps in providing a clearer picture of the fluid flow. There will be more on this later. For consistency, we ought to look at the Cartesian tensor form for $\nabla \times \mathbf{F}$. To do this involves introducing the alternating tensor that might be too much of a distraction for some. If it is, skip this bit and move to the next section. The alternating tensor is a third-order tensor with $3^3 = 27$ components. It is called ϵ_{ijk} and defined as follows:

$$\epsilon_{ijk} = \begin{cases} 1, & \text{if } i,j,k \text{ cycle in this order,} \\ -1, & \text{if } i,j,k \text{ cycle in the order } k,j,i, \\ 0, & \text{if any two indices are the same.} \end{cases}$$

Having defined this tensor, direct calculation reveals that only 6 of its components are non-zero, 3 have the value 1 and three have the value -1, and the rest are zero. The Cartesian tensor version of $\nabla \times \mathbf{F}$ can now be written, in fact $\nabla \times \mathbf{F} = \epsilon_{ijk} F_{j,k}$. Summation occurs over the indices j and k and most (21) terms vanish leaving just the six required for curl (see Equation 6.1). On the face of it, this might seem clumsy to do, but it is here for completeness. The clumsiness disappears if tensors are pursued.

6.5 Vector Identities

Having introduced three operators, grad., div. and curl, it is a good idea to see how they behave and to derive results involving them. The first thing to realise is that grad., div. and curl are differential operators, so you might expect rules of addition and some kinds of product rules will apply. The additional complication is the involvement of vector quantities and that is why care is needed before boldly stating rules. Linearity of all the operators helps. Therefore, rather than just to list lots of identities, which is what will eventually happen, let us start with a couple with implications that make them more interesting, and that are easy enough to prove. They are presented as an example:

Example 6.5 Show that $\nabla \times \nabla \phi \equiv 0$ for any scalar field $\phi = \phi(x, y, z)$ and also that $\nabla.\nabla \times \mathbf{F} \equiv 0$ for any vector field $\mathbf{F} = \mathbf{F}(x, y, z)$.

Solution: First

$$\nabla \times \nabla \phi = \begin{vmatrix} \mathbf{i} & \mathbf{j} & \mathbf{k} \\ \dfrac{\partial}{\partial x} & \dfrac{\partial}{\partial y} & \dfrac{\partial}{\partial z} \\ \dfrac{\partial \phi}{\partial x} & \dfrac{\partial \phi}{\partial y} & \dfrac{\partial \phi}{\partial z} \end{vmatrix}.$$

As the second and third rows of the determinant are more or less the same, it should come as no surprise that the determinant multiplies to zero, here are the terms explicitly:

$$\left(\frac{\partial}{\partial y} \frac{\partial \phi}{\partial z} - \frac{\partial}{\partial z} \frac{\partial \phi}{\partial y} \right) \mathbf{i} + \left(\frac{\partial}{\partial z} \frac{\partial \phi}{\partial x} - \frac{\partial}{\partial x} \frac{\partial \phi}{\partial z} \right) \mathbf{j}$$

$$+ \left(\frac{\partial}{\partial x} \frac{\partial \phi}{\partial y} - \frac{\partial}{\partial y} \frac{\partial \phi}{\partial x} \right) \mathbf{k} \equiv \mathbf{0}.$$

The second identity is now expanded

$$\nabla \cdot (\nabla \times \mathbf{F}) = \frac{\partial}{\partial x} \left(\frac{\partial F_3}{\partial y} - \frac{\partial F_2}{\partial z} \right) + \frac{\partial}{\partial y} \left(\frac{\partial F_1}{\partial z} - \frac{\partial F_3}{\partial x} \right)$$

$$+ \frac{\partial}{\partial z} \left(\frac{\partial F_2}{\partial x} - \frac{\partial F_1}{\partial y} \right) \equiv 0$$

as each second-order derivative has a cancelling equivalent. □

What is more difficult to prove are the converse two statements, formalised in the following two theorems:

Theorem 6.1 If for a vector field $\mathbf{F}(x, y, z)$ and it is true that $\nabla \times \mathbf{F} = 0$, then there exists a scalar field $\phi(x, y, z)$ called the *scalar potential* such that $\mathbf{F} = \nabla \phi$.

Theorem 6.2 If for a vector field $\mathbf{F}(x, y, z)$, it is true that $\nabla.\mathbf{F} = 0$ then there exists a vector field $\mathbf{A}(x, y, z)$ called the *vector potential* such that $\mathbf{F} = \nabla \times \mathbf{A}$.

These will be proved, but need vector integration hence the postponement for a few chapters. It is the first of these that is particularly useful, as it is easier to solve a problem to find a scalar function $\phi(x, y, z)$ than having to find a vector function $\mathbf{F}(x, y, z)$. On the face of it, the second theorem has merely exchanged finding $\mathbf{F}(x, y, z)$ with finding another vector function $\mathbf{A}(x, y, z)$ seemingly an equally tricky problem. Nonetheless, the concept of a vector potential is very useful in electrodynamics where Maxwell's equations are solved. A vector field $\mathbf{F}$ for which $\nabla \times \mathbf{F} = \mathbf{0}$ is called *irrotational* and a vector field $\mathbf{F}$ for which $\nabla \cdot \mathbf{F} = 0$ is called *solenoidal*. Here is a partial list of vector identities, some of the more obvious have been omitted, and those that remain are extensions or applications of the product rule. Those with four terms on the right are not obvious at all.

1) $\nabla \cdot (\mathbf{A} + \mathbf{B}) = \nabla \cdot \mathbf{A} + \nabla \cdot \mathbf{B}$.
2) $\nabla \times (\mathbf{A} + \mathbf{B}) = \nabla \times \mathbf{A} + \nabla \times \mathbf{B}$.
3) $\nabla \cdot (\phi \mathbf{A}) = \nabla \phi \cdot \mathbf{A} + \phi(\nabla \cdot \mathbf{A})$.
4) $\nabla \times (\phi \mathbf{A}) = \nabla \phi \times \mathbf{A} + \phi(\nabla \times \mathbf{A})$.
5) $\nabla \cdot (\mathbf{A} \times \mathbf{B}) = \mathbf{B} \cdot \nabla \times \mathbf{A} - \mathbf{A} \cdot \nabla \times \mathbf{B}$.
6) $\nabla \times (\mathbf{A} \times \mathbf{B}) = (\mathbf{B} \cdot \nabla)\mathbf{A} - (\mathbf{A} \cdot \nabla)\mathbf{B} - \mathbf{B}(\nabla \cdot \mathbf{A}) + \mathbf{A}(\nabla \cdot \mathbf{B})$.
7) $\nabla(\mathbf{A} \cdot \mathbf{B}) = (\mathbf{B} \cdot \nabla)\mathbf{A} + (\mathbf{A} \cdot \nabla)\mathbf{B} + \mathbf{B} \times (\nabla \times \mathbf{A}) + \mathbf{A} \times (\nabla \times \mathbf{B})$.
8) $\mathbf{A} \cdot ((\mathbf{B} \times \nabla) \times \mathbf{C}) = ((\mathbf{A} \cdot \nabla)\mathbf{C}) \cdot \mathbf{B} - (\mathbf{A} \cdot \mathbf{B})(\nabla \cdot \mathbf{C})$.
9) $\nabla^2(fg) = f\nabla^2 g + g\nabla^2 f + 2(\nabla f \cdot \nabla g)$.
10) $\nabla \cdot (\nabla f \times \nabla g) = 0$.
11) $\nabla \cdot (f\nabla g - g\nabla f) = f\nabla^2 g - g\nabla^2 f$.

In this list, it is number 7 that is used in fluid mechanics, and so this is proved now. The proof is tedious while not being enlightening, but it is worth doing at least one or two proofs.

Example 6.6 Prove the identity:

$$\nabla(\mathbf{A} \cdot \mathbf{B}) = (\mathbf{B} \cdot \nabla)\mathbf{A} + (\mathbf{A} \cdot \nabla)\mathbf{B} + \mathbf{B} \times (\nabla \times \mathbf{A}) + \mathbf{A} \times (\nabla \times \mathbf{B}).$$

Solution: Without loss of generality, it is possible to align the x axis with $\mathbf{A}$ and let the x, y plane contain the vector $\mathbf{B}$. Hence set $\mathbf{A} = a\mathbf{i}$ and $\mathbf{B} = b\mathbf{i} + c\mathbf{j}$. The proof merely computes both sides of the identity and shows that they are the same; doing it with the simple representations of the vectors in Cartesian co-ordinates is really worthwhile and helps reduce cumbersome algebra. Some might be tempted to put general expressions for $\mathbf{A}$ and $\mathbf{B}$ but this only works when there is a lot of symmetry, and there isn't enough here to really help. Therefore, the terms on the right are: $(\mathbf{B} \cdot \nabla)\mathbf{A} = (ba_x + ca_y)\mathbf{i}$; $(\mathbf{A} \cdot \nabla)\mathbf{B} = ab_x\mathbf{i} + ac_x\mathbf{j}$ for the first two. The third and fourth terms are

$$\mathbf{B} \times (\nabla \times \mathbf{A}) = \mathbf{B} \times \begin{vmatrix} \mathbf{i} & \mathbf{j} & \mathbf{k} \\ \dfrac{\partial}{\partial x} & \dfrac{\partial}{\partial y} & \dfrac{\partial}{\partial z} \\ a & 0 & 0 \end{vmatrix} = (b\mathbf{i} + c\mathbf{j}) \times (a_z\mathbf{j} - a_y\mathbf{k})$$

$$= -\mathbf{i}ca_y + \mathbf{j}ba_y + \mathbf{k}ba_z$$

and

$$
\mathbf{A} \times (\nabla \times \mathbf{B}) = \mathbf{A} \times \begin{vmatrix} \mathbf{i} & \mathbf{j} & \mathbf{k} \\ \dfrac{\partial}{\partial x} & \dfrac{\partial}{\partial y} & \dfrac{\partial}{\partial z} \\ b & c & 0 \end{vmatrix} = a\mathbf{i} \times (-\mathbf{i}c_z + \mathbf{j}b_z + \mathbf{k}(c_x - b_y))
$$

$$
= -\mathbf{j}(ac_x - ab_y) + \mathbf{k}ab_z.
$$

Summing up the components of the four terms thus gives

$$
\text{R.H.S.} = \mathbf{i}(ba_x + ab_x + ca_y - ca_y) + \mathbf{j}(ac_x + ba_y - ac_x + ab_y)
$$
$$
+ \mathbf{k}(ba_z + ab_z)
$$
$$
= \mathbf{i}(ab)_x + \mathbf{j}(ab)_y + \mathbf{k}(ab)_z = \nabla(\mathbf{A} \cdot \mathbf{B}) = \text{L.H.S.},
$$

which completes the proof. □

Example 6.7 Prove that

$$
\nabla \cdot (\mathbf{A} \times \mathbf{B}) = \mathbf{B} \cdot \nabla \times \mathbf{A} - \mathbf{A} \cdot \nabla \times \mathbf{B}.
$$

Solution: Proceed as with the previous example and set $\mathbf{A} = a\mathbf{i}$ and $\mathbf{B} = b\mathbf{i} + c\mathbf{j}$. Then proceed to evaluate both sides of the expression that needs to be proved. Direct calculation gives:

$$
\mathbf{B} \cdot \nabla \times \mathbf{A} = (\mathbf{i}b + \mathbf{j}c) \begin{vmatrix} \mathbf{i} & \mathbf{j} & \mathbf{k} \\ \dfrac{\partial}{\partial x} & \dfrac{\partial}{\partial y} & \dfrac{\partial}{\partial z} \\ a & 0 & 0 \end{vmatrix} = ca_z
$$

and

$$
\mathbf{A} \cdot \nabla \times \mathbf{B} = \mathbf{i}a \begin{vmatrix} \mathbf{i} & \mathbf{j} & \mathbf{k} \\ \dfrac{\partial}{\partial x} & \dfrac{\partial}{\partial y} & \dfrac{\partial}{\partial z} \\ b & c & 0 \end{vmatrix} = -ac_z.
$$

Finally,

$$
\nabla \cdot \mathbf{A} \times \mathbf{B} = \nabla \begin{vmatrix} \mathbf{i} & \mathbf{j} & \mathbf{k} \\ a & 0 & 0 \\ b & c & 0 \end{vmatrix} = \nabla \cdot (\mathbf{k}ac) = (ac)_z
$$

and this proves the identity. □

As the proofs of most of the identities follow similarly, no more of these identities will be proved here. The later of these identities involves the second-order operator $\nabla \cdot \nabla$ that gets the symbol ∇^2 and is called the *Laplacian*. Therefore, for a scalar field, $\mathbf{\nabla \cdot \nabla}\phi = \nabla^2\phi$ and the equation $\nabla^2\phi = 0$ is called Laplace's equation. In words, the Laplacian is 'div. grad.'. In Cartesian co-ordinates,

$$\nabla^2\phi = \mathbf{\nabla \cdot \nabla}\phi = \mathbf{\nabla} \cdot \left(\mathbf{i}\frac{\partial\phi}{\partial x} + \mathbf{j}\frac{\partial\phi}{\partial y} + \mathbf{k}\frac{\partial\phi}{\partial z} \right) = \frac{\partial^2\phi}{\partial x^2} + \frac{\partial^2\phi}{\partial y^2} + \frac{\partial^2\phi}{\partial z^2}.$$

The equation

$$\nabla^2\phi = 0$$

is perhaps the most well-known partial differential equation in mathematical physics (used in its widest sense, not just particle physics). It should be noted that equations written using ∇ will be co-ordinate free and as such we are free to use any co-ordinate system to solve them. That we used Cartesian co-ordinates to define grad., div. and curl does not matter which will be obvious once co-ordinate free versions are given later. A reasonable question is therefore is there a vector form of Laplacian? The direct answer is 'yes', but the details are only straightforward in Cartesian co-ordinates, otherwise the vector Laplacian is defined by:

$$\nabla^2\mathbf{A} = \mathbf{\nabla}(\mathbf{\nabla} \cdot \mathbf{A}) - \mathbf{\nabla} \times \mathbf{\nabla} \times \mathbf{A}.$$

As the right-hand side is a grad. plus a curl, it certainly gives a vector field as required. In Cartesian co-ordinates:

$$\nabla^2\mathbf{A} = \mathbf{i}\frac{\partial^2 A_1}{\partial x^2} + \mathbf{j}\frac{\partial^2 A_2}{\partial y^2} + \mathbf{k}\frac{\partial^2 A_3}{\partial z^2}$$

as might be expected, where $\mathbf{A} = A_1\mathbf{i} + A_2\mathbf{j} + A_3\mathbf{k}$. This does look a strange result, so it is worth running through an example.

Example 6.8 Calculate the vector Laplacian for the vector $\mathbf{A} = (0, 0, f(x, y, z))$ using the notation for components where $f(x, y, z)$ is a function with continuous second-order derivatives.

Solution: Proceeding with the calculation,

$$\mathbf{\nabla} \times \mathbf{A} = \begin{vmatrix} \mathbf{i} & \mathbf{j} & \mathbf{k} \\ \dfrac{\partial}{\partial x} & \dfrac{\partial}{\partial y} & \dfrac{\partial}{\partial z} \\ 0 & 0 & f \end{vmatrix} = (f_y, -f_x, 0),$$

$$\mathbf{\nabla} \times \mathbf{\nabla} \times \mathbf{A} = \begin{vmatrix} \mathbf{i} & \mathbf{j} & \mathbf{k} \\ \dfrac{\partial}{\partial x} & \dfrac{\partial}{\partial y} & \dfrac{\partial}{\partial z} \\ f_y & -f_x & 0 \end{vmatrix} = (f_{xz}, f_{yz}, -\nabla^2 f),$$

$$\mathbf{\nabla} \cdot \mathbf{A} = f_z \quad \text{and so} \quad \mathbf{\nabla}(\mathbf{\nabla} \cdot \mathbf{A}) = (f_{zx}, f_{zy}, f_{zz}),$$

and so

$$\nabla(\nabla \cdot \mathbf{A}) - \nabla \times \nabla \times \mathbf{A} = (f_{zx}, f_{zy}, f_{zz}) - (f_{xz}, f_{yz}, -\nabla^2 f) = \nabla^2 f \mathbf{k} = \nabla^2 \mathbf{A}.$$

Hence, the vector Laplacian formula is verified. □

Those with stamina can now go ahead and prove it in general; just multiply out the determinants and be careful with minus signs. It does depend on the continuity of all second derivatives so set $\mathbf{A} = (f, g, h)$ and use $f_{xy} = f_{yx}; f_{zx} = f_{xz}; f_{yz} = f_{zy}$ and similarly for the functions g and h.

In older books, dyadics would now be introduced as being compatible with vector notation. These are now a little obscure and have largely been superseded by covariant and contravariant second-order tensor notation, and the use of the suffix and superfix that generalises more easily to higher dimensions where they are used, for example in the study of general relativity. To say these are never used elsewhere would be false as they do appear both in turbulence theory and elasticity, but they are not used in this text. Instead, there is one more topic that is useful in applications the concept of conjugate functions.

6.6 Conjugate Functions

Consider a vector function $\mathbf{F}$ that satisfies both

$$\nabla \cdot \mathbf{F} = 0 \quad \text{and} \quad \nabla \times \mathbf{F} = \mathbf{0}.$$

From the second of these, it is the case that exists a scalar function ϕ such that $\mathbf{F} = \nabla\phi$. Insert this into the first expression and we see that

$$\nabla \cdot \nabla\phi = 0 \quad \text{or} \quad \nabla^2 \phi = 0$$

and we say that ϕ is a *harmonic function*, this simply means that it satisfies Laplace's partial differential equation. This topic can now break off into the general theory of harmonic functions, but although there is some of this once vector integration is covered, here discussion is now restricted to functions of two variables. This is limited but very useful in fluid mechanics. Therefore, if $\phi = \phi(x, y)$ and $\mathbf{F} = \mathbf{i}F_1(x, y) + \mathbf{j}F_2(x, y)$, then zero divergence leads to

$$\nabla \cdot \mathbf{F} = \frac{\partial F_1}{\partial x} + \frac{\partial F_2}{\partial y} = 0,$$

and this means that it is possible to set

$$F_1 = \frac{\partial \psi}{\partial y} \quad F_2 = -\frac{\partial \psi}{\partial x}$$

for some suitably well-behaved function $\psi(x, y)$ as this satisfies $\nabla \cdot \mathbf{F} = 0$ identically. Since $\mathbf{F} = \nabla\phi$, it must be true that:

$$\mathbf{F} = \mathbf{i}F_1(x, y) + \mathbf{j}F_2(x, y) = \mathbf{i}\frac{\partial \psi}{\partial y} - \mathbf{j}\frac{\partial \psi}{\partial x} = \mathbf{i}\frac{\partial \phi}{\partial x} + \mathbf{j}\frac{\partial \phi}{\partial y},$$

from which, equating coefficients of **i** and **j**

$$\frac{\partial \psi}{\partial y} = \frac{\partial \phi}{\partial x} \quad \text{and} \quad -\frac{\partial \psi}{\partial x} = \frac{\partial \phi}{\partial y}.$$

These are called the *Cauchy–Riemann equations* and the functions $\phi(x, y)$ and $\psi(x, y)$ are called *conjugate functions* (see also section 2.8 on page 52). There are a host of mathematically neat and physically useful results that follow. To complete the general two-dimensional picture, substitute

$$\mathbf{F} = \mathbf{i}\frac{\partial \psi}{\partial y} - \mathbf{j}\frac{\partial \psi}{\partial x} \quad \text{into} \quad \nabla \times \mathbf{F} = \mathbf{0}.$$

Adopting the suffix derivative notation for convenience, this gives

$$\begin{vmatrix} \mathbf{i} & \mathbf{j} & \mathbf{k} \\ \dfrac{\partial}{\partial x} & \dfrac{\partial}{\partial y} & \dfrac{\partial}{\partial z} \\ \psi_y & -\psi_x & 0 \end{vmatrix} = \mathbf{0} \quad \text{or} \quad \mathbf{k}(-\psi_{xx} - \psi_{yy}) = \mathbf{0},$$

so

$$\nabla^2 \psi = 0$$

and ψ is also a harmonic function. Now consider the product $\nabla\phi \cdot \nabla\psi$:

$$\nabla\phi \cdot \nabla\psi = \left(\mathbf{i}\frac{\partial \phi}{\partial x} + \mathbf{j}\frac{\partial \phi}{\partial y}\right) \cdot \left(\mathbf{i}\frac{\partial \psi}{\partial x} + \mathbf{j}\frac{\partial \psi}{\partial y}\right) = \frac{\partial \phi}{\partial x}\frac{\partial \psi}{\partial x} + \frac{\partial \phi}{\partial y}\frac{\partial \psi}{\partial y}$$

or using the suffix derivative notation, this equals $\phi_x\psi_x + \phi_y\psi_y$. The Cauchy–Riemann equations tell us that $\psi_y = \phi_x$ and $\psi_x = -\phi_y$, so $\phi_x\psi_x + \phi_y\psi_y = -\phi_x\phi_y + \phi_y\phi_x = 0$. Thus $\nabla\phi \cdot \nabla\psi = 0$. In two dimensions, the two sets of curves $\phi = $ constant and $\psi = $ constant represent sets of curves or contours in the x, y plane. Now $\nabla\phi$ is normal to the contours $\phi = $ constant and $\nabla\psi$ in turn is normal to the contours $\psi = $ constant. Therefore, these two families of curves are orthogonal, that is, they intersect at right angles. In fluid mechanics, $\phi = $ constant represents the potential and $\psi = $ constant the streamlines, but more of that later in the book. The question arises how to find conjugate functions; there is an infinite source to those who know about complex analysis. If $z = x + iy$ where $i = \sqrt{-1}$, then z is a complex number, x is the real part and y the imaginary part. Those of you struggling with this either skip this bit or come back to it after reading about complex variable theory. $f(z)$ denotes a function of the complex number z and so is a function of a complex variable. If we demand that the derivative of such a function exists in the form of the uniqueness of the limit

$$\lim_{\Delta z \to 0} \left\{ \frac{f(z + \Delta z) - f(z)}{\Delta z} \right\},$$

then this is a bigger deal than it is in real variables because the way Δz can approach zero is two dimensional. By splitting $f(z) = \phi(x, y) + i\psi(x, y)$ into real and imaginary parts, it turns out that demanding that $f(z)$ is differentiable forces the functions ϕ and ψ to obey the Cauchy–Riemann equations. They can be proved by choosing the directions of the x and y axes as separate paths for the limit as $\Delta z \to 0$ then demanding they give the same unique limit as must be the case. As has been said this is a big deal and the names *analytic* and *regular* are preserved for differentiability of complex valued functions. A separate word to differentiable is needed because if differentiating once is fine, so is differentiating any number of times. The consequence that helps us here is that the real and imaginary part of a regular complex function are a pair of conjugate functions.

Example 6.9 Determine the conjugate functions associated with the two complex functions:

a) z^2;
b) $-ie^z$.

Solution: For part (*a*), put $f(z) = z^2$ splitting this into real and imaginary parts gives

$$f(z) = \phi_1(x, y) + i\psi_1(x, y)$$
$$= z^2 = (x + iy)^2 = x^2 + 2ixy - y^2 = (x^2 - y^2) + i(2xy)$$

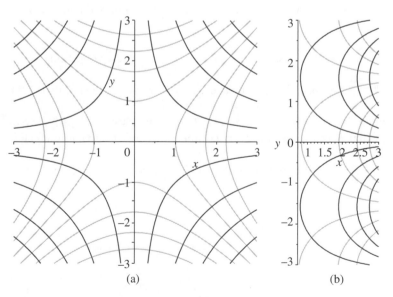

(a)   (b)

Figure 6.1 (a) The contours $x^2 - y^2 = \text{constant}$ (light) and $xy = \text{constant}$ (dark) (b) The contours $e^x \cos y = \text{constant}$ (light) and $e^x \sin y = \text{constant}$ (dark).

so equating real and imaginary parts gives $\phi_1 = x^2 - y^2, \psi_1 = 2xy$ as a pair of conjugate functions. This one gives two sets of rectangular hyperbolae $xy =$ constant and $x^2 - y^2 =$ constant tilted at $\pi/4$ to each other that is pictured in Figure 6.1a. See how they intersect at right angles. For part (b), put $g(z) = -ie^z =$ and split into real and imaginary parts as before. This gives

$$g(z) = e^x \sin y - ie^x \cos y = \phi_2(x, y) + i\psi_2(x, y),$$

so equating real and imaginary parts gives $\phi_2(x, y) = e^x \sin y$ and $\psi_2 = -e^x \cos y$. These curves are displayed in Figure 6.1b and once again of course they are orthogonal as the theory dictates. Figure 6.1b only displays the values for $x > 0$ as the presence of the exponential renders values for $x < 0$ very small.

In the theory of conjugate functions, the quantity $\mathbf{k}\psi$ can be identified with the vector potential defined earlier.

Exercises

6.1 If $\phi(x, y, z) = 2x^4y - xyz$, find $\nabla\phi$ at the point $(1, 1, -2)$.

6.2 Prove the product rule for grad.: $\nabla(\phi\psi) = \phi\nabla\psi + \psi\nabla\phi$.

6.3 Determine the value of ∇r^n where n is any real number, and $r^2 = x^2 + y^2 + z^2$.

6.4 Determine $\nabla(\ln r)$ where r is as defined in the previous question.

6.5 Find the unit normal to the surface $x^4 + y^4 + z^4 = 3$ at the point $(1, 1, 1)$.

6.6 In what direction from the point $(1, 2, 3)$ is the directional derivative of the scalar field $\phi = xy^2 + xyz$ a maximum, and find this maximum value.

6.7 If $F(x, y, z, t)$ is a differentiable scalar field with respect to all the variables (x, y, z, t), show that

$$\frac{dF}{dt} = \frac{\partial F}{\partial t} + \nabla F \cdot \frac{d\mathbf{r}}{dt}$$

and comment on the case $F = F(x, y, z)$ if x, y, z vary with t.

6.8 Show that $\nabla u \cdot \nabla v \times \nabla w$ is the Jacobian

$$\frac{\partial(u, v, w)}{\partial(x, y, z)} \quad \text{[see Chapter 2]},$$

where $u = u(x, y, z)$, $v = v(x, y, z)$ and $w = w(x, y, z)$. What is the consequence of this Jacobian being zero?

6.9 If **a** is a constant vector, show that
a) $\nabla(\mathbf{a} \cdot \mathbf{r}) = \mathbf{a}$.
b) $\nabla|\mathbf{a} \times \mathbf{r}|^n = n|\mathbf{a} \times \mathbf{r}|^{n-2}\mathbf{a} \times (\mathbf{r} \times \mathbf{a})$.
c) $\mathbf{a} \cdot \nabla\left(\mathbf{a} \cdot \nabla\dfrac{1}{r}\right) = \dfrac{3(\mathbf{a} \cdot \mathbf{r})^2 - a^2 r^2}{r^5}$.

6.10 Find the divergence of the vector field
$$\frac{-x\mathbf{i} - y\mathbf{j}}{\sqrt{x^2 + y^2}}$$
and discuss its physical significance.

6.11 If $\mathbf{F} = x^2yz\mathbf{i} + xy^2z\mathbf{i} + xyz^2\mathbf{k}$ determine $\nabla \cdot \mathbf{F}$ and find its value at the point $(1, 1, -1)$.

6.12 If ϕ and ψ are differentiable scalar fields, show that $\nabla\phi \times \nabla\psi$ is solenoidal.

6.13 Determine $\nabla \times \mathbf{F}$ if $\mathbf{F} = x^2yz\mathbf{i} + xy^2z\mathbf{i} + xyz^2\mathbf{k}$ and find its value at $(1, 1, -1)$.

6.14 Find the values of (*a*) $\mathbf{A} \times (\nabla \times \mathbf{B})$ and (*b*) $(\mathbf{A} \times \nabla) \times \mathbf{B}$ at the point $(1, 1, 1)$ given that $\mathbf{A} = x^2y\mathbf{i} + xy^2\mathbf{j} + xyz^2\mathbf{k}$ and $\mathbf{B} = yz\mathbf{i} + xz\mathbf{j} + xy\mathbf{k}$.

6.15 Show that $\phi(x, y) = \ln(x^2 + y^2)$ and $\psi(x, y) = \tan^{-1}(y/x)$ are conjugate harmonic functions.

7

Curvilinear Co-ordinates

7.1 Introduction

It should come as no surprise that Cartesian co-ordinates are sometimes not the best co-ordinates to use. Even in two dimensions, plane polar co-ordinates are often a very useful alternative. The behaviour of vectors and the various quantities defined in the previous chapters such as div., grad. and curl are independent of choice of co-ordinate as has been made clear; however, it is useful to have expressions that hold for other systems, hence this chapter. The special feature of Cartesian co-ordinates is that the three planes $x = $ constant, $y = $ constant and $z = $ constant intersect in three straight lines that give the co-ordinate system. There is not a curve in sight. In all other co-ordinate systems, there is at least one curve, and it is this that makes the generalisation into such systems not obvious. The section on differential geometry helps here. A curvilinear co-ordinate system is one that uses curved and not straight lines as reference axes. In this chapter, attention will be restricted to curved co-ordinate systems that are orthogonal, that is at every point the three curved axes intersect mutually at right angles. Different distinct notation is called for, so name the unit vectors tangential to the three intersecting curves at the origin $\hat{e}_1, \hat{e}_2$ and $\hat{e}_3$ and the curves themselves u, v and w so that $x = x(u, v, w), y = y(u, v, w)$ and $z = z(u, v, w)$ expresses Cartesian co-ordinates in terms of these new orthogonal curvilinear ones.

7.2 Curved Axes and Scale Factors

First, before doing specific calculations, there are some general aspects of behaviour and notation that need to be covered. The general curvilinear co-ordinates $\hat{e}_1, \hat{e}_2$ and $\hat{e}_3$ are constant in magnitude as they are unit vectors, but they are not constant in direction, so they should be written

Two and Three Dimensional Calculus: With Applications in Science and Engineering, First Edition. Phil Dyke.
© 2018 John Wiley & Sons Ltd. Published 2018 by John Wiley & Sons Ltd.

$\hat{e}_1(u, v, w), \hat{e}_2(u, v, w), \hat{e}_3(u, v, w)$. Any vector quantity $\mathbf{F}$ can be referred to these orthogonal axes in the usual way:

$$\mathbf{F}(u, v, w) =$$
$$F_1(u, v, w)\hat{e}_1(u, v, w) + F_2(u, v, w)\hat{e}_2(u, v, w) + F_3(u, v, w)\hat{e}_3(u, v, w),$$

so that any partial derivative has to be calculated through the product rule. For example,

$$\frac{\partial \mathbf{F}}{\partial u} = \frac{\partial F_1}{\partial u}\hat{e}_1 + \frac{\partial F_2}{\partial u}\hat{e}_2 + \frac{\partial F_3}{\partial u}\hat{e}_3 + F_1\frac{\partial \hat{e}_1}{\partial u} + F_2\frac{\partial \hat{e}_2}{\partial u} + F_3\frac{\partial \hat{e}_3}{\partial u}.$$

Another more pure mathematical point is the acceptance of the three vectors $\hat{e}_1(u, v, w), \hat{e}_2(u, v, w)$ and $\hat{e}_3(u, v, w)$ as a basis in the same way that $\mathbf{i}, \mathbf{j}$ and $\mathbf{k}$ are. This is more subtle than you might think. In fact, it is true 'almost everywhere' as mathematicians are fond of saying. In curvilinear co-ordinates, there are isolated points (e.g. the origin in polar co-ordinates) where technically this fails, but as they *are* isolated, they can be easily dealt with as special cases and are no trouble in calculations. In terms of notation, $\mathbf{F} = (F_1, F_2, F_3)$ is kept for Cartesian co-ordinates and is avoided here as the chosen co-ordinate system is not explicitly shown.

Let $\mathbf{r}$ be expressed in terms of the curvilinear co-ordinates $\hat{e}_1, \hat{e}_2$ and $\hat{e}_3$. The way to do this is to hold two of u, v, w constant so that $\mathbf{r}(u, v_0, w_0)$ is the 'u axis', $\mathbf{r}(u_0, v, w_0)$ is the 'v axis' and $\mathbf{r}(u_0, v_0, w)$ is the 'w axis'. As mentioned above there may be isolated points, the origin in polar co-ordinates is one, where this description fails. As a single parameter (u) expression, $\mathbf{r}(u, v_0, w_0)$ represents a curve in space, see Chapter 5, so the partial derivative

$$\frac{\partial \mathbf{r}}{\partial u}$$

will be a tangent to this curve in the local (osculating) plane that contains the curve. This enables the *scale factors* h_1, h_2 and h_3 to be defined as follows:

$$h_1 = \left|\frac{\partial \mathbf{r}}{\partial u}\right|, h_2 = \left|\frac{\partial \mathbf{r}}{\partial v}\right| \quad \text{and} \quad h_3 = \left|\frac{\partial \mathbf{r}}{\partial w}\right|.$$

The scale factors are not simply local radii of curvature, but the ratio of the infinitesimal length of $\mathbf{r}$ divided by the change in the direction of the three co-ordinates taken in turn. It is possible using the language of manifolds and tangent bundles to put all this far more rigorously, but this would be out of place here. Our three scale factors turn out to be the diagonal elements of a 3×3 matrix or second-order tensor that is only diagonal for orthogonal curvilinear co-ordinate systems. As attention is actually restricted to these orthogonal systems here, such excursions into generality would be an unnecessary complication. Hence, the scale factors appear simply as ratios of infinitesimals; they do not have a consistent dimension though to consider them as lengths related to radius of curvature might be useful for the polar examples encountered later. Here, we have

$$\hat{\mathbf{e}}_1 = \frac{\partial \mathbf{r}/\partial u}{|\partial \mathbf{r}/\partial u|}, \hat{\mathbf{e}}_2 = \frac{\partial \mathbf{r}/\partial v}{|\partial \mathbf{r}/\partial v|} \quad \text{and} \quad \hat{\mathbf{e}}_3 = \frac{\partial \mathbf{r}/\partial w}{|\partial \mathbf{r}/\partial w|}$$

so that

$$d\mathbf{r} = \hat{\mathbf{e}}_1 h_1 du + \hat{\mathbf{e}}_2 h_2 dv + \hat{\mathbf{e}}_3 h_3 dw.$$

Thus

$$\hat{\mathbf{e}}_1 \cdot d\mathbf{r} = h_1 du, \quad \hat{\mathbf{e}}_2 \cdot d\mathbf{r} = h_2 dv \quad \text{and} \quad \hat{\mathbf{e}}_3 \cdot d\mathbf{r} = h_3 dw.$$

Bearing in mind the practical nature of this text, to fix ideas, let us derive what they look like in spherical polar co-ordinates; see Figure 7.1. The three curvilinear co-ordinates are r, θ and ϕ with $x = x(r, \phi, \theta) = r \sin \phi \cos \theta, y = y(r, \phi, \theta) = r \sin \phi \sin \theta$ and $z = z(r, \phi, \theta) = r \cos \phi$. A caution here: in some books and websites, λ is used instead of ϕ and in others the roles of θ and ϕ are reversed. The choice here is to keep θ in the same role as in two-dimensional polar co-ordinates. Sadly, written in the what seems the natural order (r, θ, ϕ) the co-ordinates are left handed; however, expressing spherical polars as (r, ϕ, θ) without a diagram close by could be confusing alongside cylindrical

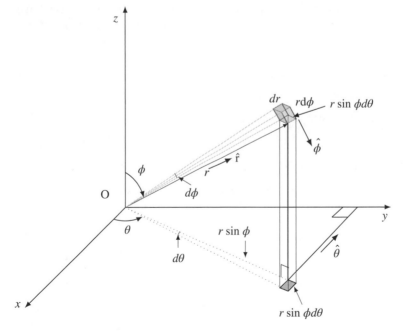

Figure 7.1 Spherical polar co-ordinates (r, ϕ, θ), showing the infinitesimal volume element of sides $dr, rd\phi$ and $r \sin \phi d\theta$.

polars expressed as (R, θ, z) with θ in the second position. The key fact is that in this text θ is the same angle in both systems. The unit vectors $\hat{e}_1 = \hat{r}, \hat{e}_2 = \hat{\phi}$ and $\hat{e}_3 = \hat{\theta}$ are shown in Figure 7.1; they are mutually orthogonal and right handed in this order. Using the general expressions above gives in spherical polar co-ordinates:

$$h_1 = \left|\frac{\partial \mathbf{r}}{\partial u}\right| = \left|\frac{\partial \mathbf{r}}{\partial r}\right| = \left|\frac{\partial}{\partial r}(r \sin \phi \cos \theta \mathbf{i} + r \sin \phi \sin \theta \mathbf{j} + r \cos \phi \mathbf{k})\right|$$

$$= |\sin \phi \cos \theta \mathbf{i} + \sin \phi \sin \theta \mathbf{j} + \cos \phi \mathbf{k}|$$

$$= \sqrt{(\sin \phi \cos \theta)^2 + (\sin \phi \sin \theta)^2 + (\cos \phi)^2} = 1,$$

$$h_2 = \left|\frac{\partial \mathbf{r}}{\partial w}\right| = \left|\frac{\partial \mathbf{r}}{\partial \phi}\right| = \left|\frac{\partial}{\partial \phi}(r \sin \phi \cos \theta \mathbf{i} + r \sin \phi \sin \theta \mathbf{j} + r \cos \phi \mathbf{k})\right|$$

$$= |r \cos \phi \cos \theta \mathbf{i} + r \cos \phi \sin \theta \mathbf{j} - r \sin \phi \mathbf{k}|$$

$$= \sqrt{(r \cos \phi \cos \theta)^2 + (r \cos \phi \sin \theta)^2 + (r \sin \phi)^2} = r,$$

$$h_3 = \left|\frac{\partial \mathbf{r}}{\partial v}\right| = \left|\frac{\partial \mathbf{r}}{\partial \theta}\right| = \left|\frac{\partial}{\partial \theta}(r \sin \phi \cos \theta \mathbf{i} + r \sin \phi \sin \theta \mathbf{j} + r \cos \phi \mathbf{k})\right|$$

$$= |-r \sin \phi \sin \theta \mathbf{i} + r \sin \phi \cos \theta \mathbf{j}|$$

$$= \sqrt{(r \sin \phi \sin \theta)^2 + (r \sin \phi \cos \theta)^2} = r \sin \phi.$$

given the scale factors as $1, r, r \sin \phi$. A glance at Figure 7.1 and it is seen that the r axis is straight with $h_1 = 1$, the θ axis is circular with radius $r \sin \phi$ and ϕ axis is circular with radius r so $h_2 = r$ and $h_3 = r \sin \phi$. To get to grips with the meaning of the scale factors, some differential calculus is used:

$$d\mathbf{r} = \frac{\partial \mathbf{r}}{\partial u}du + \frac{\partial \mathbf{r}}{\partial v}dv + \frac{\partial \mathbf{r}}{\partial w}dw,$$

which can be written

$$d\mathbf{r} = \hat{e}_1 h_1 du + \hat{e}_2 h_2 dv + \hat{e}_3 h_3 dw.$$

Thus,

$$\hat{e}_1 \cdot d\mathbf{r} = h_1 du, \quad \hat{e}_2 \cdot d\mathbf{r} = h_2 dv \quad \text{and} \quad \hat{e}_3 \cdot d\mathbf{r} = h_3 dw,$$

and so it is seen that the arc length along each orthogonal curvilinear co-ordinate is h times the differential. If the co-ordinate is straight, then obviously $h = 1$ otherwise h tells us the factor by which the differential needs to be multiplied to convert the product to arc length in the direction of the tangent to the particular co-ordinate, hence the name *scale factor* is a good one.

Example 7.1   Show that the following co-ordinate system is orthogonal:

$$x = \frac{1}{2}(u^2 - v^2), \quad y = uv \quad \text{and} \quad z = z$$

and determine the scale factors. This is the *parabolic cylinder co-ordinate system*.

Figure 7.2 The x, y plane in parabolic cylinder co-ordinates.

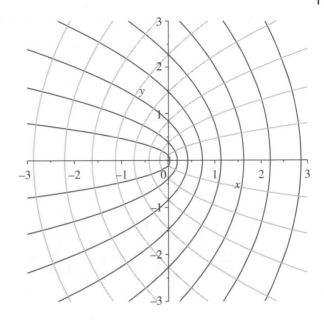

Solution: Figure 7.2 shows the co-ordinate system in the x, y plane. The curves are two systems of intersecting parabolas each set having the same focus. The algebra follows. First of all, let us find the derivatives:

$$\frac{\partial \mathbf{r}}{\partial u} = u\mathbf{i} + v\mathbf{j}, \quad \frac{\partial \mathbf{r}}{\partial v} = -v\mathbf{i} + u\mathbf{j} \quad \text{and} \quad \frac{\partial \mathbf{r}}{\partial z} = \mathbf{k}.$$

Taking the dot products of each of these shows orthogonality. The modulus of each of these vectors gives the three scale factors:

$$h_1 = h_2 = \sqrt{u^2 + v^2} \quad \text{and} \quad h_3 = 1.$$

Most of these kind of examples are best worked in terms of $\mathbf{i}, \mathbf{j}$ and $\mathbf{k}$ as this is safe ground to show orthogonality. □

7.3 Curvilinear Gradient, Divergence, and Curl

7.3.1 Gradient

In this section, the general orthogonal curvilinear expressions for grad., div. and curl will be derived. It is procedural rather than mathematically challenging, but a useful reference for later applications. Let us start with

$$d\mathbf{r} = \hat{\mathbf{e}}_1 h_1 du + \hat{\mathbf{e}}_2 h_2 dv + \hat{\mathbf{e}}_3 h_3 dw$$

and note that $\nabla\phi$ can be written in component form using the generalised co-ordinates

$$\nabla\phi = \hat{\mathbf{e}}_1 f_1 + \hat{\mathbf{e}}_2 f_2 + \hat{\mathbf{e}}_3 f_3$$

so

$$d\phi = \nabla\phi \cdot d\mathbf{r} = h_1 f_1 du + h_2 f_2 dv + h_3 f_3 dw.$$

However, it is also true that

$$d\phi = \frac{\partial\phi}{\partial u} du + \frac{\partial\phi}{\partial v} dv + \frac{\partial\phi}{\partial w} dw$$

and so equating coefficients of the (independent) differentials du, dv and dw enables f_1, f_2 and f_3 to be found as

$$f_1 = \frac{1}{h_1}\frac{\partial\phi}{\partial u}, f_2 = \frac{1}{h_2}\frac{\partial\phi}{\partial v} \quad \text{and} \quad f_3 = \frac{1}{h_3}\frac{\partial\phi}{\partial w}$$

giving

$$\nabla\phi = \frac{1}{h_1}\frac{\partial\phi}{\partial u}\hat{\mathbf{e}}_1 + \frac{1}{h_2}\frac{\partial\phi}{\partial v}\hat{\mathbf{e}}_2 + \frac{1}{h_3}\frac{\partial\phi}{\partial w}\hat{\mathbf{e}}_3$$

as the general expression for grad. in orthogonal curvilinear co-ordinates.

Example 7.2   Find $\nabla\phi$ in the parabolic cylinder co-ordinate system of the last example. Show that $\nabla\phi$ is in the $\mathbf{k}$ direction if and only if $\phi = \phi(z)$.

Solution: Using the values of the scale factors derived in Example 7.1 gives the unit vectors

$$\hat{\mathbf{e}}_1 = \frac{u\mathbf{i} + v\mathbf{j}}{\sqrt{(u^2 + v^2)}}, \quad \text{and} \quad \hat{\mathbf{e}}_2 = \frac{-v\mathbf{i} + u\mathbf{j}}{\sqrt{(u^2 + v^2)}}$$

together with $\hat{\mathbf{e}}_3 = \mathbf{k}$. Using the formula for $\nabla\phi$, thus gives

$$\nabla\phi = \frac{\phi_u(u\mathbf{i} + v\mathbf{j})}{(u^2 + v^2)} + \frac{\phi_v(-v\mathbf{i} + u\mathbf{j})}{(u^2 + v^2)} + \phi_z\mathbf{k}$$

from which it is apparent that $\nabla\phi$ is in the direction of $\mathbf{k}$ if and only if both $\phi_u = 0$ and $\phi_v = 0$ that is $\phi = \phi(z)$. This last expression can be rearranged as follows

$$\nabla\phi = \frac{1}{u^2 + v^2}\left(u\frac{\partial\phi}{\partial u} - v\frac{\partial\phi}{\partial v}\right)\mathbf{i} + \frac{1}{u^2 + v^2}\left(v\frac{\partial\phi}{\partial u} + u\frac{\partial\phi}{\partial v}\right)\mathbf{j} + \frac{\partial\phi}{\partial z}\mathbf{k}.$$

□

7.3.2 Divergence

The next vector operator to tackle is divergence or 'div.' for short. The expression for this in curvilinear co-ordinates turns out to be a bit awkward to derive. First of all, using the expression for $\nabla\phi$ derived above with ϕ equalling, in turn, u, v then w it follows that

$$\nabla u = \frac{\hat{\mathbf{e}}_1}{h_1}, \quad \nabla v = \frac{\hat{\mathbf{e}}_2}{h_2} \quad \text{and} \quad \nabla w = \frac{\hat{\mathbf{e}}_3}{h_3}.$$

As the three vectors form a right-handed triad in which $\hat{\mathbf{e}}_1 = \hat{\mathbf{e}}_2 \times \hat{\mathbf{e}}_3, \hat{\mathbf{e}}_2 = \hat{\mathbf{e}}_3 \times \hat{\mathbf{e}}_1$ and $\hat{\mathbf{e}}_3 = \hat{\mathbf{e}}_1 \times \hat{\mathbf{e}}_2$ substituting for the unit vectors in terms of these cross products of grad. gives

$$\hat{\mathbf{e}}_1 = h_2 h_3 \nabla v \times \nabla w, \quad \hat{\mathbf{e}}_2 = h_3 h_1 \nabla w \times \nabla u \quad \text{and} \quad \hat{\mathbf{e}}_3 = h_1 h_2 \nabla u \times \nabla v.$$

Now consider $\nabla \cdot (F_1 \hat{\mathbf{e}}_1)$ and use the above so that

$$\nabla \cdot (F_1 \hat{\mathbf{e}}_1) = \nabla \cdot (F_1 h_2 h_3 \nabla v \times \nabla w).$$

Now, use $\nabla \cdot (\phi \mathbf{A}) = \nabla\phi \cdot \mathbf{A} + \phi(\nabla \cdot \mathbf{A})$, number three in our list of identities, with $\phi = F_1 h_2 h_3$ and $\mathbf{A} = \nabla v \times \nabla w$ to expand this as follows

$$\nabla \cdot (F_1 h_2 h_3 \nabla v \times \nabla w) = \nabla(F_1 h_2 h_3) \cdot (\nabla v \times \nabla w) + F_1 h_2 h_3 \nabla \cdot (\nabla v \times \nabla w).$$

The second term here is zero, as the fifth of the identities $\nabla \cdot (\mathbf{A} \times \mathbf{B}) = \mathbf{B} \cdot \nabla \times \mathbf{A} - \mathbf{A} \cdot \nabla \times \mathbf{B}$ tells us that this consists of two quantities that are of the form '$\nabla \times \nabla$' that is curl of grad. which is identically zero. Hence,

$$\nabla \cdot (F_1 \hat{\mathbf{e}}_1) = \nabla(F_1 h_2 h_3) \cdot (\nabla v \times \nabla w).$$

A bit more patience, it is almost finished. Now, write $\nabla v \times \nabla w$ back in terms of $\hat{\mathbf{e}}_1$ so that

$$\nabla v \times \nabla w = \frac{\hat{\mathbf{e}}_1}{h_2 h_3}$$

and use our new expansion of grad.:

$$\nabla(F_1 h_2 h_3) = \frac{1}{h_1} \frac{\partial}{\partial u}(F_1 h_2 h_3)\hat{\mathbf{e}}_1 + \frac{1}{h_2} \frac{\partial}{\partial v}(F_1 h_2 h_3)\hat{\mathbf{e}}_2 + \frac{1}{h_3} \frac{\partial}{\partial w}(F_1 h_2 h_3)\hat{\mathbf{e}}_3.$$

Taking the scalar product of these last two expressions and using the orthogonality of $\hat{\mathbf{e}}_1$, $\hat{\mathbf{e}}_2$ and $\hat{\mathbf{e}}_3$ means that

$$\nabla \cdot (F_1 \hat{\mathbf{e}}_1) = \frac{1}{h_1 h_2 h_3} \frac{\partial}{\partial u}(F_1 h_2 h_3).$$

Therefore, the expression for divergence follows by adding similar terms:

$$\nabla \cdot \mathbf{F} = \nabla \cdot (F_1 \hat{\mathbf{e}}_1 + F_2 \hat{\mathbf{e}}_2 + F_3 \hat{\mathbf{e}}_3)$$
$$= \frac{1}{h_1 h_2 h_3} \left[\frac{\partial}{\partial u}(F_1 h_2 h_3) + \frac{\partial}{\partial v}(F_2 h_3 h_1) + \frac{\partial}{\partial w}(F_3 h_1 h_2) \right].$$

In addition, it is important to remember that in this expression $\mathbf{F} = F_1 \hat{\mathbf{e}}_1 + F_2 \hat{\mathbf{e}}_2 + F_3 \hat{\mathbf{e}}_3$ are the components in the curvilinear co-ordinate system, and not in Cartesian co-ordinates. Here is an example.

Example 7.3 Show that the following equations define an orthogonal curvilinear co-ordinate system

$$x = uv \cos \phi \quad y = uv \sin \phi \quad \text{and} \quad z = \frac{1}{2}(u^2 - v^2).$$

Describe the geometry and calculate $\nabla.\mathbf{F}$, where

$$\mathbf{F} = \frac{v}{\sqrt{u^2 + v^2}}\hat{\mathbf{e}}_1 + \frac{u}{\sqrt{u^2 + v^2}}\hat{\mathbf{e}}_2.$$

Solution: This co-ordinate system is based on parabolas, as in the parabolic cylinder co-ordinate system, but the parabolas are in the x, z plane and rotated about the z-axis. See Figure 7.2; but replace the x-axis by z then y can be either x or y. When rotated, all points on the parabolas describe circles $x^2 + y^2 = r^2$ with $r = uv$ so that given $z = \frac{1}{2}(u^2 - v^2)$ all planes through the origin containing the z-axis exhibit the parabolic cylinder co-ordinates of Example 7.1 as shown in Figure 7.2. This co-ordinate system is called paraboloidal. Calculating the curvilinear unit vectors and scale factors follows as before:

$$\frac{\partial \mathbf{r}}{\partial u} = v \cos \phi \mathbf{i} + v \sin \phi \mathbf{j} + u \mathbf{k}, \quad \frac{\partial \mathbf{r}}{\partial v} = u \cos \phi \mathbf{i} + u \sin \phi \mathbf{j} - v \mathbf{k}$$

$$\text{and} \quad \frac{\partial \mathbf{r}}{\partial \phi} = -uv \sin \phi \mathbf{i} + uv \cos \phi \mathbf{j}$$

reveals that they are mutually orthogonal. Taking the modulus yields $h_1 = h_2 = \sqrt{u^2 + v^2}$ and $h_3 = uv$. Dividing the partial derivatives by these scale factors is straightforward and gives the three unit curvilinear vectors:

$$\hat{\mathbf{e}}_1 = \frac{1}{\sqrt{u^2 + v^2}}(v \cos \phi \mathbf{i} + v \sin \phi \mathbf{j} + u \mathbf{k}),$$

$$\hat{\mathbf{e}}_2 = \frac{1}{\sqrt{u^2 + v^2}}(u \cos \phi \mathbf{i} + u \sin \phi \mathbf{j} - v \mathbf{k})$$

$$\text{and} \quad \hat{\mathbf{e}}_3 = -\sin \phi \mathbf{i} + \cos \phi \mathbf{j}.$$

In general, the divergence in this co-ordinate system is

$$\nabla \cdot \mathbf{F} = \frac{1}{uv(u^2 + v^2)}$$

$$\times \left\{ \frac{\partial}{\partial u}[uv\sqrt{u^2 + v^2}F_1] + \frac{\partial}{\partial v}[uv\sqrt{u^2 + v^2}F_2] + \frac{\partial}{\partial \phi}\sqrt{u^2 + v^2}F_3 \right\}.$$

With

$$\mathbf{F} = \frac{v}{\sqrt{u^2 + v^2}}\hat{\mathbf{e}}_1 + \frac{u}{\sqrt{u^2 + v^2}}\hat{\mathbf{e}}_2$$

the formula for divergence gives

$$\nabla \cdot \mathbf{F} = \frac{1}{uv(u^2 + v^2)} \left\{ \frac{\partial}{\partial u}(uv^2) + \frac{\partial}{\partial v}(u^2 v) \right\} = \frac{1}{uv}. \qquad \square$$

7.3.3 Curl

The third and last vector differential operator to consider is curl. This time curl is not an abbreviation. Deriving the expression for curl in terms of curvilinear co-ordinates is not easy, but as it follows along similar lines to the derivation of div., it is left as an exercise for the reader and the result is stated as follows:

$$\nabla \times \mathbf{F} = \frac{1}{h_1 h_2 h_3} \begin{vmatrix} h_1 \hat{\mathbf{e}}_1 & h_2 \hat{\mathbf{e}}_2 & h_3 \hat{\mathbf{e}}_3 \\ \dfrac{\partial}{\partial u} & \dfrac{\partial}{\partial v} & \dfrac{\partial}{\partial w} \\ h_1 F_1 & h_2 F_2 & h_3 F_3 \end{vmatrix}.$$

Here is a computational example:

Example 7.4   For the vector field $\mathbf{F}$ defined in the previous example

$$\mathbf{F} = \frac{v}{\sqrt{u^2 + v^2}} \hat{\mathbf{e}}_1 + \frac{u}{\sqrt{u^2 + v^2}} \hat{\mathbf{e}}_2,$$

show that $\mathbf{F}$ is irrotational using paraboloidal co-ordinates and find the scalar potential Φ such that $\mathbf{F} = \nabla \Phi$.

Solution: Using direct calculation,

$$\nabla \times \mathbf{F} = \frac{1}{h_1 h_2 h_3} \begin{vmatrix} h_1 \hat{\mathbf{e}}_1 & h_2 \hat{\mathbf{e}}_2 & h_3 \hat{\mathbf{e}}_3 \\ \dfrac{\partial}{\partial u} & \dfrac{\partial}{\partial v} & \dfrac{\partial}{\partial \phi} \\ \dfrac{h_1 v}{\sqrt{u^2 + v^2}} & \dfrac{h_2 u}{\sqrt{u^2 + v^2}} & 0 \end{vmatrix},$$

and since $h_1 = h_2 = \sqrt{u^2 + v^2}$ the square roots cancel and

$$\nabla \times \mathbf{F} = \frac{\hat{\mathbf{e}}_3}{h_1 h_2} \left\{ \frac{\partial}{\partial u}(u) - \frac{\partial}{\partial v}(v) \right\} = \mathbf{0}$$

hence $\mathbf{F}$ is irrotational. $\mathbf{F} = \nabla \Phi$ implies

$$\frac{1}{h_1} \frac{\partial \Phi}{\partial u} = \frac{v}{\sqrt{u^2 + v^2}} \quad \text{and} \quad \frac{1}{h_2} \frac{\partial \Phi}{\partial v} = \frac{u}{\sqrt{u^2 + v^2}}$$

from which $\Phi = uv$ neglecting the arbitrary constant. $\qquad \square$

A comment on this last example is appropriate here. Whether or not a vector field is irrotational is not dependent upon the co-ordinates chosen, so although the last example was completed using paraboloidal co-ordinates, the result stands in any chosen co-ordinate system. Indeed, in another co-ordinate system, the algebra would probably be a lot more challenging.

7.4 Further Results and Tensors

Combining the two formulae for div. and grad. gives the curvilinear co-ordinate version of Laplacian:

$$\nabla^2 \psi = \frac{1}{h_1 h_2 h_3} \left[\frac{\partial}{\partial u} \left(\frac{h_2 h_3}{h_1} \frac{\partial \psi}{\partial u} \right) + \frac{\partial}{\partial v} \left(\frac{h_3 h_1}{h_2} \frac{\partial \psi}{\partial v} \right) + \frac{\partial}{\partial w} \left(\frac{h_1 h_2}{h_3} \frac{\partial \psi}{\partial w} \right) \right].$$

In practice, there are only three co-ordinate systems commonly used:

Cartesian Co-ordinates: x, y, z where $h_1 = 1, h_2 = 1$, and $h_3 = 1$;
Cylindrical Polar Co-ordinates: r, θ, z where $h_1 = 1, h_2 = r$, and $h_3 = 1$;
Spherical Polar Co-ordinates: r, θ, ϕ where $h_1 = 1, h_2 = r \sin \phi$, and $h_3 = r$.

Inserting the appropriate values for h_1, h_2 and h_3 give the following two Laplace's equations for the general function ψ. In cylindrical polar co-ordinates, we have

$$\nabla^2 \psi = \frac{1}{r} \frac{\partial}{\partial r} \left(r \frac{\partial \psi}{\partial r} \right) + \frac{1}{r^2} \frac{\partial^2 \psi}{\partial \theta^2} + \frac{\partial^2 \psi}{\partial z^2} = 0$$

and in spherical polar co-ordinates, we have

$$\nabla^2 \psi = \frac{1}{r^2} \frac{\partial}{\partial r^2} \left(r^2 \frac{\partial \psi}{\partial r} \right) + \frac{1}{r^2 \sin^2 \phi} \frac{\partial^2 \psi}{\partial \theta^2} + \frac{1}{r^2 \sin \phi} \frac{\partial}{\partial \phi} \left(\sin \phi \frac{\partial \psi}{\partial \phi} \right) = 0.$$

7.4.1 Tensor Notation

In some circumstances, the tensor notation is useful. Of course, academic pure mathematicians prefer to use it as do most theoretical physicists as for them vector notation is too limited. Tensors are a generalisation of the notion of scalar (zero-order tensor) and vector (first-order tensor) to an arbitrary number of orders. An nth-order tensor in ordinary three-dimensional space will have 3^n components. Of course, in m-dimensional space an nth-order tensor will have m^n components. In Chapter 6, as grad., div. and curl were introduced, so were the Cartesian tensors $\delta_{i,j}$ called Kronecker's delta and ϵ_{ijk} the alternating tensor. Here they both are as a reminder: first Kronecker delta

$$\delta_{ij} = 1 \quad \text{if} \quad i = j, \quad \delta_{ij} = 0 \quad \text{otherwise;}$$

and the alternating tensor:

$$\epsilon_{ijk} = \begin{cases} 1, & \text{if } i,j,k \quad \text{are all different and permute cyclically,} \\ -1, & \text{if } i,j,k \quad \text{are all different and permute anti-cyclically,} \\ 0, & \text{in all other cases.} \end{cases}$$

Therefore, $\epsilon_{123} = \epsilon_{231} = \epsilon_{312} = 1, \epsilon_{132} = \epsilon_{321} = \epsilon_{213} = -1$ while every other ϵ_{ijk} for example ϵ_{112} or $\epsilon_{222} = 0$. In fact all the zero members of the alternating tensor have one or more indices equal. There are 21 zero members altogether; try listing them. Therefore, in the case of this particular 3^3 component tensor, only 6 are non-zero, and they are 1 or -1. The alternating tensor appears where tensor forms of cross product or curl occur. Scalars are zero-order tensors and remain as single letters. However, the vector **a** is replaced by a_i where the index $i = 1, 2$ or 3. Therefore, a_i is a vector or first-order tensor. The quantity a_{ij} where i and j can independently assume one of the values 1, 2 or 3 is a second-order tensor. This would have 9 components and the whole can be represented by a square matrix. Third-order tensors a_{ijk} would have 27 components and representing them any other way is challenging. Tensors of higher order are met only rarely, though a sixth-order tensor does make an appearance in the theory of general relativity. This brings us to a revision of differentiation. As mentioned in Chapter 6, a derivative is denoted by a comma in the suffix. Now, consider a scalar variable $\phi = \phi(x, y, z)$. The grad. of ϕ in tensor notation is written $\phi_{,i}$; this indicates that it is a first-order tensor (vector) as it has one free index i. Moreover, the comma means that ϕ is differentiated and as there are three possible derivatives, there are three components. The expression $\phi_{i,j}$ would indicate a second-order tensor with 9 components as might be expected if the grad. of a vector (what used to be called a dyadic) was computed. The vector function **F** would now be written F_i, so how do we write the div. of this function? The answer is not $F_{i,j}$ as that is the grad., instead the index is repeated $F_{i,i}$ where the *summation convention* is obeyed whereby if an index is repeated it is summed over and is no longer a free index. The quantity $F_{i,i} = F_{1,1} + F_{2,2} + F_{3,3}$ and is a scalar. Finally, there is curl. This is not quite as neat using tensors, and the alternating tensor is required. In fact, the curl of F_i is $\epsilon_{ijk}F_{j,i}$ which is a first-order tensor as k is the only free index (i and j are summed over as they are repeated). Recalling the definition of ϵ_{ijk}, try writing out the six non-zero components and verifying that this is indeed curl. The Laplacian of the scalar quantity ϕ is $\phi_{,ii}$, another scalar quantity, of course. The notation requires a comment here. The comma denotes a derivative and the repeated index a summation, but the double index means differentiation is performed twice. The notation $\phi_{,i,i}$ could also be used, but never is in practice. Here's an example to get the flavour of working in Cartesian tensors.

Example 7.5   Establish the following three results involving the Kronecker delta and the alternating tensor:

$$\delta_{ij}\epsilon_{ijk} = 0, \quad \epsilon_{ijk}\epsilon_{rjk} = 2\delta_{ir}, \quad \text{and} \quad \epsilon_{ijk}\epsilon_{ijk} = 6.$$

Solution: The way to tackle these three problems is simply to enumerate the left-hand sides. Most of the terms are, after all, zero. Remember that repeated indices are summed over. Expression 1 is $\delta_{ij}\epsilon_{ijk}$; k is the only free index so this is a first-order tensor.

Only if $i = j$ is Kronecker delta non-zero, so

$$\delta_{ij}\epsilon_{ijk} = \delta_{11}\epsilon_{11k} + \delta_{22}\epsilon_{22k} + \delta_{33}\epsilon_{33k} = 0$$

since $\epsilon_{ijk} = 0$ if $i = j$. This establishes that $\delta_{ij}\epsilon_{ijk} = 0$. Expression 2 is $\epsilon_{ijk}\epsilon_{rjk} = F_{ir}$, a second-order tensor. First, consider the case $i = r$. If $i = 1$, then

$$\epsilon_{1jk}\epsilon_{1jk} = \epsilon_{123}\epsilon_{123} + \epsilon_{132}\epsilon_{132}$$

the rest of the terms being zero by the properties of the alternating tensor. The right-hand side is $1 \times 1 + (-1) \times (-1) = 2$. The same is true for the cases $i = 2$ and $i = 3$. For the case, $i \neq r$ consider, say, $i = 1, r = 2$ then

$$\epsilon_{1jk}\epsilon_{2jk} = 0$$

because either $j = k$, or at least one of j or k has to equal 1 or 2, in all cases there are repeating indices in at least one of the alternating tensors thus rendering the product always zero. The same follows for any different pairs of j and k so

$$\epsilon_{ijk}\epsilon_{rjk} = 0$$

for $i \neq r$. Thus, we have derived that $\epsilon_{ijk}\epsilon_{rjk} = 2$ if $i = r$ and 0 otherwise. This means that the right-hand side fits the definition $2\delta_{ir}$ and completes the proof. Expression 3 is the product $\epsilon_{ijk}\epsilon_{ijk}$, a scalar as there are three repeated indices. There are only six non-zero members of the alternating tensor, three have the value 1 and three the value -1. The sum of the squares are thus $1 \times 1 + 1 \times 1 + 1 \times 1 + (-1) \times (-1) + (-1) \times (-1) + (-1) \times (-1) = 6$ as required.   □

7.4.2   Covariance and Contravariance

Consider the velocity of a particle in mechanics, $\mathbf{v}$. This is, straightforwardly the derivative of the position vector $\mathbf{r}$

$$\mathbf{v} = \frac{d\mathbf{r}}{dt} = \left(\frac{dx}{dt}, \frac{dy}{dt}, \frac{dz}{dt} \right).$$

In contrast, consider the velocity of a fluid $\mathbf{v}$ in an irrotational flow field $\nabla \times \mathbf{v} = \mathbf{0}$ that arises from a potential ϕ with

$$\mathbf{v} = \nabla\phi = \left(\frac{\partial\phi}{\partial x}, \frac{\partial\phi}{\partial y}, \frac{\partial\phi}{\partial z} \right).$$

Apart from the obvious difference that the first is a total derivative and the second partial, note that in the first the positions x, y and z are in the numerator whereas in the second they are in the denominator, yet they are both velocities and both derivatives. The essential distinction is the one between locality and non-locality. A particle being tracked like a meteor across the sky is a local object oblivious to its surroundings and its velocity vector is the rate of change of its position. The velocity of a fluid however is not. It is a vector of a path through a field. In mathematical terms, this resembles the tangent vectors discussed in Chapter 6: The field is a differentiable manifold and the streamlines or path-lines are the tangent bundle. In Chapter 6, we called the tangent lines a covariant derivative, and it is the same here. The fluid velocity is a covariant derivative of the potential, whereas the velocity of a particle in mechanics is a contravariant derivative of its position vector. The covariant derivative is the Eulerian point of view in fluid mechanics; the contravariant derivative is the Lagrangian point of view or differentiation following the fluid.

Technically, in order to distinguish between covariance and contravariance; it is necessary to consider a transformation of co-ordinates. At this stage, given the title of the book restricting attention to three variables could be justified. However, this is pointless, so let $X^1, X^2, \ldots, X^n$ be n variables, and define n associated quantities, $A^1, A^2, \ldots, A^n$ related to associated variables $\overline{X}^1, \overline{X}^2 \ldots, \overline{X}^n$ and quantities $\overline{A}^1, \overline{A}^2, \ldots, \overline{A}^n$ via

$$\overline{A}^p = \sum_{q=1}^{n} \frac{\partial X^p}{\partial \overline{X}^q} A^q \quad (p = 1, 2, \ldots, n) \quad \text{so} \quad \overline{A}^p = \frac{\partial X^p}{\partial \overline{X}^q} A^q$$

using the summation convention. This is a transformation of a *contravariant tensor*. If, on the other hand, the transformation takes the form

$$\overline{A}_p = \sum_{q=1}^{n} \frac{\partial \overline{X}^p}{\partial X^q} A_q \quad (p = 1, 2, \ldots, n) \quad \text{so} \quad \overline{A}_p = \frac{\partial \overline{X}^p}{\partial X^q} A_q$$

again using the summation convention, we have a *covariant tensor*. Note that the index for covariance is a suffix, whereas the index for contravariance is a superfix. Even this approach is looked down upon by some mathematicians who prefer to define covariant and contravariant through multilinear mappings, relying on the algebra of vector spaces, eliminating all direct reference to calculus. Given the title of the book, this time this is unacceptable even though it does allow a cleaner approach to mixed tensors, some components of which are covariant, others contravariant.

Example 7.6   Write down the transformation for the mixed fifth-order tensor $\overline{A}^{qst}_{kl}$.

Solution: Given the above rules and using the summation convention, this can be written immediately as

$$A_{ij}^{prm} = \frac{\partial \overline{X}^p}{\partial X^q} \frac{\partial \overline{X}^r}{\partial X^s} \frac{\partial \overline{X}^m}{\partial X^t} \frac{\partial X^k}{\partial \overline{X}^i} \frac{\partial X^l}{\partial \overline{X}^j} A_{kl}^{qst}.$$

The summation convention suppresses six summations showing the power of the notation. Using more traditional notation would be dire and probably unworkable. □

As we have moved outside three dimensions for a short while, the summation convention also allows us to define an n-dimensional space:

Definition 7.1 If $(x^1, x^2, \ldots, x^n)$ are co-ordinates in an n-dimensional space, then the *metric form* or *metric* for this space is $g_{pq} dx^p dx^q$ using the summation convention, and the space is called *Riemannian*, named after Bernhard Riemann (1826–1866) surely one of the most famous of all mathematicians if only for the as-yet-unproven Riemann hypothesis.

The second-order covariant tensor g_{pq} called the *metric tensor* is a generalisation of the scale factors introduced as h_i earlier. In the special case, where the co-ordinate x^j transforms to the co-ordinate $\overline{x}^k$ such that the metric is transformed into the sum of squares

$$(d\overline{x}^1)^2 + (d\overline{x}^2)^2 + (d\overline{x}^3)^2 + \cdots + (d\overline{x}^n)^2 = (d\overline{x}^k)(d\overline{x}^k)$$

then the n-dimensional space is called Euclidean and is a direct extension to our three dimensional Cartesian world with three straight-line axes.

It is a short step from here to define a Riemannian manifold, a space that has an associated tangent space with a metric that makes it a Riemannian space. There is also an inner product that leads to g_{pq} a Riemannian metric. The whole gives rise to Riemannian geometry that generalises the curvilinear co-ordinates of the early parts of this chapter and is the starting point for the study of general relativity. The curvatures are expressed as third-order tensors called *Christoffel symbols*, a mixed tensor these days denoted as Γ^i_{jk} rather than the less convenient

$$\left\{ \begin{matrix} i \\ j\,k \end{matrix} \right\}$$

found in older texts. There are some exercises for those who want to take this further, but in a text on three-dimensional calculus primarily for second-stage undergraduates, let us leave matters here.

Exercises

7.1 Express the following surfaces in both cylindrical polar and spherical polar co-ordinates:
a) $y = x$,
b) $z = x^2 + y^2$,
c) $z^2 = 4(x^2 + y^2)$,
d) $z = 0$,
e) $x^2 + y^2 + z^2 = 16$.

7.2 Writing cylindrical polar co-ordinates as R, θ, z and spherical polar co-ordinates as r, θ, ϕ (θ is the same angle of course), describe the following surfaces:
a) $R = 3$,
b) $\phi = \pi/3$,
c) $R = \cos \theta$,
d) $R^2 = 9z$,
e) $\theta = \pi/4$
For part (e), give both the cylindrical polar and spherical polar answers.

7.3 Represent the vector field $\mathbf{F} = x^2\mathbf{i} + y\mathbf{j} + z^3\mathbf{k}$ in terms of both cylindrical and spherical polar co-ordinates, giving both answers in component form.

7.4 Show that the co-ordinate system u, v, z given by the transformation

$$x = \frac{a \sinh v}{\cosh v - \cos u}, \quad y = \frac{a \sin u}{\cosh v - \cos u} \quad \text{and} \quad z = z,$$

where $0 \le u < 2\pi, -\infty < v < \infty$ and $-\infty < z < \infty$ is orthogonal, and determine the scale factors h_u, h_v and h_z.

7.5 Find expressions for the elements of area and the element of volume in general curvilinear co-ordinates.

7.6 Show that the product

$$\left\{ \frac{\partial \mathbf{r}}{\partial u} \cdot \frac{\partial \mathbf{r}}{\partial v} \times \frac{\partial \mathbf{r}}{\partial w} \right\} \{\nabla u \cdot \nabla v \times \nabla w\} = 1$$

where u, v and w are general curvilinear co-ordinates.

7.7 Determine $\nabla \psi$, $\nabla \cdot \mathbf{F}$ and $\nabla \times \mathbf{F}$ where ψ and $\mathbf{F}$ are arbitrary scalar and vector fields using the paraboloidal co-ordinates u, v and ϕ of

Example 7.3:

$$x = uv \cos \phi, \quad y = uv \sin \phi \quad \text{and} \quad z = \frac{1}{2}(u^2 - v^2).$$

7.8 A tensor a_{ij} is symmetric if $a_{ij} = a_{ji}$ and anti-symmetric if $a_{ij} = -a_{ji}$. Show that an arbitrary tensor is always the sum of a symmetric and anti-symmetric tensor.

7.9 Prove the result $\epsilon_{ijk}\epsilon_{rsk} = \delta_{ir}\delta_{js} - \delta_{is}\delta_{jr}$.

7.10 Find Cartesian tensor forms of the scalar and vector products of two vectors $\mathbf{a} = (a_1, a_2, a_3)$ and $\mathbf{b} = (b_1, b_2, b_3)$.

7.11 Write down the laws of transformation for the tensors:
a) A^i_{jk};
b) B^n;
c) C^{ijk}_{lmn}.

7.12 Show that Kronecker delta is mixed and not a covariant tensor. How should it be written?

7.13 Generalise the alternating tensor ϵ_{ijk} to n dimensions as the Levi-Civita tensor $\epsilon_{i_1 i_2 \ldots i_n}$ and suggest its definition given it is sometimes called the *permutation* tensor.

8

Path Integrals

8.1 Introduction

Integrals have certainly been encountered before; in Chapter 1, there was an albeit brief reminder of some of those tricks needed to evaluate standard one-dimensional scalar integrals. Substitution, integration by parts and use of partial fractions were all given an airing, but the emphasis here is not on these methods more how to define and then utilise integration of a vector quantity. There is no doubt that the evaluation will, from time to time, involve these techniques but that's not the point. Here, the procedure of integrating vectors is defined, and it transpires that the definition of integration has to be different. In order to take full advantage of the power of integration, its definition has to start from a different place. The idea of the integral expressing the area under a curve is jettisoned in favour of something new, the line or path integral. This may seem a big deal, but it isn't. The rigorous definition of integration as the limit of a sum still holds, therefore our new interpretation is still a classical Riemann integral and their evaluation proceeds as before including if necessary using specialist methods.

8.2 Integration Along a Curve

Consider the integral

$$\int \mathbf{F}(t)dt.$$

This is easily evaluated by splitting $\mathbf{F}(t)$ into its components $F_1(t)\mathbf{i} + F_2(t)\mathbf{j} + F_3(t)\mathbf{k}$ and then integrating the three scalar integrals as follows

$$\int \mathbf{F}(t)dt = \int F_1(t)dt\mathbf{i} + \int F_2(t)dt\mathbf{j} + \int F_3(t)dt\mathbf{k}.$$

Two and Three Dimensional Calculus: With Applications in Science and Engineering, First Edition. Phil Dyke.
© 2018 John Wiley & Sons Ltd. Published 2018 by John Wiley & Sons Ltd.

This is straightforward and devoid of interest. Slightly more interesting is the integral of the scalar function vectorially

$$\int f(t)d\mathbf{r},$$

which is also quite easily evaluated in theory. Now, however, it is necessary to use $d\mathbf{r} = \dot{x}dt\mathbf{i} + \dot{y}dt\mathbf{j} + \dot{z}dt\mathbf{k}$, where the dot denotes differentiation with respect to t so that

$$\int f(t)d\mathbf{r} = \int f(t)\dot{x}dt\mathbf{i} + \int f(t)\dot{y}dt\mathbf{j} + \int f(t)\dot{z}dt\mathbf{k}.$$

The interpretation of these integrals is worth a close look. The integrals are with respect to the parameter t, so it is implicitly assumed that $x = x(t), y = y(t)$ and $z = z(t)$, which is a curve in three-dimensional space. Hence, the original integral is an integral *along* a curve. There are no limits specified but if there were these would correspond to two points on the curve representing the start and end points of integration. The actual evaluation proceeds as normal calculus, but the interpretation is certainly no longer an area under a curve. Instead, the integral can be interpreted as a different limit of a summation. Although the above component form is useful for computational purposes, for a better derivation let us return to the vector form:

$$\int f(t)d\mathbf{r} = \int f(t)\dot{\mathbf{r}}dt,$$

where $\mathbf{r}(t)$ traces position vectors on a curve in space. Figure 8.1 depicts a typical curve in space, here $f(t)\dot{\mathbf{r}}$ is written as a single symbol $\sigma(t)$ for convenience. The integral indicated in the last equation is thus the limit of a sum of the form

$$\sum_{n=1}^{N} \sigma(t_n^*)\Delta t_n,$$

where Δt_n is the interval $[t_n, t_{n+1}]$ and t_n^* denotes a value of t somewhere in this interval. This kind of summation gels with the standard ideas of Riemann integration. Without turning this into a rerun of all the details, the idea is that the summation can be bounded above or below by making appropriate choices of t_n^*, that either maximise or minimise the value of $\sigma(t)$ in each interval $[t_n, t_{n+1}]$. The condition that such upper and lower bounds exist is quite mild. Continuous functions will do, but so will functions that have jumps provided they are finite. The sandwiching of the summation by these bounds then, as the limit is taken as $N \to \infty$, means that its value is uniquely defined. Specifically, if the greatest lower bound of the summation is the same as the least upper bound these are the definition of the integral, the Riemann integral. It is the same as the integral defined as an area under a curve. All the mathematical steps can be found in analysis texts, the book *A Course of Analysis* by Phillips [5] remains

Figure 8.1 The curve $\sigma(t) = f(t)\hat{r}$ is displayed, together with a series of dots $\sigma_n(t)$ indicating the representation of the curve by discrete points. The exact number of points is immaterial, but the more there are the better the representation.

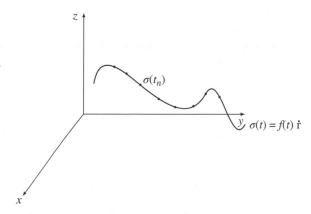

my favourite for ease of explanation and clarity, but the above visual and verbal account tells the underlying principle. Other integrals that involve scalar and vector products will succumb to the Riemann definition with a bit of work. An obvious question to ask is what does such an integral represent as it is no longer an area? This does depend on the details. Assuming no vectors, the path integral

$$\int_C \rho(x, y, z)ds$$

can be expressed as the standard integral

$$\int_{t_a}^{t_b} \rho(x(t), y(t), z(t))\dot{s}dt,$$

where s is arc length, with $ds^2 = dx^2 + dy^2 + dz^2$ so $\dot{s}^2 = \dot{x}^2 + \dot{y}^2 + \dot{z}^2$. This integral can be interpreted as the mass of a wire of uniform cross section but variable density ρ that has the shape of the curve C, a curve parameterised by $x = x(t), y = y(t)$ and $z = z(t)$ and whose endpoints correspond to the values $t = t_a$ and $t = t_b$. Here is a multiple example.

Example 8.1 Find the values of the integral

$$\int_C \rho(x, y, z)ds$$

for the following three cases:

a) C is the circle $x^2 + y^2 = a^2, z = b$, a and b constant, and $\rho(x, y, z) = $ constant.
b) C is the helix $x = a \cos t, y = a \sin t, z = bt$, a and b constant, and $\rho(x, y, z) = $ constant.
c) C is the helix $x = a \cos t, y = a \sin t, z = bt$, a and b constant, and $\rho(x, y, z) = \dfrac{\rho_0}{2\pi}\left(1 + x^2 + y^2 + \dfrac{3b^2}{4\pi^2}z^2\right)$.

Solution:

a) The circle $x^2 + y^2 = a^2$ is on the plane $z = b$ and so has the parametric form $x = a\cos t, y = a\sin t, z = b$. Thus $ds^2 = dx^2 + dy^2 + dz^2 = a^2\sin^2 t dt^2 + a^2\cos^2 t dt^2 = a^2 dt^2$ whence $ds = adt$. Thus

$$\int_C \rho(x, y, z)ds = \rho \int_0^{2\pi} adt = 2\pi a\rho.$$

b) Parametrising the helix has already been done, $x = a\cos t, y = a\sin t, z = bt$. so differentiating and squaring gives $ds^2 = a^2\sin^2 t dt^2 + a^2\cos^2 t dt^2 + b^2 dt^2 = (a^2 + b^2)dt^2$. Proceeding as before thus gives

$$\int_C \rho(x, y, z)ds = \rho \int_0^{2\pi} \sqrt{a^2 + b^2}dt = 2\pi\rho\sqrt{a^2 + b^2}.$$

c) For this last part, the path remains the same helix as for part (b) however the density is now variable. Thus, ds is the same so going straight to the integral gives

$$\int_C \rho(x, y, z)ds = \int_0^{2\pi} \sqrt{a^2 + b^2}\rho(t)dt.$$

Therefore, all that is required is to write ρ in terms of t, viz. $\rho = \frac{\rho_0}{2\pi}\left(1 + a^2 + \frac{3b^2}{4\pi^2}t^2\right)$. Hence,

$$\int_C \rho(x, y, z)ds = \int_0^{2\pi} \sqrt{a^2 + b^2}\frac{\rho_0}{2\pi}\left(1 + a^2 + \frac{3b^2}{4\pi^2}t^2\right)dt.$$

Performing the elementary integral and cancelling various factors gives

$$\int_C \rho(x, y, z)ds = \rho_0(1 + a^2 + b^2)\sqrt{a^2 + b^2}.$$

$\square$

The interpretation of what an integral means is an important aspect of vector integration, and this aspect is usually emphasised over the computational aspects. They will all still have to be evaluated of course, but that is why you do calculus before vector calculus. Here is another kind of *path integral*. This time vectors play an important part and are not a side issue as they might seem to have been above.

$$\int_C \mathbf{F} \cdot d\mathbf{r}.$$

This again an integral along a curve or path C and it remains a scalar, so the answer is still simply a number, but the interpretation is key. If $\mathbf{F}$ represents a force, then it will interest those who know about mechanics that the integral is in fact the work done by the force in moving along the curve specified by the path C. This brings in questions about whether the path is a closed one, and whether the work done moving around the curve and back to its original place is zero or not. This will be returned to later. At once, it is apparent that if $\mathbf{F}$ is perpendicular to $d\mathbf{r}$ then the work done is zero. The force being at right angles to the path has no component along the path and thus no work is done. A purely computational example involving evaluating this integral now follows.

Example 8.2   Evaluate the following integral

$$\int_C \mathbf{F} \cdot d\mathbf{r},$$

where C is the unit circle $x^2 + y^2 = 1$ with $z = 0$ and $\mathbf{F} = -y\mathbf{i} + x\mathbf{j}$.

Solution: In this question, which is two dimensional, C is parameterised by $x = \cos t, y = \sin t, 0 \le t < 2\pi$, and on C therefore $\mathbf{F} = -\sin t\mathbf{i} + \cos t\mathbf{j}$. In addition, $d\mathbf{r} = d(x\mathbf{i} + y\mathbf{j}) = -\sin t dt\mathbf{i} + \cos t dt\mathbf{j}$. Hence, $\mathbf{F} \cdot d\mathbf{r} = (\sin^2 t + \cos^2 t)dt = dt$ and the integral limits correspond to a single circuit of the unit circle starting at $t = 0$ and finishing at $t = 2\pi$, hence

$$\int_C \mathbf{F} \cdot d\mathbf{r} = \int_0^{2\pi} dt = 2\pi.$$

□

The integration certainly was easy here, and that was on purpose. The important part was getting everything in terms of the parameter t. Sometimes these path integrals throw up definite real integrals that are challenging, but in all cases put all variables in terms of the single parameter (t above) and compute the real integral that results. Here is a much meatier example:

Example 8.3   A vector field is given in component form by $\mathbf{F}(x, y, z) = 2xyz\mathbf{i} + x^2 z\mathbf{j} + x^2 y\mathbf{k}$. Evaluate

$$\int_C \mathbf{F} \cdot d\mathbf{r}$$

along the two paths, both connecting the points $(0, 0, 0)$ and $(2, 4, 6)$:

a) $x = t, y = t^2, z = 3t$.
b) the straight line segments connecting the points $(0, 0, 0)$ to $(2, 0, 0)$; $(2, 0, 0)$ to $(2, 4, 0)$ and $(2, 4, 0)$ to $(2, 4, 6)$.

Solution:
a) For the first part, the parameters are already prescribed $x = t, y = t^2, z = 3t$ so, on this curve, $d\mathbf{r} = dt\mathbf{i} + 2t dt\mathbf{j} + 3dt\mathbf{k}$ and $\mathbf{F}(x, y, z) = 2xyz\mathbf{i} + x^2 z\mathbf{j} + $

$x^2 y \mathbf{k} = 6t^4 \mathbf{i} + 3t^3 \mathbf{j} + t^4 \mathbf{k}$. Taking the scalar product of these two quantities thus gives

$$\mathbf{F} \cdot d\mathbf{r} = 15t^4 dt.$$

The point $(0, 0, 0)$ corresponds to $t = 0$, and the point $(2, 4, 6)$ to the value $t = 2$. Of course, there has to be a value of t that, given the parameterisation $x = t, y = t^2, z = 3t$, correspond to each of the points, otherwise the points do not lie on this curve (a parabola $y = x^2$ restricted to lie on the plane $z = 3x$). The required integral is thus

$$\int_C \mathbf{F} \cdot d\mathbf{r} = \int_0^2 15t^4 dt.$$

This is a straightforward integral

$$\int_0^2 15t^4 dt = [3t^5]_0^2 = 96.$$

b) For this part, there are three separate integrals:

  i) Both y and z are zero and $0 \le x \le 2$ so $d\mathbf{r} = \mathbf{i} dx$ so only the first component of $\mathbf{F}$ matters. But this is zero because y and for that matter z are zero. The contribution to the integral is thus 0.

  ii) For this part, $x = 2, z = 0$ but $0 \le y \le 2$, so here $d\mathbf{r} = \mathbf{j} dy$ hence only the second component of $\mathbf{F}$ matters this time. Again, as this is $x^2 z$ and as $z = 0$ this contribution to the integral is also zero.

  iii) Having so far had two zeros, ploughing on, here $x = 2, y = 4$ and $0 \le z \le 6$ hence $d\mathbf{r} = \mathbf{k} dz$ and only the third component of $\mathbf{F} = x^2 y \mathbf{k}$ matters. This is not zero this time as $\mathbf{F} = 16\mathbf{k}$ and the integral

$$\int_C \mathbf{F} \cdot d\mathbf{r} = \int_0^6 16 dz = [16z]_0^6 = 16 \times 6 = 96.$$

Therefore, the three integrals together are $0 + 0 + 96 = 96$. The two paths are shown in Figure 8.2, the most obvious point being that the results are the same. This is taken up next. □

There are a few more similar computational problems in the end of chapter exercises. An important special case arises if the contour C is a closed, and a special notation

$$\oint_C \mathbf{F} \cdot d\mathbf{r}$$

is sometimes used. Now, the relation $\nabla \phi \cdot d\mathbf{r} = d\phi$ follows from the chain rule. Therefore, suppose that the vector field $\mathbf{F}$ can be derived from a potential such

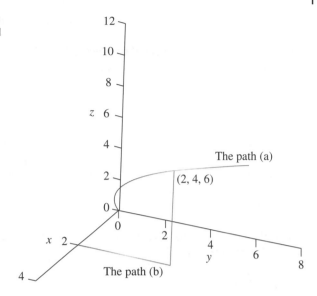

Figure 8.2 The two paths: (a) the parabolic curve and (b) a set of straight lines.

that $\mathbf{F} = \nabla\phi$ is the case, then $\mathbf{F} \cdot d\mathbf{r} = \nabla\phi \cdot d\mathbf{r} = d\phi$. Integration of this scalar product now becomes possible:

$$\oint_C \mathbf{F} \cdot d\mathbf{r} = \oint_C d\phi = [\phi]_C,$$

where the right-hand side denotes the change in the value of ϕ as the curve C is traversed. In all but a few cases this has to be zero. Hence if $\mathbf{F} = \nabla\phi$, then

$$\oint_C \mathbf{F} \cdot d\mathbf{r} = 0.$$

The converse is also true; if

$$\oint_C \mathbf{F} \cdot d\mathbf{r} = 0,$$

then there exists a scalar field ϕ such that $\mathbf{F} = \nabla\phi$. This is important in all field theory, especially fluid mechanics and electromagnetic theory. If the vector field $\mathbf{F}$ is such that $\nabla \times \mathbf{F} = \mathbf{0}$, then this guarantees the existence of such a scalar potential ϕ with $\mathbf{F} = \nabla\phi$ and in mechanics the vector field $\mathbf{F}$ is called *conservative*. In fluid mechanics, it is called *irrotational*. It means that the value of the integral

$$\int_C \mathbf{F} \cdot d\mathbf{r}$$

is only dependent upon the value of ϕ at the end points, and if these are the same and the contour C is closed its value has to be zero. In the last example, it is apparent that the integral of the vector field $\mathbf{F}(x, y, z) = 2xyz\mathbf{i} + x^2z\mathbf{j} + x^2y\mathbf{k}$ was the same between the two chosen paths. Now this could be a coincidence, but if $\mathbf{F}$ is a conservative vector field, then the value of the integral

$$\int_C \mathbf{F} \cdot d\mathbf{r}$$

will be independent of the path taken. In everyday life, the gravitational field of the Earth is conservative so the work done moving between two fixed points will be the same no matter the path taken. A point of issue for mountaineers perhaps, but then their definition of work is not quite the same as the precise mechanical one.

Example 8.4   Using the vector field $\mathbf{F}(x, y, z) = 2xyz\mathbf{i} + x^2z\mathbf{j} + x^2y\mathbf{k}$, determine whether the integral

$$\int_C \mathbf{F} \cdot d\mathbf{r}$$

is independent of path.

The best way to do this is to determine whether the field is conservative, this can be done by taking the curl

$$\nabla \times \mathbf{F}$$

$$= \begin{vmatrix} \mathbf{i} & \mathbf{j} & \mathbf{k} \\ \dfrac{\partial}{\partial x} & \dfrac{\partial}{\partial y} & \dfrac{\partial}{\partial z} \\ 2xyz & x^2z & x^2y \end{vmatrix} = (x^2 - x^2)\mathbf{i} + (2xy - 2xy)\mathbf{j} + (2xz - 2xz)\mathbf{k} = 0$$

so the theory says that there is a $\phi(x, y, z)$ and this example could stop here. However, finding ϕ is not difficult and can be done by integration of the three equations

$$\frac{\partial \phi}{\partial x} = 2xyz \quad \frac{\partial \phi}{\partial y} = x^2z \quad \frac{\partial \phi}{\partial z} = x^2y,$$

which immediately yields $\phi(x, y, z) = x^2yz$ ignoring the arbitrary constant of integration.   □

In fluid mechanics, if $\mathbf{u}$ is fluid velocity, then

$$\int_C \mathbf{u} \cdot d\mathbf{r}$$

is the *circulation*, and this is zero around a closed contour if the velocity can be derived from a potential, or is *irrotational*. This is approximately the case for either air or water flowing around an object, except that there is friction that stops this being exactly so. If the body is an aerofoil, the shedding of eddies and the non-zero circulation is vital for the lift that makes flight possible, but that belongs in a different textbook.

8.3 Practical Applications

The phrase 'practical applications' can imply calculating actual integrals that mean something, or it can mean the use of path integrals to describe something practical, or indeed both. The first two examples are of the former type, while the third is definitely the latter. Finding the work done by a mass moving in a field is a reasonably straightforward application; here is an example of it:

Example 8.5   Find the work done moving a mass of 2 kg from the point $(0, 1, -1)$ to the point $(\pi/2, -1, -2)$ in the vector field $\mathbf{F} = (y^2 \cos x + z^3)\mathbf{i} + (2y \sin x - 4)\mathbf{j} + (3xz^2 + 2)\mathbf{k}$.

Solution: The question does not specify the path between the two points, which should lead one to ask whether the result must therefore be independent of the path taken. To see whether this is true, let us find the curl:

$$\nabla \times \mathbf{F} = \begin{vmatrix} \mathbf{i} & \mathbf{j} & \mathbf{k} \\ \dfrac{\partial}{\partial x} & \dfrac{\partial}{\partial y} & \dfrac{\partial}{\partial z} \\ y^2 \cos x + z^3 & 2y \sin x - 4 & 3xz^2 + 2 \end{vmatrix}$$

$$= (0 - 0)\mathbf{i} + (3z^2 - 3z^2)\mathbf{j} + (2y \cos x - 2y \cos x)\mathbf{k} = \mathbf{0}.$$

Therefore, the vector field is conservative and the integral

$$\int_C \mathbf{F} \cdot d\mathbf{r}$$

is independent of the path. It is worth finding the ϕ such that $\mathbf{F} = \nabla \phi$. To do this, solve the equations

$$\frac{\partial \phi}{\partial x} = y^2 \cos x + z^3,$$

$$\frac{\partial \phi}{\partial y} = 2y \sin x - 4,$$

$$\frac{\partial \phi}{\partial z} = 3xz^2 + 2.$$

Integrating each gives

$$\phi(x, y, z) = y^2 \sin x + xz^3 + f_1(y, z),$$

$$\phi(x, y, z) = y^2 \sin x - 4y + f_2(x, z),$$
$$\phi(x, y, z) = xz^3 + 2z + f_3(x, y),$$

where the functions $f_1(y, z), f_2(x, z)$ and $f_3(x, y)$ are arbitrary and determined by what follows. Comparing the three expressions for $\phi(x, y, z)$ leads to

$$\phi(x, y, z) = y^2 \sin x - 4y + xz^3 + 2z + k,$$

where k is an arbitrary constant. Let us now determine the values of $\phi(0, 1, -1)$ and $\phi(\pi/2, -1, -2)$. Inserting the values of x, y and z into the formula just obtained gives:

$$\phi(0, 1, -1) = -4 + 2 + k = k - 2$$

and

$$\phi(\pi/2, -1, -2) = 1 + 4 + 8\pi - 2 + k = 8\pi - 3 + k,$$

hence, the value of the integral

$$\int_C \mathbf{F} \cdot d\mathbf{r} = \phi(\pi/2, -1, -2) - \phi(0, 1, -1) = 8\pi - 3 + k - k + 2 = 8\pi - 1.$$

Notice how the unknown k cancels. This confirms the unimportance of constants in the formula for a potential. The work done moving a mass of 2 kg would be double this:

$$(16\pi - 2)\text{kg.m.}$$

□

The other obvious species of path integral is the fully vectorial:

$$\int_C \mathbf{F} \times d\mathbf{r},$$

and this has application to electromagnetism. This is evaluated in the same way (put everything in terms of a parameter and then integrate as usual) so does not tell us anything particularly new in mathematical terms, but demonstrates a different application.

Example 8.6 A coil of wire is in the shape of a circle $x^2 + y^2 = a^2, z = 0$. Calculate the magnetic induction $\mathbf{B}$ at an arbitrary point on the z-axis due to a current I flowing in the coil if

$$\mathbf{B} = I \int_C \frac{d\mathbf{s} \times \mathbf{p}}{|\mathbf{p}|^3},$$

where $\mathbf{p}$ is a vector joining an arbitrary point on the z-axis to a point on the coil and C is the circle $x^2 + y^2 = a^2, z = 0$.

Solution: This is a result known as *Biot–Savart Law* and although this problem is securely an electromagnetism problem, it is a good example of evaluating

Figure 8.3 The coil
$x^2 + y^2 = a^2$ on the plane
$z = 0$ together with the
vector **p**.

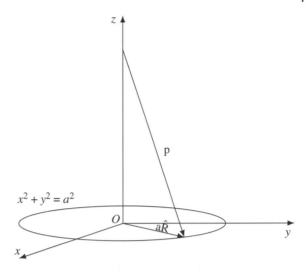

a vector integral in cylindrical polar co-ordinates R, θ and z. The set up is shown in Figure 8.3 and shows the vector $a\hat{\mathbf{R}}$. Some elementary vector algebra gives

$$\mathbf{p} = -z\mathbf{k} + a\hat{\mathbf{R}}.$$

The element of arc length is $d\mathbf{s} = ad\theta\hat{\boldsymbol{\theta}}$ so

$$\frac{d\mathbf{s} \times \mathbf{p}}{p^3} = \frac{-az\hat{\mathbf{R}} - a^2\mathbf{k}}{(z^2 + a^2)^{3/2}} d\theta$$

using that $\hat{\mathbf{R}}, \hat{\boldsymbol{\theta}}$ and $\mathbf{k}$ is a mutually orthogonally and right-handed system of unit vectors. Finally, before integration can take place $\hat{\mathbf{R}} = \mathbf{i}\cos\theta + \mathbf{j}\sin\theta$, so

$$\mathbf{B} = I \int_C \frac{d\mathbf{s} \times \mathbf{p}}{|\mathbf{p}|^3} = I \int_0^{2\pi} \frac{-az\cos\theta\mathbf{i} - az\sin\theta\mathbf{j} - a^2\mathbf{k}}{(z^2 + a^2)^{3/2}} d\theta.$$

Both the **i** and **j** terms integrate to zero so the only non-zero term is the **k** term, and the integration is trivial; simply multiply by 2π to obtain

$$\mathbf{B} = -\frac{2\pi a^2 I}{(z^2 + a^2)^{3/2}}\mathbf{k}.$$

This result confirms that the induced magnetic field **B** is in the $-z$ direction as might be expected by Faraday's Law of electromagnetic induction for those who know about such things. For more complex geometry, numerical methods would have to be used to evaluate the Biot–Savart Law integral. □

For the last example, we turn to thermodynamics, a subject introduced in Section 2.9. The application of integration in thermodynamics can be computational, but it is most often conceptual. Take for example the Carnot cycle shown in Figure 8.4. In this figure, the axes are pressure p and volume V.

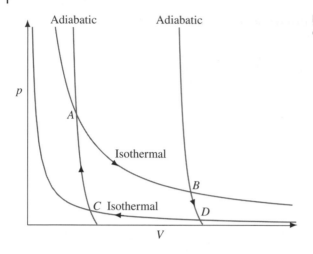

Figure 8.4 The Carnot cycle.

An isothermal process is one where the temperature does not change, it usually happens quickly. Since $pV = RT$ for an ideal gas, if T does not vary and since R is a constant for a given gas, pV is a constant (Boyle's law) and an isotherm in Figure 8.4 is a rectangular hyperbola. In this figure, AB and CD are two different isotherms corresponding to the gas at different temperatures. The adjoining curves DA and BC are adiabatic curves. An adiabatic curve represents a process that happens so quickly that there is no loss of heat. It was shown in Section 2.9 that for an adiabatic process the first law of thermodynamics implies that $pV^\gamma = $ constant, where $\gamma = c_p/c_V$ the ratio of the specific heats. Therefore, the curves DA and BC in Figure 8.4 represent paths where the temperature changes and the cycle $ABCD$ can be completed. It is called a Carnot cycle. Now there is neither the space nor the inclination to give a detailed account of thermodynamics, but it's worth explaining the cycle in terms of energy. The cycle shown can be related directly to what happens as a piston moves, and this will (finally) get us into representation with integrals. The starting point is A at a particular pressure and volume, and temperature too of course. The isotherm AB represents a change of state from A to state B without any change in temperature. Physically, the gas undergoes an expansion in volume and a decrease in pressure in accordance with Boyle's law. In terms of line integral, the work done going from A to B is W_{AB} where

$$W_{AB} = \int_{V_A}^{V_B} p\,dV = \int_{V_A}^{V_B} n\frac{RT_1}{V}\,dV = NRT_1 \ln\left(\frac{V_B}{V_A}\right),$$

so $W_{AB} = Q_{AB}$ where Q_{AB} is the heat absorbed as the temperature is not changed. In this expression, R is the gas constant and n a number representing the number of moles of the gas as $pV = nRT$ so $n = m/M$ with m the mass and M the molar mass. In the next phase, the gas experiences adiabatic expansion from B to C. This means that on $BC, pV^\gamma = p_A V_A^\gamma = p_B V_B^\gamma$, so

$$W_{BC} = \int_{V_B}^{V_C} pdV = \int_{V_B}^{V_C} \frac{p_B V_B^\gamma}{V^\gamma} dV = \frac{p_C V_C^\gamma V_C^{1-\gamma} - p_B V_B^\gamma V_B^{1-\gamma}}{1 - \gamma}$$

$$= \frac{p_C V_C - p_B V_B}{\gamma - 1}.$$

This is more usefully put in terms of temperature using $pV = nRT$ as

$$W_{BC} = \frac{nR(T_1 - T_2)}{\gamma - 1}.$$

Along the path CD, this is a different isotherm but the calculation follows as for AB:

$$W_{CD} = nRT_2 \ln \left(\frac{V_D}{V_C} \right),$$

and as before this is also Q_{CD} the heat *absorbed* this time. Finally for the adiabatic compression DA, this follows along similar lines to BC and

$$W_{DA} = \frac{nR(T_2 - T_1)}{\gamma - 1}.$$

The total work done will be the sum of all the above, which can be written as the path integral

$$\oint_C pdV = W = W_{AB} + W_{BC} + W_{CD} + W_{DA}.$$

Before any of the mathematics was done, the original work by Carnot from the early nineteenth century was thought by some to be a way to produce infinite energy simply by going around this cycle infinitely many times. However, having done the integration, the total work done can be found. Noting that the two adiabatic contributions cancel, the others sum to

$$W = nRT_1 \ln \left(\frac{V_B}{V_A} \right) - nRT_2 \ln \left(\frac{V_C}{V_D} \right)$$

upon reversing the second. This can be written in terms of efficiency θ as

$$\theta = \frac{W}{Q_{AB}} = 1 - \frac{T_2 \ln(V_C/V_D)}{T_1 \ln(V_B/V_A)}.$$

It can easily be deduced that the logarithm terms cancel since $V_C/V_D = V_B/V_A$ hence the efficiency is not infinity but

$$\theta = 1 - \frac{T_2}{T_1}.$$

Therefore, here is an example of the use of a path integral to explain physics rather than just to calculate. The original Carnot cycle was an account with no mathematics, sadly Sadi Carnot (1796–1832) died of cholera following a largely military career and did not get the chance to publish results that Joule (1818–1889) duplicated and was credited with later. It was Lord Kelvin, Sir William Thompson (1824–1907), who did the calculation repeated above, about 50 years after Carnot's death.

The concept of entropy emerged from the Carnot cycle. By generalising the above arguments, Clausius (1822–1888) proposed that closed path integral

$$\oint \frac{dQ}{T} = 0,$$

where remember Q is heat and of course T is temperature. This leads to defining a quantity S such that

$$dS = \frac{dQ}{T}$$

and S is the *entropy*. It turns out that entropy is as fundamental quantity as energy, and $dS = 0$ in a reversible system but $dS \geq 0$, in general, which is the second law of thermodynamics and stated loosely says that the quantity of disorder in a real system always increases. This in turn led to the development of the Kelvin absolute temperature scale, and statements such as $S = 0$ only when $T = 0$ in degrees Kelvin. Let's leave thermodynamics at this point.

Exercises

8.1 Evaluate the integrals

a)

$$\int_C x^2 y\, ds,$$

where C is the semi-circle $x^2 + y^2 = a^2, z = 0, y \geq 0$.

b)

$$\int_C xy^2\, ds,$$

where C is the semi-circle $x^2 + y^2 = a^2, z = 0, y \geq 0$.

c)

$$\int_C (x^2 + y^4 + z^6)\, dx,$$

where C is that part of the curve $x = t^3, y = t^2, z = t$ between the points $(0, 0, 0)$ and $(1, 1, 1)$.

d)

$$\int_C \sqrt{a^2 + b^2 - r^2} ds,$$

where C is the ellipse $x = a \cos \theta, y = b \sin \theta, 0 \le \theta \le 2\pi$.

8.2 Consider the integral

$$\int_C \mathbf{F} \cdot d\mathbf{r}$$

evaluated along the following three paths, each connecting the point $(-1, 0)$ to the point $(1, 0)$ in the x, y plane:
a) the x-axis from $(-1, 0)$ to $(1, 0)$,
b) the semi-circle $x^2 + y^2 = 1$ with $y \ge 0$,
c) three straight lines: $(-1, 0)$ to $(-1, 1), (-1, 1)$ to $(1, 1)$ and $(1, 1)$ to $(1, 0)$,

where the vector field $\mathbf{F} = x^2 \mathbf{i} + y \mathbf{j}$. What do your answers tell you about the vector field $\mathbf{F}$?

8.3 Re-visit the previous question using $\mathbf{F} = (z - 1)\mathbf{i}$. Evaluate the work done on the x, y plane between the points $(-1, 0)$ and $(1, 0)$ along the same three paths and discuss the results. What happens if the paths are now between $(-1, 0, 1)$ and $(1, 0, 1)$ on the plane $z = 1$ instead of the x, y plane?

8.4 Evaluate the integral

$$\int_C \mathbf{F} \cdot d\mathbf{r}$$

a) for $\mathbf{F} = x\mathbf{i}$, where C is the line $x = y, z = 0$ from $(0, 0, 0)$ to $(2, 2, 0)$,
b) for $\mathbf{F} = \mathbf{r}$, where C is the helix $(\sin t, \cos t, t)$ from $(0, 1, 0)$ to $(0, 1, 2\pi)$,
c) for $\mathbf{F} = \mathbf{r}$, where C is the straight line $x = 0, y = 1, 0 \le z \le 2\pi$.
Why are the solutions to part (b) and (c) the same?

8.5 If $\mathbf{F}$ is the field $x^2 \mathbf{i} + y^2 \mathbf{j} + z^2 \mathbf{k}$ and C is the twisted cubic

$$x = t, y = \frac{t^2}{\sqrt{2}}, z = \frac{t^3}{3}$$

between the points $(0, 0, 0)$ and $\left(1, \dfrac{1}{\sqrt{2}}, \dfrac{1}{3}\right)$, evaluate the three integrals:

a) $\displaystyle\int_C \mathbf{F}ds,$

b) $\displaystyle\int_C \mathbf{F}\cdot d\mathbf{r},$

c) $\displaystyle\int_C \mathbf{F}\times d\mathbf{r}.$

8.6 Show that the *circulation* of any constant vector field $\mathbf{U}$

$$\oint_C \mathbf{U}\cdot d\mathbf{r} = 0.$$

8.7 Show that the work done by a unit mass of 1 kg moving along a curve C under the action of a force $\mathbf{F}$ is equal to the integral

$$\int_C \mathbf{F}\cdot \hat{\mathbf{T}}ds,$$

where $\hat{\mathbf{T}}$ is the unit tangent to the curve C. Hence or otherwise find the work done moving a unit mass along the following curves C under the stated forces $\mathbf{F}$.

a) $\mathbf{F} = \hat{\mathbf{T}}\sqrt{x^2+y^2}$, C is the semi-circle $x = 1 - \cos\theta, y = \sin\theta, z = 0, 0 \le \theta < 2\pi$.

b) $\mathbf{F} = \mathbf{k}z$, C is the curve $x = f_1(t), y = f_2(t), z = \sqrt{1+t^2}, 0 \le t \le 1$.

c) $\mathbf{F} = x^3\mathbf{i} + \mathbf{j} + \sinh y\mathbf{k}$, C is the curve $x = 1, y = \sinh^{-1}t, z = \sqrt{1+t^2}$, $0 \le t \le 1$.

d) $\mathbf{F} = mg\mathbf{k}$ along the same path as in part (c).

8.8 Calculate the work done by a unit mass 1 kg in the vector field $\mathbf{F} = (x - 3y)\mathbf{i} + (y - 2x)\mathbf{j}$ twice around the ellipse

$$\frac{x^2}{9} + \frac{y^2}{4} = 1$$

in a counter clockwise direction.

8.9 The force due to gravity is $-k\mathbf{r}/|\mathbf{r}|^3$. Calculate the work done falling from a radius $r = a$ to a radius $r = b$.

8.10 Revisit Example 8.6 and use the Biot–Savart Law to show that the magnetic field induced by the coil at $x^2 + y^2 = a^2, z = 0$ at a point with position vector $\mathbf{r}_0$ is given by the integral

$$Ia\int_0^{2\pi} \frac{z_0\mathbf{R} - (\mathbf{R}\cdot\mathbf{r}_0)\mathbf{k}}{[R^2 + r_0^2 - 2\mathbf{R}\cdot\mathbf{r}_0]^{3/2}}d\theta,$$

where **R** is the position vector of an arbitrary point on the coil that carries a current I. Verify that the result of Example 8.6 is regained if $\mathbf{r}_0$ represents any point on the z axis. [Note: Those familiar with MAPLE or similar software might like to experiment with the numerical evaluation of this integral, plotting the solution for different values of $\mathbf{r}_0$.]

9

Multiple Integrals

9.1 Introduction

Integration, as introduced previously, is the inverse of differentiation. Usually, the integral is interpreted as the area under a curve. For example,

$$\int_a^b f(x)dx$$

is the area under the curve $y = f(x)$ between the values $x = a$ and $x = b$. This works well provided $f(x) > 0$ in the range $a \leq x \leq b$. In the last chapter, the path or line integral was introduced. Although the integration of this chapter is still the inverse of differentiation, it is not restricted to either finding areas under curves or indeed path integrals. As a starting point, it is best to think of a pixel on a screen moving first one way to describe a line, then the line moving at right angles to encompass an area. This way a domain is covered, and the magnitude of areas are calculated. Move in three perpendicular directions and volumes can also be calculated. Importantly, so much more than areas and volumes can be found in this way. For those more interested in mechanics, quantities such as the centre of mass and the moment of inertia can be found. For the statistically inclined, there are probability distributions defined over an area that occur in experimental design, for example. It is also possible to use co-ordinate systems other than Cartesian: plane polars in two dimensions and cylindrical and spherical polars in three dimensions. Given the new material, even though vectors could be involved here, a decision has been made to keep this chapter clear of vectors. We start with double integrals.

9.2 The Double Integral

Figure 9.1 shows a typical region D in the x, y plane. The small black rectangular area $dx \times dy$ is shown and the idea is that the expression

$$\int_{y_1}^{y_2} \int_{x=f_1(y)}^{x=f_2(y)} f(x, y)dx \, dy$$

Two and Three Dimensional Calculus: With Applications in Science and Engineering, First Edition. Phil Dyke.
© 2018 John Wiley & Sons Ltd. Published 2018 by John Wiley & Sons Ltd.

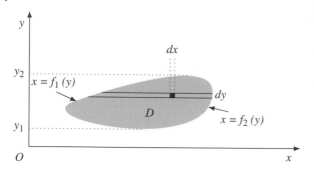

Figure 9.1 The x, y plane showing a convex region D and the pixel $dx \times dy$; the left-hand half is $x = f_1(y)$ and the right-hand half is $x = f_2(y)$.

should be uniquely defined. First of all, the integrand $f(x, y)$ is an integrable function of both x and y. Functions that are piecewise continuous throughout and on the domain D will suffice. The limits are defined to indicate the domain D. Take a look at Figure 9.1. The small black rectangle has sides dx and dy. The way the integral is written, first integration with respect to x takes place. This means that y is held constant at some arbitrary value indicated by the tramlines of width dy. The bottom limit is labelled $x = f_1(y)$ and $f_1(y)$ is the function that describes the left-hand side of the border of the domain D. The top limit is labelled $x = f_2(y)$ and this function describes the right-hand side of the domain. They have to be different functions otherwise a horizontal line through the domain D at a fixed y would have to produce the same value ($f_1(y)$, say) at both ends, which is clearly impossible. Performing this first integration is tantamount to adding up all the local values of $f(x, y)dx\, dy$ along the line through D from the left-hand side at $f_1(y)$ to the right at $f_2(y)$. As y is constant throughout this integration, it could be termed *partial* integration analogous to partial differentiation, but the term is never used. Once the integration along this line has been performed, the second integration sums all lines from the bottom of D at y_1 to the top at y_2. In this way, the values of $f(x, y)dx\, dy$ are summed over the entire domain D. There is of course no reason why the integration cannot be done in the other order with the stripes along the lines $x = $ constant first. This would be written

$$\int_{x_1}^{x_2} \int_{y=g_1(x)}^{y=g_2(x)} f(x, y)dy\, dx$$

and the same domain integrated in this reverse order is shown in Figure 9.2. The reader is left to relate this double integral to Figure 9.2. Sometimes, due to the technicalities of performing the integration, it is only possible to evaluate a double integral one way round. Changing the order is then mandatory and is something that is tackled in Section 9.2.2. Note that the region is assumed to be convex, that is a straight line passing through the region only enters and leaves it once. Regions that are not convex can be managed by splitting the domain,

Figure 9.2 The same domain as Figure 9.1 but reversing the order of integration.

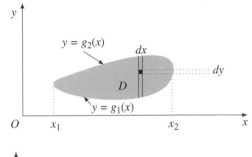

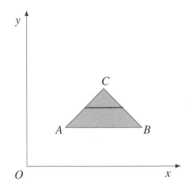

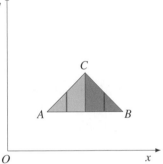

Figure 9.3 The triangular domain *ABC*. The left graph indicates integration with respect to *x* first (the horizontal black line) where the entire triangle can be covered with one double integral, whereas the right graph indicates integration with respect to *y* first (the two black vertical lines) where the triangle has to be integrated in two halves, indicated by the different shading involving two, separate, double integrals.

it is inconvenient and the details are not particularly rewarding, so domains are convex here. Sometimes, the border of a domain has corners; a triangle, for example. Using a change of order can halve the algebra in these circumstances; see Figure 9.3. It is worth emphasising that the function that is being integrated, $f(x, y)$, has nothing whatsoever to do with the limits of integration. If $f(x, y) = 1$, then the double integral is the area of *D*; in fact, finding areas in this way is a common use of double integration. Letting $f(x, y)$ be other simple functions leads to finding centres of mass or moments of inertia in mechanics, or evaluating probabilities in statistics. These applications come later. One final point on notation. In this text, the integrals are evaluated from the inside out, so the if the order is $dx\ dy$, then the inner integral sign refers to integration with respect to *x* and the outer integral sign integration with respect to *y*. Some books do not use this notation, instead of calculating from inside out, others calculate from left to right; so using this alternative notation the integral signs would be the other way round. The advice, as always, is to be careful when consulting different sources. Let us now do some examples, the first is introductory.

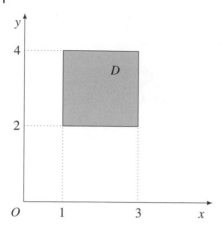

Figure 9.4 The domain of integration: a square.

Example 9.1   Evaluate the two double integrals:

a) $\displaystyle\int_{y=2}^{y=4}\int_{x=1}^{x=3}(40-2xy)dx\,dy;$

b) $\displaystyle\int_{x=1}^{x=3}\int_{y=2}^{y=4}(40-2xy)dy\,dx.$

Solution: The advice for all multiple integration problems is to draw a diagram, so let us do this.

a) Figure 9.4 shows the region of integration for the first integral. Integration proceeds as follows:

$$\int_{y=2}^{y=4}\int_{x=1}^{x=3}(40-2xy)dx\,dy = \int_{y=2}^{y=4}[40x-x^2y]_{x=1}^{x=3}dy$$

$$= \int_{y=2}^{y=4}[120-9y-(40-y)]dy$$

$$= \int_{y=2}^{y=4}(80-8y)dy$$

$$= [80y-4y^2]_{y=2}^{y=4}$$

$$= (320-64)-(160-16) = 112.$$

b) The second integral proceeds similarly, the region remains the same (Figure 9.4) so we have

$$\int_{x=1}^{x=3}\int_{y=2}^{y=4}(40-2xy)dx\,dy = \int_{x=1}^{x=3}[40y-y^2x]_{y=2}^{y=4}dx$$

$$= \int_{x=1}^{x=3} [(160 - 16x) - (80 - 4x)]dx$$

$$= \int_{x=1}^{x=3} (80 - 12x)dx$$

$$= [80x - 6x^2]_{x=1}^{x=3}$$

$$= (240 - 54) - (80 - 6) = 112$$

and it is confirmed that both results are the same. ☐

There is a lot of detail in the above example as it is the first. Do not come away with the impression that to change the order of integration, one merely has to change the limits. This only works if the limits are constant, that is the domain D is a rectangle. This next example is more complex, and it is essential to draw a diagram as the limits are not given.

Example 9.2 Evaluate the double integral

$$\iint_D (2x - y^2)dx \, dy,$$

where D is the triangular region enclosed by the lines $y = 0, y = 2x$ and $x = 3$.

Solution: The triangular region is shown in Figure 9.5. In this figure, the vertical tramline shows the decision is to integrate with respect to y first. The upper

Figure 9.5 The triangular domain.

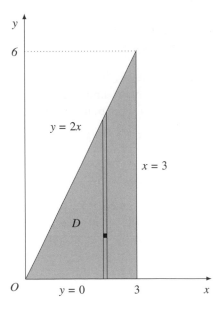

limit is the line $y = 2x$ and the lower limit is $y = 0$, then the line is taken from $x = 0$ on the left to the line $x = 3$ on the right. This leads to the double integral

$$\int_0^3 \int_0^{2x} (2x - y^2) dy \, dx,$$

where the limits are written just as limits without $y = 0, y = 2x$, and so on in front of each. The calculation proceeds as follows:

$$\int_0^3 \int_0^{2x} (2x - y^2) dy \, dx = \int_0^3 \left[2xy - \frac{1}{3}y^3 \right]_0^{2x} dx$$

$$= \int_0^3 4x^2 - \frac{8}{3}x^3 \, dx$$

$$= \left[\frac{4}{3}x^3 - \frac{2}{3}x^4 \right]_0^3$$

$$= 36 - 54 = -18. \qquad \square$$

As an exercise, try reversing the order from Figure 9.5. The horizontal line through the black rectangle $dx \times dy$ is at the left-most point of the same straight line $y = 2x$; only this time as x is varying; it needs to be called $x = y/2$, the right-hand limit will be $x = 3$, and then this straight line is taken from the bottom at $y = 0$ to the top point at $y = 6$. The integral when evaluated should still give -18 as before. Note how the limits are very different this time and need to be created from knowledge of the shape of the domain D. There's more practice for doing this in the next few pages.

Example 9.3 Find the area between the parabolas $y = x^2 - 4x + 6$ and $y = -x^2 + 6x - 2$ using double integration.

Solution: Figure 9.6 shows the required area. There are no hints here, so there are some points to find and decisions to be made. To define the area properly we need to find the points where the two parabolas intersect so we solve

$$x^2 - 4x + 6 = -x^2 + 6x - 2 \quad \text{or} \quad 2x^2 - 10x + 8 = 0,$$

which is

$$2(x - 4)(x - 1) = 0 \quad \text{so} \quad x = 1, 4 \quad \text{giving} \quad y = 3, 6.$$

The choice of whether to integrate with respect to x or y first is reasonably straightforward to make after a little thought. Figure 9.6 shows the area and also shows that y is the correct (only) choice as the direction in which to integrate

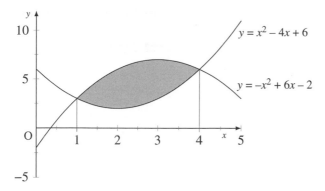

Figure 9.6 The two parabolas are shown, and the domain of the integral is shaded.

first. If x is chosen, the first integral would be in four parts, and putting x in terms of y is not easy and involves square roots which then have to be integrated. Therefore, the integral to be evaluated is

$$\int_1^4 \int_{x^2-4x+6}^{-x^2+6x-2} dy\ dx.$$

Evaluating gives

$$\int_1^4 \int_{x^2-4x+6}^{-x^2+6x-2} dy\ dx = \int_1^4 [-x^2 + 6x - 2 - (x^2 - 4x + 6)]dx$$

$$= \int_1^4 -2x^2 + 10x - 8\ dx$$

$$= \left[-\frac{2}{3}x^3 + 5x^2 - 8x \right]$$

$$= -\frac{128}{3} + 80 - 32 - \left(-\frac{2}{3} + 5 - 8 \right)$$

$$= -\frac{126}{3} + 88 - 37 = 9. \qquad \square$$

The enthusiastic amongst you could reverse the order of integration to check this answer. (The author used MAPLE.) There is another option using elementary calculus; some of you might have spotted this. Using the area under $y = f(x)$

$$\int_a^b f(x)dx,$$

the shaded area of Figure 9.6 is given by the difference

$$\int_1^4 (-x^2 + 6x - 2)dx - \int_1^4 (x^2 - 4x + 6)dx$$

after which the calculation proceeds along the previous lines. Here is an example where the domain has to be split.

Example 9.4 Evaluate the integral

$$\iint_D xy \, dy \, dx,$$

where D is the rectangle with vertices $(1, 1), (2, 0), (4, 2)$ and $(3, 3)$.

Solution: For this example, no matter which is chosen as the initial direction to integrate along, x or y, the rectangle has to be split into three, see Figure 9.7. Here, y is chosen as the first integration variable. Each place where the vertical line representing integration in the y direction meets a corner, the limit has to change corresponding to the equation of the new line. Integration proceeds

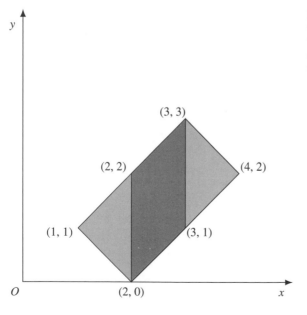

Figure 9.7 The rectangular domain shown split into three; each region corresponding to a different double integral.

as follows:

$$\iint_D xy\, dy\, dx = \int_1^2 \int_{2-x}^x xy\, dy\, dx + \int_2^3 \int_{x-2}^x xy\, dy\, dx$$

$$+ \int_3^4 \int_{x-2}^{6-x} xy\, dy\, dx.$$

These integrals are routine if a little long:

$$\int_1^2 \int_{2-x}^x xy\, dy\, dx = \frac{1}{2} \int_1^2 x(4x - 4)dx = 2\left[\frac{1}{3}x^3 - \frac{1}{2}x^2\right]_1^2$$

$$= 2\left(\frac{8}{3} - 2 - \frac{1}{3} + \frac{1}{2}\right) = \frac{14}{3} - 3 = \frac{5}{3},$$

$$\int_2^3 \int_{x-2}^x xy\, dy\, dx = \frac{1}{2} \int_2^3 x(4x - 4)dx = 13 - \frac{16}{3} = \frac{23}{3},$$

$$\int_3^4 \int_{x-2}^{6-x} xy\, dy\, dx = \int_3^4 x(16 - 4x)dx = \left[8x^2 - \frac{4}{3}x^3\right]_3^4$$

$$= 4\left(23 - \frac{64}{3}\right) = \frac{20}{3}.$$

The sum of these is 16, the required answer. □

9.2.1 Rotation and Translation

In the last example, the domain, though a simple rectangle, was awkwardly placed. Often, integrals over such a domain can be simplified by moving the domain to somewhere more convenient, and this is usually achieved through a rotation and translation. The new co-ordinates (X, Y) are related to (x, y) via the equations

$$\begin{pmatrix} x \\ y \end{pmatrix} = \begin{pmatrix} \cos\theta & \sin\theta \\ -\sin\theta & \cos\theta \end{pmatrix} \begin{pmatrix} X \\ Y \end{pmatrix} + \begin{pmatrix} x_0 \\ y_0 \end{pmatrix},$$

where θ is the positive (anti-clockwise) rotation of the axes and through translation the point (x_0, y_0) becomes the origin $(0, 0)$ in the new (X, Y) co-ordinates. Without matrices, the equations are

$$x = X\cos\theta + Y\sin\theta + x_0,$$

$$y = -X\sin\theta + Y\cos\theta + y_0.$$

The last example can thus be simplified through use of this transformation with $\theta = -\pi/4$ and $(x_0, y_0) = (2, 0)$ hence

$$x = \frac{X - Y}{\sqrt{2}} + 2,$$

$$y = \frac{X + Y}{\sqrt{2}}$$

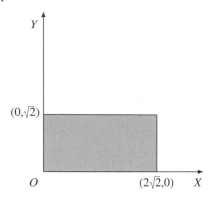

Figure 9.8 The transformed rectangle in the X, Y plane.

takes the rectangle in the x, y plane, and rotates and translates it to be in the position indicated in Figure 9.8.

In the x, y plane, the vertices of the rectangle are $(2, 0), (1, 1), (4, 2)$ and $(3, 3)$ and they transform to $(0, 0), (0, \sqrt{2}), (2\sqrt{2}, 0)$ and $(2\sqrt{2}, \sqrt{2})$ respectively in the new X, Y co-ordinates. The double integral in this new plane is

$$\int_0^{\sqrt{2}} \int_0^{2\sqrt{2}} \left[\frac{1}{2}(X^2 - Y^2) + (X + Y)\sqrt{2} \right] dx \, dy$$

and this is much more easily evaluated and has the same value as before: 16. The double integral

$$\iint_D (x^2 + y^2) dy \, dx$$

has meaning: It is the moment of inertia of a lamina the shape of D about an axis perpendicular to the x, y plane through the origin in the x, y plane. It was tempting to set this as the last example, but the three integrals simply take too long to compute. However, now that the equations that transform the rectangle to that in Figure 9.8 have been found, the required double integral can be stated as

$$I = \int_0^{2\sqrt{2}} \int_0^{\sqrt{2}} \left(\frac{X - Y}{\sqrt{2}} + 2 \right)^2 + \left(\frac{X + Y}{\sqrt{2}} \right)^2 dY \, dX$$

and this can be evaluated quite easily. Multiplying out this expression:

$$I = \int_0^{2\sqrt{2}} \int_0^{\sqrt{2}} [X^2 + Y^2 + 2\sqrt{2}(X - Y) + 4] dY \, dX$$

$$= \int_0^{2\sqrt{2}} \left[X^2 Y + \frac{1}{3} Y^3 + 2\sqrt{2} \left(XY - \frac{1}{2} Y^2 \right) + 4Y \right]_0^{\sqrt{2}} dX$$

$$= \int_0^{2\sqrt{2}} \left[X^2\sqrt{2} + \frac{2}{3}\sqrt{2} + 2\sqrt{2}(X\sqrt{2} - 1) + 4\sqrt{2} \right] dX$$

$$= \left[\frac{1}{3}X^3\sqrt{2} + 2X^2 + \frac{8}{3}\sqrt{2}X \right]_0^{2\sqrt{2}}$$

$$= \frac{1}{3}16\sqrt{2} \cdot \sqrt{2} + 16 + \frac{8}{3} \cdot 4 = \frac{112}{3}.$$

This value can be calculated from the three integrals in the x, y co-ordinate system by those with more stamina than the author. The author was content to verify the above result using MAPLE.

9.2.2 Change of Order of Integration

In this subsection, the formal process of changing the order of integration is done. The process is not a complex one and is best explained through an example.

Example 9.5   Reverse the order and hence evaluate the double integral

$$\int_{x=0}^{x=3} \int_{y=\frac{1}{3}x}^{y=1} y(x + y)^{1/2} dy \, dx.$$

Solution: Figure 9.9 shows the triangular area, drawn from examination of the limits $y = \frac{1}{3}x$ and $y = 1$ with $x = 0$ to $x = 3$. As written, the order is integrated with respect to y first. Although this is possible, it is not easy and reversing the order means looking at the triangle re-calculating the limits. Integrating with respect to x first means that we look at the extremities of a horizontal line through the grey area of Figure 9.9. The left is the y-axis or $x = 0$ and right is the line $y = \frac{1}{3}x$ now written as $x = 3y$. Looking now at the y direction, this line

Figure 9.9 The triangular area.

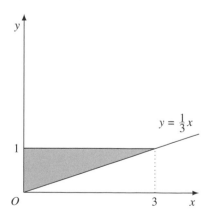

starts at the origin $y = 0$ and ends at the line $y = 1$. This gives rise to the double integral

$$\int_{y=0}^{y=1} \int_{x=0}^{x=3y} y(x+y)^{1/2} dx \, dy$$

and now the integral is with respect to x and as y is held constant, it is a simple square root that integrates at once to

$$\int_{y=0}^{y=1} y \left[\frac{2}{3}(x+y)^{3/2} \right]_{x=0}^{x=3y} dy$$

and once the limits for x are inserted, we get the much simpler

$$\int_0^1 \frac{2}{3} y [8y^{3/2} - y^{3/2}] dy = \frac{14}{3} \int_0^1 y^{5/2} \, dy$$

a power law integral that gives

$$\frac{14}{3} \left[\frac{2}{7} y^{7/2} \right]_0^1 = \frac{14}{3} \cdot \frac{2}{7} = \frac{4}{3}. \qquad \square$$

It is the nature of this kind of problem that the given integral has limits that need to be interpreted as a domain. The task then is to reverse the order of integration given this knowledge. Here is an example where reversing the order has to be done.

Example 9.6   Evaluate the integral

$$\int_{y=0}^{y=\ln 2} \int_{x=e^y}^{x=2} \frac{y}{[\ln(x)]^2} dx \, dy.$$

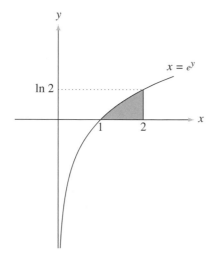

Figure 9.10 The curved triangular domain is shaded.

Solution: This integral is not possible unless the order of integration is reversed. The inner limits tell us that the integration is between $x = e^y$ (or $y = \ln x$) and $x = 2$. This line is then taken from $y = 0$ (the x-axis) to the point where $y = \ln 2$ which is where $2 = e^y$. As this coincides with $x = 2$ (as $x = e^y$), the domain has to be as shown in Figure 9.10. Reversing the order of integration using this figure gives:

$$\int_{y=0}^{y=\ln 2} \int_{x=e^y}^{x=2} \frac{y}{[\ln(x)]^2} dx\, dy = \int_1^2 \int_0^{\ln x} \frac{y}{[\ln x]^2} dy\, dx$$

the right-hand side of which is easily evaluated as follows:

$$\int_1^2 \int_0^{\ln x} \frac{y}{[\ln x]^2} dy\, dx = \int_1^2 \left[\frac{1}{2} \frac{y^2}{[\ln x]^2} \right]_0^{\ln x} dx$$
$$= \int_1^2 \frac{1}{2} \frac{[\ln x]^2}{[\ln x]^2} dx$$
$$= \int_1^2 \frac{1}{2} dx = \left[\frac{1}{2} x \right]_1^2 = \frac{1}{2}. \qquad \Box$$

9.2.3   Plane Polar Co-ordinates

The next extension to the toolkit for solving problems is to examine different co-ordinate systems, in particular, the use of polar co-ordinates. It is also useful to look at general features of changing the co-ordinate system; but actual problems will almost always use either Cartesian or plane polar co-ordinates. In double integration, one has to deal with the element of area $dx\, dy$ in Cartesian co-ordinates. In a general co-ordinate system ξ, η, this will be $d\xi\, d\eta$ and the relationship between these is

$$d\xi\, d\eta = \frac{\partial(\xi, \eta)}{\partial(x, y)} dx\, dy$$

or

$$dx\, dy = \frac{\partial(x, y)}{\partial(\xi, \eta)} d\xi\, d\eta.$$

In plane polars $\xi = r$ and $\eta = \theta$

$$x = r \cos\theta \quad \text{and} \quad y = r \sin\theta$$

so

$$\frac{\partial(x, y)}{\partial(r, \theta)} = \begin{pmatrix} \dfrac{\partial r}{\partial x} & \dfrac{\partial r}{\partial y} \\[2mm] \dfrac{\partial \theta}{\partial x} & \dfrac{\partial \theta}{\partial y} \end{pmatrix} = \frac{\partial r}{\partial x} \frac{\partial \theta}{\partial y} - \frac{\partial r}{\partial y} \frac{\partial \theta}{\partial x} = r$$

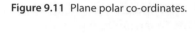

Figure 9.11 Plane polar co-ordinates.

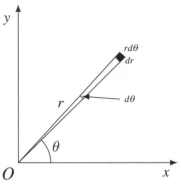

and we arrive at the relationship

$$dx \, dy = r \, dr \, d\theta$$

purely algebraically. Figure 9.11 does the same thing graphically. The small black area is to first order a rectangle of sides dr and $r \, d\theta$ so the area of the element is $r \, dr \, d\theta$ as derived earlier. Here is a simple example, but one that can only be done using polar co-ordinates.

Example 9.7   Evaluate the two integrals

$$\iint_D x^2 \, dx \, dy \quad \text{and} \quad \iint_D y^2 \, dx \, dy,$$

where D is the annular domain $4 \leq x^2 + y^2 \leq 9$.

Solution: The reason polar co-ordinates really have to be used here is that the domain D is multiply connected. Put simply, D has a hole in it. Figure 9.12 shows the domain of the double integrals. Using $x = r \cos \theta$ and $y = r \sin \theta$ together with $dx \, dy = r \, dr \, d\theta$ in plane polar co-ordinates, they become

$$\int_0^{2\pi} \int_2^3 r^3 \cos^2\theta \, dr \, d\theta \quad \text{and} \quad \int_0^{2\pi} \int_2^3 r^3 \sin^2\theta \, dr \, d\theta;$$

details of the integration are left to the reader. Both answers are

$$\frac{65}{4}\pi.$$

□

The area A of a domain D in plane polar co-ordinates is

$$A = \iint_D r \, dr \, d\theta.$$

Figure 9.12 The annular domain D (shaded).

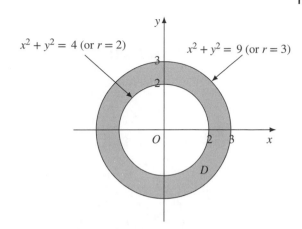

$x^2 + y^2 = 4$ (or $r = 2$)

$x^2 + y^2 = 9$ (or $r = 3$)

However, it is quite common in polar co-ordinates for a curve to be in the form $r = r(\theta)$, and so the domain D is captured by the limits as

$$A = \int_{\theta_1}^{\theta_2} \int_0^{r(\theta)} r \, dr \, d\theta$$

and it is not unusual to find that $\theta_1 = 0$ and $\theta_2 = 2\pi$. The selection of plane polar co-ordinates typically means that the curve $r = r(\theta)$ in some sense surrounds the origin. It could be a circle of course but need not be. An ellipse or other closed curve would still have this form. The point here is that whatever the form $r = r(\theta)$ takes, one integration can take place and that the area is

$$A = \frac{1}{2} \int_{\theta_1}^{\theta_2} r^2(\theta) d\theta.$$

For $r = a$, a circle centred at the origin we have the expected result

$$A = \frac{1}{2} \int_0^{2\pi} a^2 \, d\theta = \frac{1}{2} \cdot 2\pi \cdot a^2 = \pi a^2.$$

Here is a slightly more involved example.

Example 9.8 Find the area enclosed by the curve $r = 2 + \sin(5\theta)$.

Solution: The domain is shown in Figure 9.13 and the calculation, using the convenient formula just derived proceeds as follows:

$$
\begin{aligned}
\text{Area} &= \frac{1}{2} \int_0^{2\pi} r^2(\theta) d\theta = \frac{1}{2} \int_0^{2\pi} (2 + \sin 5\theta)^2 d\theta \\
&= \frac{1}{2} \int_0^{2\pi} \left[\frac{9}{2} + 4\sin 5\theta - \frac{1}{2} \cos 10\theta \right] d\theta \\
&= \frac{1}{2} 9\pi;
\end{aligned}
$$

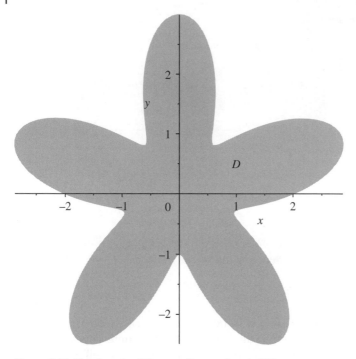

Figure 9.13 The domain of the equation $r = 2 + \sin(5\theta)$.

all the trigonometric terms integrate to zero as is typical in these kind of examples. The answer is therefore $9\pi/2$.   □

Example 9.9   Determine the area enclosed by the limaçon $r = 2 + 4\cos\theta$ but excluding the small loop. This area is shaded in Figure 9.14.

Solution: This is not as straightforward a problem as the last one. The area as a double integral in plane polar co-ordinates is still

$$\iint_D r \, dr \, d\theta$$

but the problem is finding the limits. The limaçon traditionally has equation

$$r = 2 + 4\cos\theta$$

but it is multivalued for some values of θ and r is negative, hence technically has no value, for other values. It is helpful therefore to plot special values of r and θ; even sophisticated polar-plot software needs to be used with caution. When $\theta = 0, r = 6$ and this corresponds to the point $(6, 0)$ on the x-axis.

Figure 9.14 The domain is the area within the limaçon $r = 2 + 4\cos\theta$ and is shaded and labelled D.

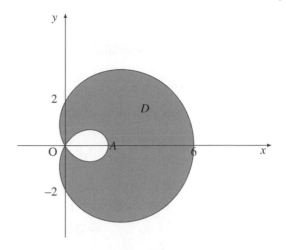

As θ increases the outer loop is described anti-clockwise, when $\theta = \pi/2$ the point $(2, 0)$ labelled as 2 in Figure 9.14 is reached. The origin is then reached where $r = 0$ and $\theta = 2\pi/3$. For the range $2\pi/3 \leq \theta \leq 4\pi/3, r \leq 0$ so there is no curve. The tangents to the curve through the origin are $\theta = 2\pi/3$ and $\theta = 4\pi/3$ and between these two values on the left of the y-axis in Figure 9.14 there is no curve. The limaçon path followed for $\theta > 4\pi/3$ is then the lower arc from the origin to $(0, -2)$ where $\theta = 3\pi/2$, then on to $(6, 0)$ where $\theta = 2\pi$. The inner loop is a representation of the curve $r = -(2 + 4\cos\theta)$ for $2\pi/3 \leq \theta \leq 4\pi/3$ so the real equation is better written either as

$$r^2 = (2 + 4\cos\theta)^2 \quad \text{or} \quad r = |2 + 4\cos\theta|, \tag{9.1}$$

in most textbooks or websites that contain the geometry of this curve either a parametric representation

$$x = 2(1 + 2\cos t)\cos t; \quad y = 2(1 + 2\cos t)\sin t$$

or Cartesian co-ordinates are used. The area of the domain is given by the expression derived earlier

$$\int_{\theta_1}^{\theta_2} r^2 \, d\theta;$$

it is just a question of finding θ_1 and θ_2. For this problem, it is best to calculate the entire area with the inner loop filled, then subtract the area of the inner loop. It was seen above that where $\theta = \pi$ in the true representation of the curve, Equation 9.1, this corresponded to the point $A(2, 0)$ on the x-axis at the right-hand extremity of the inner loop, and $2\pi/3 \leq \theta \leq \pi$ corresponded to the upper arc of inner loop and $\pi \leq \theta \leq 4\pi/3$ to the lower arc of inner loop.

Thus, the area of this inner loop is

$$\frac{1}{2}\int_{2\pi/3}^{4\pi/3} (2 + 4\cos\theta)^2 d\theta.$$

This is evaluated as follows:

$$\frac{1}{2}\int_{2\pi/3}^{4\pi/3} (2 + 4\cos\theta)^2 d\theta = 2\int_{2\pi/3}^{4\pi/3} [3 + 4\cos\theta + \sin 2\theta]d\theta$$

$$= 2[3\theta + 4\sin\theta + \sin 2\theta]_{2\pi/3}^{4\pi/3}$$

$$= 4\left[\pi - 2\sqrt{3} + \frac{\sqrt{3}}{2}\right]$$

$$= 2[2\pi - 3\sqrt{3}].$$

The whole area is a bit easier: as the entire area is required, simply use the two tangents at the origin as limits: $\theta = -2\pi/3$ and $\theta = 2\pi/3$ as these exclude that range of θ where there are no positive values of r.

$$\frac{1}{2}\int_{-2\pi/3}^{2\pi/3} (2 + 4\cos\theta)^2 d\theta = 2\int_{-2\pi/3}^{2\pi/3} [3 + 4\cos\theta + \sin 2\theta]d\theta$$

$$= 2[3\theta + 4\sin\theta + \sin 2\theta]_{-2\pi/3}^{2\pi/3}$$

$$= 4\left[2\pi + 2\sqrt{3} - \frac{\sqrt{3}}{2}\right]$$

$$= 2[4\pi + 3\sqrt{3}].$$

Hence, the shaded area is $2[4\pi + 3\sqrt{3}] - 2[2\pi - 3\sqrt{3}] = 4[\pi + 3\sqrt{3}]$. $\square$

9.2.4 Applications of Double Integration

In this subsection, double integration will be used to solve problems that arise in mechanics and probability. It is not the purpose of this book to derive expressions in either mechanics or statistics; the formulas required will be stated and described, but the background details belong in other texts or can be found on the internet (with the usual health warning; the internet is full of glamorous nonsense).

In mechanics, the centre of mass of a rigid body is the point through which the mass appears to act. Put another way, if the mass is supported directly under its centre of mass, it balances. In two dimensions, double integration is used to calculate the centre of mass of laminas. Suppose a lamina is in the shape of a domain D, then the mass of a small rectangle of this domain is $\rho\, dx\, dy$ and the mass of the entire domain (lamina) will be the double integral

$$\iint_D \rho(x, y)dx\, dy,$$

where the dependence of the density $\rho(x, y)$ on x and y has been explicitly stated, but it is usually constant for a given material. If the position of the centre of mass of the lamina is at the point $(\bar{x}, \bar{y})$, then

$$(\bar{x}, \bar{y}) = \left(\frac{\iint_D x\rho\, dx\, dy}{\iint_D \rho\, dx\, dy}, \frac{\iint_D y\rho\, dx\, dy}{\iint_D \rho\, dx\, dy} \right).$$

There are similar formulas in plane polar co-ordinates. In vector notation and in complete generality the formula for the centre of mass of a body of volume (three-dimensional domain) V and density ρ is given by $\bar{\mathbf{r}}$ where

$$\bar{\mathbf{r}} = \frac{\iiint_V \mathbf{r}\rho\, dV}{\iiint_V \rho\, dV}$$

and this will be used once three-dimensional or triple integration has been covered. Here is an example of finding the centre of mass of a lamina.

Example 9.10   Find the centre of mass of a uniform lamina shaped as the area under the parabola $a\, y = a^2 - x^2$ and over the x-axis.

Solution: First of all, let us find the area and evaluate

$$\frac{1}{a} \int_0^a \int_{-\sqrt{a^2 - ay}}^{\sqrt{a^2 - ay}} dx\, dy = \frac{1}{a} \int_0^a 2\sqrt{a^2 - ay}\, dy,$$

which succumbs to the substitution $u = a^2 - ay$ to become

$$2 \int_0^{a^2} u^{1/2}\, du = \frac{4}{3}a^3.$$

The integral

$$\frac{1}{a} \int_0^a \int_{-\sqrt{a^2 - ay}}^{\sqrt{a^2 - ay}} a^2 - x^2\, dx\, dy,$$

which, without making a song and dance about it, also succumbs to the same substitution and gives the result $8a^4/15$. Thus,

$$\bar{y} = \frac{8}{15}a^4 / \frac{4}{3}a^3 = \frac{2}{5}a.$$

By symmetry, the centre of mass must be on the y-axis; see Figure 9.15. Therefore, $\bar{x} = 0$ and the position of the centre of mass is

$$(0, 2a/5)$$

marked as G in Figure 9.15.   ☐

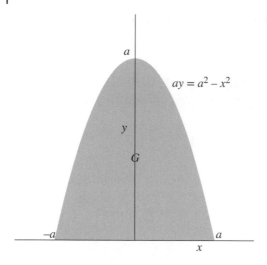

Figure 9.15 The parabolic domain.

Now, we move to probability. To give the context, suppose that there are two independent random variables X and Y. The point (X, Y) then corresponds to the outcome of a trial, and there are lots of trials. Suppose further that there is a region D of the (x, y) plane corresponding to where all the outcomes occur. The key is now to assume that if we look at a very small part of this region or domain $dx\ dy$, then there is a probability $p(x, y)dx\ dy$ that the experimental outcome is on this pixel. If all the probabilities are summed over the domain of outcomes, then we must have that

$$\iint_D p(x, y)dx\ dy = 1$$

and so the kind of problem that can be solved is different to the centre-of-mass type problem. In addition, the two quantities $E(X)$ and $E(Y)$ are called the expectation values associated with outcome X, Y. They are defined by

$$E(X) = \iint_D xp(x, y)dx\ dy \quad \text{and} \quad E(Y) = \iint_D yp(x, y)dx\ dy.$$

Here are a couple of examples that use these probabilistic notions.

Example 9.11 Suppose that the domain of the outcomes of many trials is the square $0 \le x \le 1; 0 \le y \le 1$ with the probability that the trial is at the point (x, y) is $ae^{-(x+y)}$ where a is a constant. Determine the value of a and the probability that the outcome of trials is in the square $0 \le x \le 0.5; 0 \le y \le 0.5$.

Solution: Since every outcome of the trial must lie in the square $0 \leq x \leq 1; 0 \leq y \leq 1$, the total probability must be unity. Hence,

$$\int_0^1 \int_0^1 ae^{-(x+y)} \, dx \, dy = 1$$

so

$$a \int_0^1 e^{-x} \, dx \int_0^1 e^{-y} \, dy$$
$$= a[-e^{-x}]_0^1[-e^{-y}]_0^1 = a(1 - e^{-1})(1 - e^{-1}) = 1$$

giving

$$a = \frac{e^2}{(e-1)^2} = 2.50265.$$

The required probability is given by

$$P = \int_0^{0.5} \int_0^{0.5} ae^{-(x+y)} \, dx \, dy$$
$$= a[-e^{-x}]_0^{0.5}[-e^{-y}]_0^{0.5} = a(1 - e^{-0.5})(1 - e^{-0.5}),$$

which in terms of the constant e is

$$P = e\left(\frac{\sqrt{e} - 1}{e - 1}\right)^2 = 0.38742.$$

□

This is an example of use of double integration. However, if the emphasis here was on the outcome of applying probability theory to trials, there would be a finite number of them and the probability function $p(x, y)$ would not have such a convenient closed form. The required probability would take the form of estimates and summations; nevertheless, it is useful to know that double integrals can be used here and are a good approximation when the number of trials is very large. Finally in this section, here is a more practical example that, by its very nature, involves some statistical notions that may not be familiar to some of you. If this is the case, skip it and go to the next section.

Example 9.12 In a restaurant, customers wait on average 10 min for a table. From the time they are seated to the end of the meal, it takes, on average, another 30 min. What is the probability that a customer spends over 1 h in the restaurant? You may assume that waiting time and eating time are not related. In addition, the Poisson distribution (exponential model) can be used as the model for probabilities. Calculate also the 'expected times' for waiting for a table and eating the meal.

Solution: Both 'waiting' and 'dining' are modelled with an exponential function; so from the data given, it can be assumed that the probability density functions $p_1(x)$ for waiting for a table and $p_2(y)$ for dining are as follows:

$$p_1(x) = \begin{cases} 0 & \text{if } x < 0 \\ \frac{1}{10}e^{-x/10} & \text{if } x \geq 0 \end{cases} \quad \text{and} \quad p_2(y) = \begin{cases} 0 & \text{if } y < 0, \\ \frac{1}{30}e^{-y/30} & \text{if } y \geq 0. \end{cases}$$

Since x and y are independent events, the probability for the total time of the restaurant experience is $p(x, y) = p_1(x)p_2(y)$ where

$$p(x, y) = \begin{cases} 0 & \text{if either } x < 0 \text{ or } y < 0, \\ \frac{1}{300}e^{-x/10}e^{-y/30} & \text{if } x, y \geq 0. \end{cases}$$

The total time is less than 60 min if $x + y < 60$ and this is displayed shaded in the triangle depicted in Figure 9.16. In order to answer the question, the function $p(x, y)$ has to be integrated over this shaded region. This is given by the double integral

$$P = \iint_R \frac{1}{300} e^{-x/10} e^{-y/30} \, dR = \frac{1}{300} \int_0^{60} \int_0^{60-x} e^{-x/10} e^{-y/30} \, dy \, dx$$

choosing to integrate with respect to y first. Carrying out the integration gives, without presenting all the stages,

$$P = \frac{1}{10} \int_0^{60} [e^{-x/10} - e^{-2}e^{-x/15}] dx = 1 - \frac{3}{e^2} + \frac{1}{2e^6} = 0.7982.$$

Therefore, there is a 79.82% chance of entering the restaurant and finishing the meal within 1 h. The last part of the question asks for the expectation values.

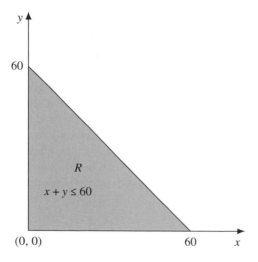

Figure 9.16 The triangular region R is shaded.

One could evaluate the two integrals:

$$E(X) = \frac{1}{300} \int_0^{60} \int_0^{60-x} xe^{-x/10}e^{-y/30} \, dy \, dx$$

and

$$E(Y) = \frac{1}{300} \int_0^{60} \int_0^{60-x} ye^{-x/10}e^{-y/30} \, dy \, dx$$

that are the statistical equivalent to the centre of mass in mechanics, but those with knowledge of this area of statistics will know that $E(X)$ is the average waiting time and $E(Y)$ the average dining time. Both of these are given in the question, so without any further calculation $E(X) = 10$ min and $E(Y) = 30$ min. You are welcome to check these using elementary integration by parts. □

The time has come to extend integration to three dimensions.

9.3   Triple Integration

The extension of double integration to triple integration is reasonably straightforward. The triple integral

$$\iiint_V f(x, y, z)dV$$

is displayed in Figure 9.17 and is written

$$\int_{x_1}^{x_2} \int_{y_1(x)}^{y_2(x)} \int_{z_1(x,y)}^{z_2(x,y)} f(x, y, z)dz \, dy \, dx.$$

With triple integration, there is even more choice of order of integration; above it is z first, then y and finally x. The technicalities are the same as for double integrals; however, it is often difficult to visualise let alone draw the three-dimensional domain. In particular, it is sometimes hard to see whether or not the region is convex. Mathematical packages such as MAPLE or MATLAB can be useful; or there is the old, tried-and-tested method of looking at slices through the solid domain. Each of these will be done in the examples that follow. If the function being integrated $f(x, y, z) = 1$, then it is the volume of the domain V that is being calculated. Integrating over a cuboid will be left to the exercises as it does not present anything very new, instead let us start with a tetrahedron shape solid with a vertex at the origin.

Example 9.13   Determine the volume of the solid bound by the planes $z + y + 2x = 4, x = 0, y = 0$ and $z = 0$ in the positive octant $x, y, z \geq 0$.

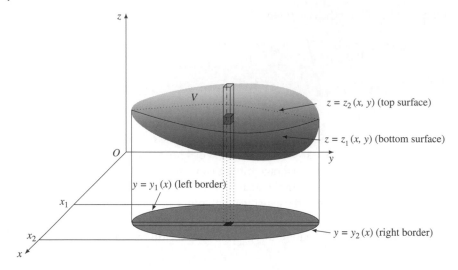

Figure 9.17 The triple integral over the volume domain V showing the limits $z_1(x, y), z_2(x, y)$ as surfaces, $y_1(x), y_2(x)$ as the left and right borders of the projection of V on to the x, y plane and, finally, x_1, x_2 as the final limits.

Solution: The plane $z + y + 2x = 4$ cuts the three co-ordinate axes x, y and z at the points $x = 2, y = 4$ and $z = 4$ respectively. This is the shape of a plane that cuts diagonally through the three axes at the points $(2, 0, 0), (0, 4, 0)$ and $(0, 0, 4)$ and is shown in Figure 9.18. The triple integral can be written

$$V = \int_0^2 \int_0^{4-2x} \int_0^{4-2x-y} dz \, dy \, dx,$$

which integrates at once with respect to z to give

$$V = \int_0^2 \int_0^{4-2x} (4 - 2x - y) dy \, dx$$

and the calculation proceeds

$$V = \int_0^2 \left[(4 - 2x)y - \frac{1}{2}y^2 \right]_0^{4-2x}$$

$$= \frac{1}{2} \int_0^2 (4 - 2x)^2 dx$$

$$= -\frac{1}{12} [(4 - 2x)^3]_0^2$$

$$= \frac{64}{12} = \frac{16}{3},$$

where some of the elementary steps have been omitted. Some of you might have difficulty with the limits of the second integration, if not in this example

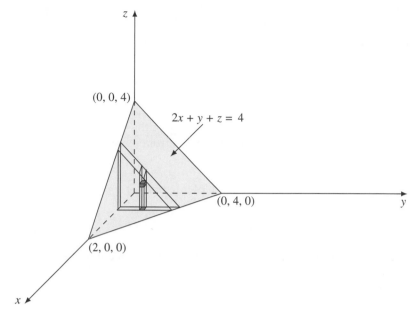

Figure 9.18 The tetrahedral region bounded by the planes $2x + y + z = 0, x = 0, y = 0$ and $z = 0$.

maybe in one that's more complex. Once the integration with respect to z has occurred, only x and y remain and visualisation can be helped by projecting the volume on to the x, y plane. Think of a beam of light parallel to the z-axis and the shadow cast by V on to the x, y plane. This can be seen in Figure 9.18 but it is shown more clearly in Figure 9.19. In fact, if there is difficulty actually drawing the equivalent of Figure 9.18 due to the complexity of the definition of the volume domain, then it is worth trying to draw the plan view (Figure 9.19)

Figure 9.19 The x, y plane for integration once the integration with respect to z has taken place. The equation for the line is $2x + y = 4$ obtained by putting $z = 0$ the equation of the x, y plane into the original equation for the plane $2x + y + z = 4$.

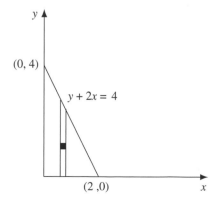

and the elevation in the y, z plane. More of this will be covered later. To finish this example, as the shape is a pyramid on its side, this result can be checked using the formula 'volume $= \frac{1}{3}$ base area $\times$ height' whence

$$V = \frac{1}{3} \cdot \frac{1}{2} \cdot 2 \times 4 \times 4 = \frac{16}{3}$$

as before. □

The next example tackles the question of changing the order of integration around as well as interpreting what the domain is from a knowledge of all the limits.

Example 9.14   Sketch the volume domain of integral

$$\int_{x=0}^{x=4} \int_{y=0}^{y=\sqrt{16-x^2}} \int_{z=(x^2+y^2)/4}^{z=4} dz \, dy \, dx$$

and reform the integral in the order y then x then z, and again in the order z then y then x.

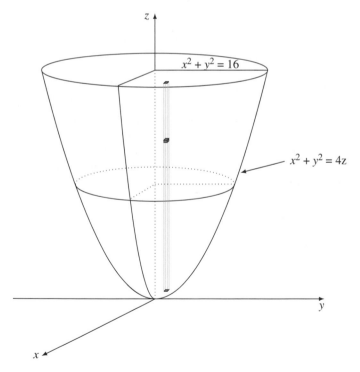

Figure 9.20 The domain is the quarter of this paraboloid that is in the first octant $(x, y, z \geq 0)$; and the first z direction integration is shown.

Solution: The domain of the triple integral is displayed in Figure 9.20. The first limits are in terms of z, and are $4z = x^2 + y^2$ at the bottom and $z = 4$ at the top. Once this has taken place, the x, y integration is over the circle $x^2 + y^2 = 16$ but restricted to $y \geq 0$ and $x \geq 0$ which means only a quarter of the circle is covered. The domain of the triple integral is thus a quarter of the paraboloid indicated in Figure 9.20. In this figure, the given order is indicated in that the z integral is done first. This leaves the integration in the y direction, and this is achieved by considering the x, y plane only where the domain is the circle $x^2 + y^2 = 16, x \geq 0, y \geq 0$ shown in Figure 9.21. The y limits are therefore $y = 0$ to $y = \sqrt{16 - x^2}$ and, finally the x limits are $x = 0$ to $x = 4$. Evaluating the integral is not asked for; instead this is an exercise in changing the order. This is easier once the domain of the integral has been found (the quarter paraboloid shown in Figure 9.20). The next order required is y then x then z. Take a look at Figure 9.20 once again. It will be seen that the path through the tiny cube in the y direction starts at $y = 0$ and ends on the positive surface of the paraboloid. This is obtained by making y the subject of the formula for the paraboloid $4z = x^2 + y^2$, that is $y = \sqrt{4z - x^2}$ so $y = 0$ and $y = \sqrt{4z - x^2}$ are the inner limits. Once y has been integrated, only x and z remain. The shape of the region is that bounded above by the parabola $x^2 = 4z$ and below by the z-axis which, as there is no longer any y in the picture, $x = 0$. Finally, the last integration with respect to z must be from $z = 0$ to $z = 4$. All this can be seen from Figure 9.20. The triple integral is thus

$$\int_0^4 \int_0^{2\sqrt{z}} \int_0^{\sqrt{4z-x^2}} dy \, dx \, dz.$$

Figure 9.21 The projection of the paraboloid $4z = x^2 + y^2$ on to the plane $z = 4$ together with indication of the order of integration.

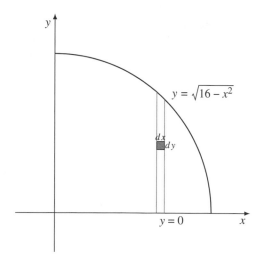

The last ordering is x then z then y. As the paraboloid is axisymmetric, integrating x first and integrating y first is the same algebra; simply exchange the roles of x and y thus obtaining the limits $x = 0$ and $x = \sqrt{4z - y^2}$. The integration with respect to z is in the y, z plane where the region is between the parabola $y^2 = 4z$ and the line $z = 4$. In Figure 9.20, it is as if the small black volume and its associated vertical tramlines have been pushed against the y, z plane (algebraically put $x = 0$). The limits are thus $z = y^2/4$ and $z = 4$. Finally, the x limits are $x = 0$ to $x = 4$ resulting in the triple integral

$$\int_0^4 \int_{y^2/4}^4 \int_0^{\sqrt{4z-y^2}} dx \, dz \, dy.$$

It is not part of the example to evaluate this triple integral, but please do have a go at it. The second is the easiest to find and the answer (to all three) is 8π. This fact can be easily verified: at an arbitrary height z the cross-section has area πy^2 (or πx^2 as $x = y$ on the surface). Therefore, the total area of the paraboloid must be

$$\int_0^4 \pi y^2 \, dz = 4\pi \int_0^4 z \, dz$$

as $y^2 = 4z$ on the surface. The right-hand integral is simply $4\pi \times 2 \times 2^2 = 32\pi$, a quarter of this is 8π which checks the above result. Finally in this section, here is an example of a triple integral where decisions have to be made on the order.

Example 9.15   Find the volume of the solid in the first octant bounded by the surfaces $z = 1 - y^2, y = 2x$ and $x = 3$.

Solution: The volume is given by the sliced cylinder shown in Figure 9.22 it is a parabolic cylinder $z = 1 - y^2$ with x arbitrary, cut at right angles by the plane $x = 3$ and at an angle from the z axis by the plane $y = 2x$. Figure 9.22 shows these details, and tells us that we have to be careful in our choice of order of integration. For example, integrating with respect to y first will not do as this would mean dividing the region, that nearest the origin would have $y = 2x$ as the upper limit, whereas the awkward $y = \sqrt{1 - z}$ would be the upper limit after the intersection of the parabolic cylinder with the plane $y = 2x$. To avoid such complications, choose to integrate with respect to z first. The limits are from $z = 0$ to $z = 1 - y^2$. The x, y area is then the projection of the solid on to the x, y plane which is a trapezium. This is also outlined in Figure 9.22. Again, to avoid a split integral, x has to be chosen next and the limits are the end lines of the trapezium, $x = y/2$ and $x = 3$. Finally, these lines are taken from $y = 0$ to $y = 1$ and the integration is complete. The triple integral in the order z then x then y

Figure 9.22 The parabolic cylinder $z = 1 - y^2$ cut by the two planes $x = 3$ and $y = 2x$. The black lined trapezium is the planar region of integration once the z integration has been completed.

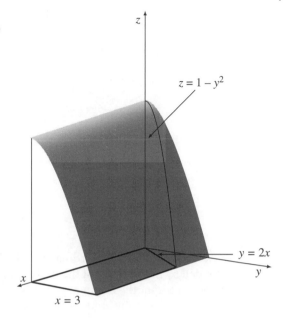

is thus

$$V = \int_0^1 \int_{y/2}^3 \int_0^{1-y^2} dz \, dx \, dy$$

and the details of the integration are not remarkable.

$$V = \int_0^1 \int_{y/2}^3 (1 - y^2) dx \, dy$$

$$= \int_0^1 [x - xy^2]_{y/2}^3 dy$$

$$= \int_0^1 \left[3 - 3y^2 - \frac{1}{2}y + \frac{1}{2}y^3 \right] dy$$

$$= \left[3y - y^3 - \frac{1}{4} + \frac{1}{6}y^4 \right]_0^1$$

$$= 3 - 1 - \frac{1}{4} + \frac{1}{6} = \frac{15}{8}.$$ □

9.3.1 Cylindrical and Spherical Polar Co-ordinates

Many three-dimensional problems that involve triple integrals are conveniently posed and subsequently solved in polar co-ordinates. In three dimensions,

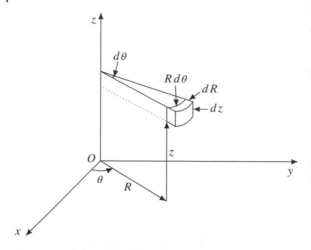

Figure 9.23 Cylindrical polar co-ordinates (R, θ, z) together with the element of volume.

there are two different polar co-ordinates: cylindrical where plane polars are simply extended by the addition of the z-axis and spherical that consists of the radial distance out from the origin together with two angles. Both of these are dealt with here; let us first introduce cylindrical polar co-ordinates. Figure 9.23 shows the co-ordinates and the infinitesimal volume. From this figure, we can see that the Cartesian $dx\,dy\,dz$ is replaced by $R\,dR\,d\theta\,dz$; but we need not take this just from the figure, as the relationship between Jacobians and differentials can be derived with the knowledge given in Section 9.2.3. The required relationship here is

$$dx\,dy\,dz = \frac{\partial(x, y, z)}{\partial(R, \theta, z)} dR\,d\theta\,dz.$$

Therefore, the relationship between (x, y, z) and (R, θ, z) must be used to find the Jacobian. Fortunately, it is easy as cylindrical polar co-ordinates are simply plane polar co-ordinates but with z added. Note that the lower-case r has been replaced by the upper-case R. This is not universally done, but is preferred here to avoid confusion with r being used as the distance from the origin in any dimension. In particular, it relates to the confusion with its use in spherical polar co-ordinates that comes next. To derive the relationship required between the infinitesimal volumes, we need the relationships between r, θ, z and x, y, z. For cylindrical polar co-ordinates, this is simply plane polar co-ordinates with the addition of $z = z$. Thus,

$$x = R\cos\theta, \quad y = R\sin\theta, \quad z = z$$

so that

$$\frac{\partial(x, y, z)}{\partial(R, \theta, z)} = \begin{vmatrix} \frac{\partial x}{\partial R} & \frac{\partial y}{\partial R} & \frac{\partial z}{\partial R} \\ \frac{\partial x}{\partial \theta} & \frac{\partial y}{\partial \theta} & \frac{\partial z}{\partial \theta} \\ \frac{\partial x}{\partial z} & \frac{\partial y}{\partial z} & \frac{\partial z}{\partial z} \end{vmatrix} = \begin{vmatrix} \cos\theta & \sin\theta & 0 \\ -R\sin\theta & R\cos\theta & 0 \\ 0 & 0 & 1 \end{vmatrix} = R$$

so establishing what is seen in Figure 9.23 that $dx\,dy\,dz = R\,dR\,d\theta\,dz$. Here is a reasonably straightforward example using cylindrical polar co-ordinates.

Example 9.16   Determine the volume common to the two intersecting cylinders $y^2 + z^2 = a^2$ and $x^2 + z^2 = a^2$.

Solution: The two cylinders $x^2 + z^2 = a^2$ and $y^2 + z^2 = a^2$ intersect along the plane $x = y$. This is seen simply through the algebra of equating the two surfaces. The actual curve of intersection is, of course, not the whole plane but an ellipse in that plane. A quarter of this ellipse is shown in Figure 9.24. The problem of finding the volume of intersection is best tackled by finding the volume in the positive octant where $x, y, z \geq 0$, in particular, in one half of this to avoid splitting the integration. Consider the left half of Figure 9.24 and integrate with respect to z first. Then, integrate from the plane $y = 0$ to the diving plane $y = x$ (the triangle is marked in Figure 9.24) and finally from $x = 0$ to $x = a$; thus,

$$\frac{V}{16} = \int_0^a \int_0^x \int_0^{\sqrt{a^2-x^2}} dz\,dy\,dx,$$

where V is the required volume. The calculation is again unremarkable:

$$\frac{V}{16} = \int_0^a \int_0^x \sqrt{a^2 - x^2}\,dy\,dx = \int_0^a x\sqrt{a^2 - x^2}\,dx = \int_0^a u^2\,du$$

writing $u^2 = a^2 - x^2$ and converting $x\,dx = -u\,du$ and limits (then swapping them around to eliminate the minus sign). This integral has the value $a^3/3$, whence

$$V = \frac{16}{3}a^3$$

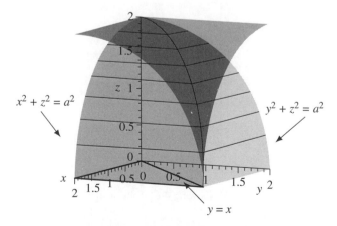

Figure 9.24 Two intersecting cylinders $x^2 + z^2 = a^2$ and $y^2 + z^2 = a^2$ drawn with $a = 2$ showing the intersection $y = x$ which is a quarter ellipse in this first octant $x, y, z \geq 0$. The plane triangular region corresponding to the domain of the x, y plane covered by the last two integrations is outlined.

is remarkable as it does not contain the transcendental number π despite being a volume seemingly contained in two circles. The volume inside three intersecting cylinders is more challenging and is in an exercise. Try it before looking at the solution online or in the solution section of the book. □

The second three-dimensional polar co-ordinate system, spherical polars, is now introduced. This is drawn in Figure 9.25 with the three co-ordinates r, ϕ, θ labelled as well as the infinitesimal volume. This volume has the magnitude $r^2 \sin \phi \, dr \, d\phi \, d\theta$ which will be proved shortly. First, a word about notation: as for cylindrical polars, there are differences. In this text, θ is retained as the angle in the x, y plane, sometimes referred to as the azimuthal angle, and ϕ is used as the angle measured down from the z-axis. This second angle, sometimes called the polar angle, gets the symbol ϕ but in alternative notation, this is often called λ or other Greek letter (sadly, θ can be used which is very confusing as this means ϕ and θ swap places). Finally, some books replace r by ρ again to distinguish it from planar uses. The notation adopted here is the most frequently used. The spherical polar system is almost the geographical system of position, latitude and longitude, except that latitude ($\pi/2 - \phi$) is replaced by co-latitude (ϕ) in order to preserve a right-handed system. Looking carefully at Figure 9.25, on the x, y plane the right-angled triangle has hypotenuse $r \sin \phi$ so

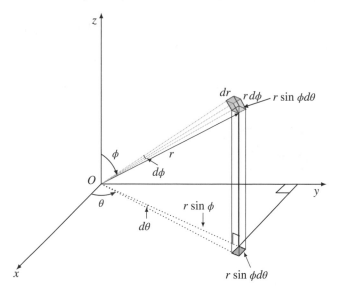

Figure 9.25 Spherical polar co-ordinates (r, ϕ, θ), showing the infinitesimal volume element of sides $dr, r \, d\phi$ and $r \sin \phi \, d\theta$.

that the sides are $r \cos \theta \sin \phi$ and $r \sin \theta \sin \phi$; hence, the relationships between (r, ϕ, θ) and (x, y, z) are

$$x = r \cos \theta \sin \phi, \quad y = r \sin \theta \sin \phi \quad \text{and} \quad z = r \cos \phi.$$

Now, we can use the Jacobian relationship

$$dx\, dy\, dz = \frac{\partial(x, y, z)}{\partial(r, \phi, \theta)} dr\, d\phi\, d\theta$$

as in the derivation of cylindrical polar's infinitesimal volume. The Jacobian is calculated first:

$$\frac{\partial(x, y, z)}{\partial(r, \phi, \theta)} = \begin{vmatrix} \frac{\partial x}{\partial r} & \frac{\partial y}{\partial r} & \frac{\partial z}{\partial r} \\ \frac{\partial x}{\partial \phi} & \frac{\partial y}{\partial \phi} & \frac{\partial z}{\partial \phi} \\ \frac{\partial x}{\partial \theta} & \frac{\partial y}{\partial \theta} & \frac{\partial z}{\partial \theta} \end{vmatrix} = \begin{vmatrix} \cos \theta \sin \phi & \sin \theta \sin \phi & \cos \phi \\ r \cos \theta \cos \phi & r \sin \theta \cos \phi & -r \sin \phi \\ -r \sin \theta \sin \phi & r \cos \theta \sin \phi & 0 \end{vmatrix}.$$

Multiplying the determinant on the right using the bottom row gives

$$\frac{\partial(x, y, z)}{\partial(r, \phi, \theta)} = -r \sin \theta \sin \phi(-r \sin \theta \sin^2\phi - r \sin \theta \cos^2\phi)$$
$$- r \cos \theta \sin \phi(-r \cos \theta \sin^2\phi - r \cos \theta \cos^2\phi),$$

so using $\cos^2\phi + \sin^2\phi = 1$ and then $\cos^2\theta + \sin^2\theta = 1$, this simplifies to

$$\frac{\partial(x, y, z)}{\partial(r, \phi, \theta)} = r^2 \sin \phi,$$

whence

$$dx\, dy\, dz = r^2 \sin \phi\, dr\, d\theta\, d\phi$$

the same as seen geometrically in Figure 9.25. We will now do a couple of examples.

Example 9.17 Find the volume bounded above by the cone $x^2 + y^2 = z^2$, below by the plane $z = 0$ and at the sides by the sphere $x^2 + y^2 + z^2 = a^2$.

Solution: Figure 9.26 shows the cone and the sphere together with spherical polar co-ordinates (r, ϕ, θ). It is for problems that are axisymmetric that spherical polar co-ordinates are ideal; make the z-axis the axis of symmetry as here. If there are any straight lines parallel to x, y or z, then it would be wise to try either Cartesian or cylindrical polars first. In this problem, the cone makes an angle of $\pi/4$ both with the plane $z = 0$ and the z-axis. Start by integrating with

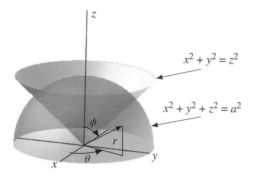

Figure 9.26 The cone and sphere together with spherical polar co-ordinates.

respect to θ first all the way round from 0 to 2π. Then, integrate radially from $r = 0$ out to the surface of the sphere at $r = a$. Finally, complete the integration over the volume by integrating with respect to ϕ from $\pi/4$ to $\pi/2$. The limits are all constant; a very good indication that the correct co-ordinate system has been chosen.

$$V = \int_{\pi/4}^{\pi/2} \int_0^a \int_0^{2\pi} r^2 \sin\phi \, d\theta \, dr \, d\phi.$$

The calculation is not difficult; just remember that the infinitesimal volume is $r^2 \sin\phi \, d\theta \, dr \, d\phi$ for spherical polar co-ordinates:

$$V = 2\pi \int_{\pi/4}^{\pi/2} \int_0^a r^2 \sin\phi \, dr \, d\phi$$

$$= 2\pi \int_{\pi/4}^{\pi/2} \left[\frac{1}{3}r^3\right]_0^a \sin\phi \, d\phi = 2\pi \frac{a^3}{3}[-\cos\phi]_{\pi/4}^{\pi/2}$$

and this gives

$$V = \frac{a^3 \pi \sqrt{2}}{3}.$$

$\square$

Here is another problem that is best solved using spherical polar co-ordinates. There is a reasonable likelihood that many of you have not been exposed to much co-ordinate geometry as this seems to be out of fashion these days. Most know that a sphere with centre at the origin $(0, 0, 0)$ and radius a has equation $x^2 + y^2 + z^2 = a^2$. Again, maybe you know that if the centre of the sphere is at (x_0, y_0, z_0) then the sphere with this centre, radius a has the equation

$$(x - x_0)^2 + (y - y_0)^2 + (z - z_0)^2 = a^2.$$

Multiplying this out gives the messier equation

$$x^2 + y^2 + z^2 - 2xx_0 - 2yy_0 - 2zz_0 = a^2 - x_0^2 - y_0^2 - z_0^2,$$

and as a guess, presented with this equation many would not immediately recognise it as representing a sphere. With the constants x_0, y_0, z_0 and a as numbers, the radius and centre would be obscured. The key is that it is quadratic, and the coefficients of x^2, y^2 and z^2 are all unity (and the same sign). This is the signature of a sphere, though in the form

$$(x - x_0)^2 + (y - y_0)^2 + (z - z_0)^2 = a^2,$$

it is much more useful as the centre and radius are explicit. With this knowledge, the equation $x^2 + y^2 + z^2 = 2az$ is recognised as a sphere, and with a slight re-arrangement into

$$x^2 + y^2 + (z - a)^2 = a^2,$$

it is seen that the sphere has radius a but is centred just along the z-axis from the origin at $(0, 0, a)$. The next example needs this information:

Example 9.18 Find using spherical polar co-ordinates and triple integration the volume and centre of mass of the uniform solid above the cone $x^2 + y^2 = z^2$ but beneath the sphere $x^2 + y^2 + z^2 = 2az$.

Solution: As the equations are in Cartesians and we need limits in spherical polar co-ordinates, let us convert. The cone $x^2 + y^2 = z^2$ is given by $\phi = \pi/4$. If this passes you by, write $x = r \cos \theta \sin \phi$ and $y = r \sin \theta \sin \phi$ so that $x^2 + y^2 = r^2 \sin^2 \phi$ and since $z^2 = r^2 \cos^2 \phi$ the cone $x^2 + y^2 = z^2$ transforms to $r^2 \sin^2 \phi = r^2 \cos^2 \phi$ which as $r \neq 0$ means $\tan^2 \phi = 1$ so $\phi = \pi/4$ or $\phi = 5\pi/4$ which is a cone. Theoretically, a cone extends indefinitely in both directions, but here it is the downward-pointing vertex at the origin and the plane at $z = 1$ that marks the horizontal diameter of the (hemi)sphere that mark its limits. See the Figure 9.27. The sphere $x^2 + y^2 + z^2 = 2az$ in spherical co-ordinates is

Figure 9.27 The hemisphere $r = 2a \cos \phi$ on the cone $\phi = \pi/4$.

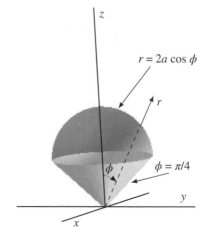

$r = 2a \cos \phi$

r

$\phi = \pi/4$

$r^2 = 2ar \cos \phi$ or $r = 2a \cos \phi$ which is a simpler equation than we had the right to expect. The volume of the solid shown in this figure can thus be written in spherical co-ordinates integrating in the order r, ϕ then θ as follows:

$$V = \int_0^{2\pi} \int_0^{\pi/4} \int_0^{2a \cos \phi} r^2 \sin \phi \, dr \, d\phi \, d\theta.$$

The steps of the integration are as follows:

$$V = \int_0^{2\pi} \int_0^{\pi/4} \left[\frac{r^3}{3} \sin \phi \right]_0^{2a \cos \phi} d\phi \, d\theta$$

$$= \int_0^{2\pi} \int_0^{\pi/4} \frac{8a^3}{3} \cos^3 \phi \sin \phi \, d\phi \, d\theta$$

$$= \int_0^{2\pi} \left[-\frac{8a^3}{12} \cos^4 \phi \right]_0^{\pi/4} d\theta$$

$$= \int_0^{2\pi} \frac{2a^3}{3} \left[1 - \frac{1}{4} \right] d\theta$$

$$= \pi a^3.$$

This result is conveniently checked; the volume of the top hemisphere (the ice cream part) is $\frac{2}{3}\pi a^3$ and that of the cone wafer is $\frac{1}{3}\pi a^3$. The sum is thus πa^3.

For the centre of mass, the limits of the integration are of course the same but the integrand contains an additional $z = r \cos \phi$ whence using the general formula

$$\bar{z} = \frac{\iiint_V z \, dV}{\iiint_V dV}$$

results in

$$\bar{z} = \frac{\int_0^{2\pi} \int_0^{\pi/4} \int_0^{2a \cos \phi} r \cos \phi r^2 \sin \phi \, dr \, d\phi \, d\theta}{\int_0^{2\pi} \int_0^{\pi/4} \int_0^{2a \cos \phi} r^2 \sin \phi \, dr \, d\phi \, d\theta}.$$

The denominator has been found as πa^3 and the calculation for the numerator now follows:

$$\int_0^{2\pi} \int_0^{\pi/4} \int_0^{2a \cos \phi} r^3 \cos \phi \sin \phi \, dr \, d\phi \, d\theta$$

$$= \int_0^{2\pi} \int_0^{\pi/4} \left[\frac{r^4}{4} \right]_0^{2a \cos \phi} \cos \phi \sin \phi \, dr \, d\phi \, d\theta$$

$$= \int_0^{2\pi} \int_0^{\pi/4} \frac{(2a)^4}{4} \sin \theta \cos^5 \phi \, d\phi \, d\theta$$

$$= 8a^4 \pi \left[-\frac{\cos^6 \phi}{6} \right]_0^{\pi/4}$$

$$= \frac{4}{3}a^4 \left[1 - \frac{1}{8}\right]$$

$$= \frac{7}{6}\pi a^4,$$

which gives

$$\bar{z} = \frac{7}{6}a,$$

so the centre of mass is at $\left(0, 0, \frac{7}{6}a\right)$ by reasons of symmetry. This is a good example of using spherical polar coordinates to find a volume, but reverting to Cartesian coordinates for finding the centre of mass. This has to be done in the majority of cases, as the definition of centre of mass depends upon moments that are essentially Cartesian. □

9.3.2 Applications of Triple Integration

As for double integrals, the applications of triple integration are in mechanics and probability. In the section on the applications of double integration (Section 9.2.4), the general formula for calculating the centre of mass

$$\bar{\mathbf{r}} = \frac{\iiint_V \mathbf{r}\rho \, dV}{\iiint_V \rho \, dV}$$

was introduced. Now that three-dimensional integration has been covered, this formula can be used properly. Let us revisit the solid of Figure 9.18; the volume has been found as 16/3 and so the denominator for this equation to find the centre of mass has already been calculated. The three co-ordinates of the centre of mass $(\bar{x}, \bar{y}, \bar{z})$ are given by the three equations:

$$\bar{x} = \frac{3}{16} \int_0^2 \int_0^{4-2x} \int_0^{4-2x-y} x \, dz \, dy \, dx,$$

$$\bar{y} = \frac{3}{16} \int_0^2 \int_0^{4-2x} \int_0^{4-2x-y} y \, dz \, dy \, dx,$$

$$\bar{z} = \frac{3}{16} \int_0^2 \int_0^{4-2x} \int_0^{4-2x-y} z \, dz \, dy \, dx,$$

these will now be calculated:

$$\int_0^2 \int_0^{4-2x} \int_0^{4-2x-y} x \, dz \, dy \, dx = \int_0^2 \int_0^{4-2x} x(4 - 2x - y)dy \, dx$$

$$= \int_0^2 \left[4xy - 2x^2y - \frac{1}{2}xy^2\right]_0^{4-2x} dx$$

$$= \int_0^2 (8x - 8x^2 + 2x^3)dx$$

$$= 4(2)^2 - \frac{8}{3}(2)^3 + \frac{1}{2}(2)^4$$

$$= 24 - \frac{64}{3} = \frac{8}{3}.$$

Therefore,

$$\bar{x} = \frac{3}{16} \times \frac{8}{3} = \frac{1}{2}.$$

Similarly,

$$\int_0^2 \int_0^{4-2x} \int_0^{4-2x-y} y \, dz \, dy \, dx = \int_0^2 \int_0^{4-2x} y(4 - 2x - y) dy \, dx$$

$$= \int_0^2 \left[2y^2 - xy^2 - \frac{1}{3}y^3 \right]_0^{4-2x} dx$$

$$= \int_0^2 (2 - x)(4 - 2x)^2 - \frac{1}{3}(4 - 2x)^3 dx.$$

This last integral is best solved by substituting $u = 4 - 2x$ that gives $du = -2 \, dx$ and the limits $u = 4, 0$. In addition, $2 - x = \frac{1}{2}u$ whence

$$\int_0^2 \int_0^{4-2x} \int_0^{4-2x-y} y \, dz \, dy \, dx = \frac{1}{2} \int_0^4 \left\{ \frac{1}{2}u^3 - \frac{1}{3}u^3 \right\} du$$

$$= \left[\frac{1}{48}u^4 \right]_0^4 = \frac{16}{3}.$$

This gives

$$\bar{y} = \frac{3}{16} \times \frac{16}{3} = 1.$$

Finally,

$$\int_0^2 \int_0^{4-2x} \int_0^{4-2x-y} z \, dz \, dy \, dx = \int_0^2 \int_0^{4-2x} \frac{1}{2}(4 - 2x - y)^2 dy \, dx$$

and these integrals also need substitution: $v = 4 - 2x - y$ gives $dv = -dy$ but the limits simply swap and cancel the minus sign, whence

$$\bar{z} = \frac{3}{16} \int_0^2 \int_0^{4-2x} \frac{1}{2}v^2 \, dv \, dx = \frac{1}{32} \int_0^2 (4 - 2x)^3 dx = \frac{1}{64} \frac{(4^4)}{4} = 1,$$

where the substitution $u = 4 - 2x$ can be used again for the final integration if desired. Check all this with MAPLE or similar software, but doing it by hand cements better understanding. The centre of mass of the tetrahedron of Figure 9.18 is thus located at $(\frac{1}{2}, 1, 1)$ verified by using geometry. The centre of mass of a homogeneous pyramid whose vertices have Cartesian co-ordinates $(x_1, y_1, z_2), (x_2, y_2, z_2), (x_3, y_3, z_3)$ and (x_4, y_4, z_4) is the point with co-ordinates

$$\left(\frac{x_1 + x_2 + x_3 + x_4}{4}, \frac{y_1 + y_2 + y_3 + y_4}{4}, \frac{z_1 + z_2 + z_3 + z_4}{4} \right).$$

One may ask whether one always has to do the integrals as was done here. There are shortcuts; the use of symmetry will come into play next, and here, since two of the vertices were $(0, 0, 4)$ and $(0, 4, 0)$, one might expect the y and z co-ordinates of the centre of mass to be the same. Since the other co-ordinate along the x-axis is $(2, 0, 0)$, the x co-ordinate of the centre of mass should be half those of the y and z co-ordinate. This way, the amount of calculation required could be reduced by two-thirds. The message is be on the sharp lookout for shortcuts or checks as we are all susceptible to making errors.

Let us now calculate the centre of mass of the solid displayed in Figure 9.20. Although the original problem involved that part of the solid in the first octant, here the use of symmetry is being emphasised and the centre of mass of the paraboloid displayed must lie on the z-axis, and all that has to be done is to find how far along this axis the centre of mass is. The volume is 32π, so the centre of mass will have co-ordinates $(0, 0, \bar{z})$ and $\bar{z}$ is given by

$$\bar{z} = \frac{1}{32\pi} \int_0^4 \int_{-2\sqrt{z}}^{2\sqrt{z}} \int_{-\sqrt{4z-x^2}}^{\sqrt{4z-x^2}} z \, dy \, dx \, dz$$

so that

$$\bar{z} = \frac{1}{32\pi} \int_0^4 \int_{-2\sqrt{z}}^{2\sqrt{z}} 2z\sqrt{4z - x^2} \, dx \, dz.$$

Substitute $x = 2\sqrt{z} \sin \theta$ to eliminate the square root $\sqrt{4z - x^2} = 2\sqrt{z} \cos \theta$ in the x integral so that $dx = 2\sqrt{z} \cos \theta \, d\theta$ and the limits are $-\pi/2$ to $\pi/2$. Thus, the above integral becomes

$$\bar{z} = \frac{1}{32\pi} \int_0^4 8z^2 \int_{-\pi/2}^{\pi/2} \cos^2\theta \, d\theta \, dz.$$

Most of you will know that the $\cos^2\theta$ integral is $\frac{1}{2}\theta - \frac{1}{4}\sin 2\theta$ using the double-angle formula, so that

$$\bar{z} = \frac{1}{32\pi} \int_0^4 4\pi z^2 \, dz = \frac{1}{32\pi} \cdot 4\pi \frac{(4)^3}{3} = \frac{256\pi}{32\pi \times 3} = \frac{8}{3}.$$

Thus $\bar{z} = \frac{8}{3}$ and the centre of mass is just over halfway up the z-axis.

Although cylindrical and spherical polar co-ordinates have been covered, in order to find the centre of mass of a three-dimensional body, Cartesian co-ordinates are invariably used. This is because these are the natural way to reference a point in space; even most professionals are hard-pressed to recognise other co-ordinate systems. Having said this, the use of cylindrical and spherical polars is common for the actual calculations. Here is an example of this.

Example 9.19 Calculate the position of the centre of mass of a hemisphere $r = a, 0 \leq \phi \leq \pi/2, 0 \leq \theta \leq 2\pi$.

Solution: By symmetry, the centre of mass is on the z-axis, but in spherical polar co-ordinates $z = r \cos \phi$. The volume of a hemisphere is $\frac{2}{3}\pi a^3$; thus, we evaluate the quantity

$$\bar{z} = \frac{3}{2\pi a^3} \int_0^{2\pi} \int_0^{\pi/2} \int_0^a (r \cos \phi) r^2 \sin \phi \, dr \, d\phi \, d\theta,$$

which is not a difficult calculation:

$$\bar{z} = \frac{3}{2\pi a^3} \int_0^{2\pi} \int_0^{\pi/2} \frac{1}{4} a^4 \cos \phi \sin \phi \, d\phi \, d\theta$$

so

$$\bar{z} = \frac{3}{2\pi a^3} \times 2\pi \times \frac{a^4}{8} = \frac{3}{8} a,$$

a well-known result. Here is a different application that involves the concept of *moment of inertia I*. □

Definition 9.1 The *moment of inertia, I* of a body about a fixed axis is defined by the integral

$$I = \int_V \rho d^2 \, dV,$$

where ρ is the density of the body and d is the perpendicular distance of the infinitesimal volume dV of the body from the fixed axis.

This definition is completely general and can apply to laminas as well as solids of varying density. The volume V does, however, have to be fixed, so plasticity and flow are disallowed.

Example 9.20 Find the moment of inertia of a solid sphere radius a about any diameter. The mass of the sphere is M.

Solution: The sphere has perfect symmetry so let us take the z-axis as our chosen diameter. The distance d is in the x, y plane or parallel to it, therefore $d = r \sin \phi$ in terms of spherical polar co-ordinates. Therefore, the integral

$$I = \int_0^{2\pi} \int_0^{\pi} \int_0^a \rho(r \sin \phi)^2 r^2 \sin \phi \, dr \, d\phi \, d\theta$$

needs to be evaluated where we have written ρ for the constant density. The key integral that's a bit tricky amongst the three here is

$$\int_0^\pi \sin^3 \phi \, d\phi = \int_0^\pi (\cos^2 \phi - 1) d(\cos \phi) = \left[\frac{1}{3} \cos^3 \phi - \cos \phi \right]_0^\pi = \frac{4}{3}.$$

Thus, as the limits are constant, we have the product

$$I = 2\pi \times \frac{a^5}{5} \times \frac{4}{3} \rho = \frac{8\pi}{15} \rho a^5.$$

There's nothing wrong with this answer, but we are given the symbol M for the mass and as mass is density times volume, we have

$$M = \rho \frac{4}{3} \pi a^3$$

so that

$$\rho = \frac{3M}{4a^3 \pi} \quad \text{giving} \quad I = \frac{8\pi}{15} a^5 \times \frac{3M}{4a^3 \pi} = \frac{2}{5} M a^2,$$

which is the standard result in mechanics. □

For double integration, a probability example involving two independent random variables was chosen. It is not hard to extend this idea to three variables, for example, a journey involving waiting for a train, the train journey and a bike ride, which is followed by the question: What is the probability that the whole journey is less than a certain time? One such question is posed in the exercises. Instead, the following slightly different one is chosen:

Example 9.21 Let $f(x) = ax^2 + bx + c$ where the constants a, b, c are random variables with values any real number between 0 and 1. What is the probability that the quadratic $f(x)$ has real roots?

Solution: The three dimensions are formed by the ranges of probability of a, b and c. Assign $z = b, x = a$ and $y = c$. The condition that $b^2 > 4ac$ therefore translates to $z^2 > 4xy$ and this surface is shown in Figure 9.28. Therefore, the simple cubic region is sliced by the surface $z^2 = 4xy$ that goes under $z = 1$ reducing our domain to the volume over the curved surface but under the plane $z = 1$. The $1 \times 1 \times 1$ cube is the total probability as it captures any choice of a, c and b (x, y or z); but the required probability is that part of the cube's volume that is over the curved surface $z^2 = 4xy$ but inside the unit cube (under the plane $z = 1$), and it is this that has to be captured by a triple integral. It turns out this one is a bit tricky. The range of z is reasonable enough and it starts at the surface $z^2 = 4xy$ and ends at the plane $z = 1$. However, the region of integration for x, y is shown in Figure 9.29 and no matter which order is taken, the region has to be split. The curved surface $z^2 = 4xy$ meets $z = 1$ where $4xy = 1$, and it is this rectangular hyperbola that is shown in Figure 9.29. The region in

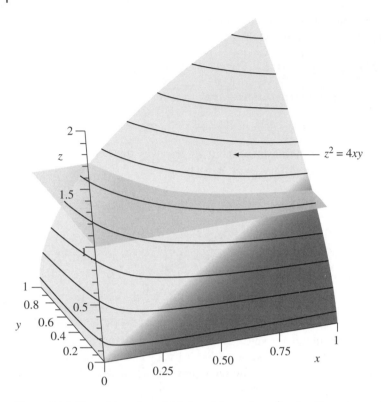

Figure 9.28 The cubic region of side 1 and the surface $z^2 = 4xy$. The volume of the domain that needs to be found is within the cube but above the surface where $z^2 \geq 4xy$.

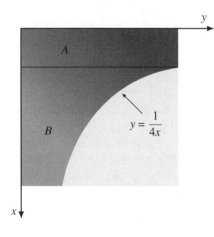

Figure 9.29 The x, y plane and the area over which the last two integrals span; it has to be split into a rectangle A and a region bordered by $y = \frac{1}{4x}$, labelled B.

the x, y plane is divided into the rectangle $0 \le x \le \frac{1}{4}; 0 \le y \le 1$ and the region $\frac{1}{4} \le x \le 1; 0 \le y \le \frac{1}{4x}$ resulting in the required probability being the sum of the two triple integrals:

$$P = \int_0^{1/4} \int_0^1 \int_{\sqrt{4xy}}^1 dz \, dy \, dx + \int_{1/4}^1 \int_0^{\frac{1}{4x}} \int_{\sqrt{4xy}}^1 dz \, dy \, dx.$$

The important steps have now been done and these integrals can be evaluated either by hand or by using software. Both are done here; the 'by-hand' evaluation follows:

$$P = \int_0^{\frac{1}{4}} \int_0^1 (1 - \sqrt{4xy}) dy \, dx + \int_{\frac{1}{4}}^1 \int_0^{\frac{1}{4x}} (1 - \sqrt{4xy}) dy \, dx$$

$$= \int_0^{\frac{1}{4}} \left[y - \frac{4}{3} y^{\frac{3}{2}} x^{\frac{1}{2}} \right]_0^1 dx + \int_{\frac{1}{4}}^1 \left[y - \frac{4}{3} y^{\frac{3}{2}} x^{\frac{1}{2}} \right]_0^{\frac{1}{4x}} dx$$

$$= \int_0^{\frac{1}{4}} 1 - \frac{4}{3} x^{\frac{1}{2}} dx + \int_{\frac{1}{4}}^1 \frac{1}{12x} dx$$

$$= \left[x - \frac{8}{9} x^{\frac{3}{2}} \right]_0^{\frac{1}{4}} + \left[\frac{1}{12} \ln x \right]_{\frac{1}{4}}^1$$

$$= \left[\frac{1}{4} - \frac{8}{9} \cdot \frac{1}{8} \right] - \frac{1}{12} \ln \left(\frac{1}{4} \right)$$

$$= \frac{5}{36} + \frac{1}{6} \ln 2 \approx 0.2544. \qquad \square$$

The next step would be to solve multiple integrals in more general curvilinear co-ordinates and to solve multiple integrals in more than three dimensions. Neither of these is pursued here.

Exercises

9.1 Evaluate the following double integrals with constant limits:

a) $\int_{y=0}^{y=3} \int_{x=1}^{x=2} (3x^2 + 2xy + 4y) dx \, dy$,

b) $\int_{x=1}^{x=2} \int_{y=0}^{y=3} (3x^2 + 2xy + 4y) dy \, dx$,

c) $\int_{x=0}^{x=2} \int_{y=0}^{y=3} y \sin x \, dy \, dx$,

d) $\int_0^1 \int_0^1 \frac{x}{(xy+1)^2} dy \, dx$,

e) $\displaystyle\int_{\pi/2}^{\pi}\int_{1}^{2} x\cos(xy)dy\ dx.$

9.2 Evaluate the following double integrals with variable limits:

a) $\displaystyle\int_{0}^{1}\int_{-x}^{x^2} y^2x\ dy\ dx,$

b) $\displaystyle\int_{0}^{\pi/3}\int_{0}^{\cos y} x\sin y\ dx\ dy.$

9.3 In the following exercises, find the limits of integration given the domains in diagrammatic form.

a) See Figure 9.30; fill in the limits in both cases and evaluate each integral, confirming that they have the same value:

$$\int_{x=}^{x=}\int_{y=}^{y=} x\ dy\ dx,$$

$$\int_{y=}^{y=}\int_{x=}^{x=} x\ dx\ dy.$$

b) See Figure 9.31; fill in the limits in both cases and evaluate each integral, confirming that they have the same value:

$$\int_{x=}^{x=}\int_{y=}^{y=} (x+y)dy\ dx,$$

$$\int_{y=}^{y=}\int_{x=}^{x=} (x+y)dx\ dy.$$

c) See Figure 9.32; fill in the limits in both cases and evaluate each integral, confirming that they have the same value:

$$\int_{x=}^{x=}\int_{y=}^{y=} x^2\ dy\ dx,$$

$$\int_{y=}^{y=}\int_{x=}^{x=} x^2\ dx\ dy.$$

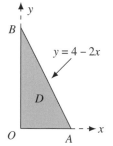

Figure 9.30 Triangle for Exercise 9.3(a).

Figure 9.31 Domain for Exercise 9.3(*b*).

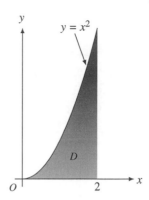

Figure 9.32 Domain for Exercise 9.3(*c*).

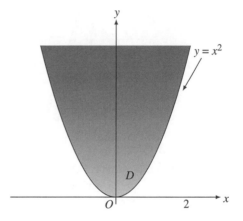

d) See Figure 9.33; fill in the limits in both cases and evaluate each integral, confirming that they have the same value:

$$\int_{x=}^{x=} \int_{y=}^{y=} y \, dy \, dx,$$

$$\int_{y=}^{y=} \int_{x=}^{x=} y \, dx \, dy.$$

e) See Figure 9.34; fill in the limits in both cases and evaluate each integral, confirming that they have the same value, taking care with the second that has to be done in two parts:

$$\int_{y=}^{y=} \int_{x=}^{x=} dx \, dy,$$

$$\int_{x=}^{x=} \int_{y=}^{y=} dy \, dx + \int_{x=}^{x=} \int_{y=}^{y=} dy \, dx.$$

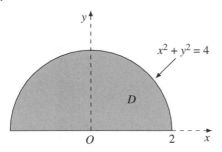

Figure 9.33 The semi-circular domain for Exercise 9.3(*d*).

$x^2 + y^2 = 4$

D

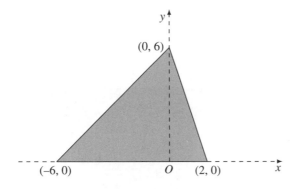

Figure 9.34 The triangular domain, shaded, for Exercise 9.3(*e*).

$(0, 6)$

$(-6, 0)$ O $(2, 0)$

9.4 Sketch the domain D that is bounded by the lines $y = 2x$, $x = 0$ and the parabola $y = 15 - x^2$ for $x \geq 0$. Hence write down the two forms for the double integral

$$\iint_D \cos x \, dD$$

and evaluate it using your choice of order of integration.

9.5 Evaluate the following double integrals in plane polar co-ordinates (r, θ):-

a) $\displaystyle\int_0^{\pi/2} \int_0^{\sin \theta} r \cos \theta r \, dr \, d\theta,$

b) $\displaystyle\int_0^{\pi} \int_0^{1+\cos \theta} r \, dr \, d\theta.$

9.6 Evaluate the following double integrals by converting them into plane polar co-ordinates (r, θ):

a) $\displaystyle\int_{-1}^{1} \int_0^{\sqrt{1-x^2}} (x^2 + y^2)^{3/2} dy \, dx,$

b)

$$\iint_R \sqrt{9 - x^2 - y^2} \, dR,$$

where R is the region in the first quadrant of a circle within the domain $x^2 + y^2 \leq 9$.

9.7 Evaluate the following integrals by first converting them into polar co-ordinates:

a) $\displaystyle\int_0^1 \int_0^{\sqrt{1-x^2}} (x^2 + y^2) dy \, dx,$

b) $\displaystyle\int_0^2 \int_0^{\sqrt{2x-x^2}} \sqrt{x^2 + y^2} \, dy \, dx,$

c) $\displaystyle\int_0^{\sqrt{2}} \int_y^{\sqrt{4-y^2}} \frac{1}{\sqrt{1 + x^2 + y^2}} \, dx \, dy.$

9.8 The following four integrals are either difficult or impossible to evaluate without reversing the order of integration. Sketch the domain, reverse the order and evaluate each of the integrals:

a) $\displaystyle\int_0^1 \int_{\sqrt{x}}^1 \sin\left(\frac{y^3 + 1}{2}\right) dy \, dx,$

b) $\displaystyle\int_0^1 \int_{x^2}^1 \frac{x^3}{\sqrt{(x^4 + y^2)}} dy \, dx,$

c) $\displaystyle\int_0^1 \int_0^{\cos^{-1} y} e^{\sin x} \, dx \, dy,$

d) $\displaystyle\int_0^1 \int_x^1 x^2 e^{y^2} \, dy \, dx.$

9.9 Use double integration to show that the area of the shaded section of Figure 9.35 is given by the expression

$$a^2 \left[\phi - \frac{1}{2} \sin(2\phi) \right].$$

9.10 Evaluate the following triple integrals:

a) $\displaystyle\int_0^1 \int_0^2 \int_1^3 x^2 yz \, dz \, dy \, dx,$

b) $\displaystyle\int_0^1 \int_{x^2}^{\sqrt{x}} \int_{x-z}^{x+z} (x + y + z) dy \, dz \, dx.$

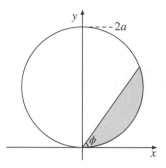

Figure 9.35 The shaded area for Exercise 9.9.

9.11 Sketch the region contained between the parabolic cylinder $y = x^2$ and the planes $z = 0$ and $x + y + z = 2$. Show that the volume can be written as the triple integral

$$\int_0^1 \int_{x^2}^{2-x} \int_0^{2-x-y} dz \, dy \, dx$$

and evaluate the volume.

9.12 Find the volume of the solid bounded by the planes $z = 0, y = 0, x = 0$ and the cylinder $x^2 + y^2 = 4$ and the hyperbolic paraboloid $z = 6 - xy$ in the positive octant.

9.13 A hole with diameter 4 cm is bored through a solid sphere of diameter 10 cm, the axis of the hole being the diameter of the sphere. Write down the triple integral for the volume of the solid remaining. Hence, determine the volume.

9.14 Evaluate the following integrals that involve cylindrical polar co-ordinates:

a) $\int_0^{2\pi} \int_0^1 \int_0^{\sqrt{1-r^2}} zr \, dz \, dr \, d\theta,$

b) $\int_0^{\pi/2} \int_0^{\cos\theta} \int_0^{r^2} r\sin\theta \, dz \, dr \, d\theta.$

9.15 Use cylindrical polar co-ordinates to evaluate the following volumes:
a) The solid enclosed by the paraboloid $z = x^2 + y^2$ and the plane $z = 9$.
b) The solid that is bounded above and below by the sphere $x^2 + y^2 + z^2 = 9$ and inside the cylinder $x^2 + y^2 = 4$.

9.16 Evaluate the following integrals that involve spherical polar co-ordinates:

a) $\displaystyle\int_0^{\pi/2} \int_0^{\pi/2} \int_0^1 r^3 \sin\phi \cos\phi \, dr \, d\phi \, d\theta,$

b) $\displaystyle\int_0^{2\pi} \int_0^{\pi/4} \int_0^{a\sec\phi} r^2 \sin\phi \, dr \, d\phi \, d\theta, \quad a > 0.$

9.17 Evaluate the following volumes using either cylindrical or spherical polar co-ordinates:

    a) The solid bounded above by the sphere $r = 4$ and below by the cone $\theta = \pi/3$.

    b) The solid enclosed by the sphere $x^2 + y^2 + z^2 = 4a^2$ and the planes $z = 0$ and $z = a$.

9.18 Use either cylindrical or spherical polar co-ordinates to evaluate the following integrals:

a) $\displaystyle\int_0^a \int_0^{\sqrt{a^2-x^2}} \int_0^{a^2-x^2-y^2} x^2 \, dz \, dy \, dx,$

b) $\displaystyle\int_{-2}^2 \int_{-\sqrt{4-x^2}}^{\sqrt{4-x^2}} \int_0^{\sqrt{4-x^2-y^2}} z^2 \sqrt{x^2 + y^2 + z^2} \, dz \, dy \, dx.$

9.19 Find the volume common to the three intersecting cylinders $x^2 + y^2 = a^2$, $x^2 + z^2 = a^2$ and $y^2 + z^2 = a^2$.

9.20 Using spherical polar co-ordinates, determine the position of the centre of mass of the cone with equation $z^2 = x^2 + y^2, z \geq 0$ with the base at $z = a$.

9.21 Find the moment of inertia of the solid of Figure 9.27 about its axis of symmetry. Use the formula

$$\iiint_V \rho d^2 \, dV,$$

where ρ is the density and d is the perpendicular distance of the element of volume dV from this axis. Express your answer in terms of the mass M of the body.

9.22 The journey home is in three parts. There is a tube journey to the main station that takes on average 10 min, followed by a train journey of

40 min, and finally a 5 min walk to home. This evening there's a trip to the theatre and lateness would be troublesome. Our city worker has to be home in an hour. Assuming each journey is independent and governed by a Poisson distribution, use a triple integral to determine the probability he or she is late.

10

Surface Integrals

10.1 Introduction

Surface integrals are integrals whose domain covers two dimensions. In the last chapter, repeated integrals were introduced, and integrals repeated twice are in fact an example of a surface integral, but the surface is a flat one. Here, we integrate over a curved surface. It can be expected however that the results for double integration covered in Section 9.2 will certainly be used and there is bound to be a little repetition, no bad thing. The new aspect is the use of vectors. In Chapter 8, the path or line integral was introduced, and surface integrals that contain vectors will have similarities to this. One of the principal kinds of surface integral involves the calculation of *flux*, which, loosely, is the calculation of the amount of stuff that goes through a surface. Figure 10.1 gives a visual interpretation of flux. It shows a curved surface S an infinitesimal patch of which is labelled $d\mathbf{S}$ and is a vector with direction $\hat{\mathbf{n}}$ its unit normal. This normal will of course change with choice of $d\mathbf{S}$. In addition, shown is a vector field $\mathbf{F}$ and once more the infinitesimal patch is so small that this field is constant in both magnitude (unimportant here) and direction (very important here) over the patch $d\mathbf{S}$. In general, there will be an angle between $\mathbf{F}$ and $d\mathbf{S}$ and this is labelled θ in Figure 10.1. This angle is of course also constant over the patch. The flux of the vector field $\mathbf{F}$ must be the scalar quantity $\mathbf{F} \cdot d\mathbf{S} = |\mathbf{F}||d\mathbf{S}| \cos\theta$ that is the scalar product of $\mathbf{F}$ with $d\mathbf{S}$ and is the amount of $\mathbf{F}$ that passes through the patch $d\mathbf{S}$. In order to find the total amount of $\mathbf{F}$ emanating from the surface S, one adds together the contributions from all the patches that comprise the surface S, which means integrating over the entire surface. This is the surface integral

$$\int_S \mathbf{F} \cdot d\mathbf{S},$$

and this is the kind of object calculated in this chapter. The vector field $\mathbf{F}$ could be the heat flow, fluid velocity, the electric field or the magnetic field, all choices lead to interesting applications of surface integrals.

Two and Three Dimensional Calculus: With Applications in Science and Engineering, First Edition. Phil Dyke.
© 2018 John Wiley & Sons Ltd. Published 2018 by John Wiley & Sons Ltd.

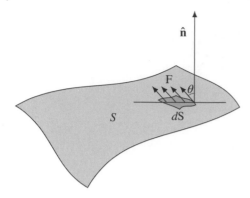

Figure 10.1 A surface S showing an infinitesimal patch $d\mathbf{S}$ that is small enough to be flat and is a vector with direction $\hat{\mathbf{n}}$ the (unit) normal to this patch.

10.2 Green's Theorem in the Plane

Before getting to surface integration proper, here is a useful result stated in the form of a theorem. It is called *Green's theorem in the plane*. It could be argued that it really belongs in the next chapter, but as the emphasis throughout this book is on results, it gets an airing a chapter early. Its 'big brother' is called Stokes' theorem, which does belong in Chapter 11.

Theorem 10.1 (Green's Theorem in the Plane) If D is a closed region of the x, y plane bounded by a simple closed curve C, and if $P(x, y)$ and $Q(x, y)$ are functions with continuous derivatives with respect to both x and y inside and on the curve C, then

$$\oint_C (Pdx + Qdy) = \iint_D \left(\frac{\partial Q}{\partial x} - \frac{\partial P}{\partial y} \right) dxdy.$$

Proof: To prove this theorem, it is best to consider first a square (see Figure 10.2).

Consider the integral

$$\iint \frac{\partial Q}{\partial x} dxdy = \int_{y_1}^{y_2} \int_{x_1}^{x_2} \frac{\partial Q}{\partial x} dxdy,$$

integrating with respect to x yields

$$\iint \frac{\partial Q}{\partial x} dxdy = \int_{y_1}^{y_2} [Q]_{x_1}^{x_2} dy$$

$$= \int_{y_1}^{y_2} [Q(x_2, y) - Q(x_1, y)] dy.$$

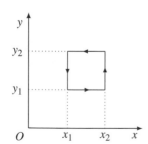

Figure 10.2 The square domain.

Now, around the same square,

$$\oint Qdy = \int_{y_1}^{y_2} [Q(x_2, y) - Q(x_1, y)]dy$$

as along the sides $y = y_1$ and $y = y_2$, x is constant and so $dx = 0$ and the four contributions to the closed integral around the square reduce to two. This is why the dy in front of the integral on the left is so important not to omit, it tells you that x is not varying. The presence of ds, for example would tell a completely different story; the theorem would be false. Hence around the square it has been proved that

$$\iint \frac{\partial Q}{\partial x}dxdy = \oint Qdy.$$

In exactly the same way, it is shown that

$$-\iint \frac{\partial P}{\partial y}dydx = \oint Pdx.$$

Adding these results, reversing the integration in the second equality gives the result for this square. Now consider Figure 10.3. The general shape can be drawn on fine resolution graph paper and

$$\oint_{C_n} (Pdx + Qdy) = \iint_{D_n} \left(\frac{\partial Q}{\partial x} - \frac{\partial P}{\partial y} \right) dxdy$$

applied to each very small square D_n with border C_n. These can then be summed:

$$\sum_n \left\{ \oint_{C_n} (Pdx + Qdy) \right\} = \sum_n \left\{ \iint_{D_n} \left(\frac{\partial Q}{\partial x} - \frac{\partial P}{\partial y} \right) dxdy \right\}.$$

Figure 10.3 A general domain D drawn over a fine grid. Green's theorem in the plane is applied to each grey square then summed over the area occupied by D.

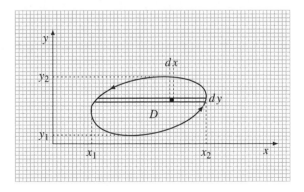

All contributions to the right-hand side will add like pixels making a picture, whereas on the left, contributions from adjacent internal sides of every square will cancel. Therefore, in the limits as the squares get infinitesimally small,

$$\oint_C (P dx + Q dy) = \iint_D \left(\frac{\partial Q}{\partial x} - \frac{\partial P}{\partial y} \right) dx dy$$

and the theorem is proved. This proof works even for shapes that are not convex, but a little more commentary is required for multi-connected regions. It is false if in the limiting process the border is non-rectifiable, that is it cannot be measured (like a fractal). Such boundaries are not considered here. Just a word on notation: if a domain D is surrounded by a smooth closed curve, it is given a symbol ∂D instead of C so Green's theorem in the plane is written

$$\oint_{\partial D} (P dx + Q dy) = \iint_D \left(\frac{\partial Q}{\partial x} - \frac{\partial P}{\partial y} \right) dx dy.$$

□

Green's theorem in the plane is mainly used in the next chapter, but it is appropriate here to do a constructive example.

Example 10.1 Use Green's theorem in the plane to show that the path integral

$$\oint_C \frac{1}{2} (x dy - y dx)$$

gives the area inside the closed curve C. Use this to find the area of an ellipse that has major axis a and minor axis b.

Solution: To solve this, start with Green's theorem in the plane

$$\oint_{\partial D} (P dx + Q dy) = \iint_D \left(\frac{\partial Q}{\partial x} - \frac{\partial P}{\partial y} \right) dx dy$$

and choose $P = -\frac{1}{2} y$ and $Q = \frac{1}{2} x$ so that Green's theorem becomes

$$\oint_C \frac{1}{2} (-y dx + x dy) = \iint_D dx dy,$$

and the right-hand side is the area enclosed by the curve C as required. The second part is using this to find the area of an ellipse that has axes a and b, and if it is assumed the centre is at the origin for convenience this has the (Cartesian) equation

$$\frac{x^2}{a^2} + \frac{y^2}{b^2} = 1.$$

The cleanest way of evaluating this kind of integral is to parametrise the curve. This ellipse is parameterised by $x = a\cos\theta, y = b\sin\theta$ where $0 \le \theta < 2\pi$. On the ellipse, therefore, we can write:

$$xdy - ydx = (a\cos\theta)(b\cos\theta d\theta) - (b\sin\theta)(-a\sin\theta d\theta)$$
$$= ab(\cos^2\theta + \sin^2\theta)d\theta = abd\theta$$

hence

$$\oint_C \frac{1}{2}(xdy - ydx) = \int_0^{2\pi} \frac{1}{2}abd\theta = \pi ab,$$

perhaps the easiest way to establish the well-known expression πab for the area enclosed by an ellipse. □

Here is a more directly computational example.

Example 10.2   Evaluate the integral

$$\oint_C (x^2 + y^2)dx + 3xy^2 dy,$$

where C is the circle centre the origin of radius 2 by using Green's theorem in the plane.

Solution: Substituting the expressions $P(x, y) = x^2 + y^2$ and $Q(x, y) = 3xy^2$ into Green's theorem in the plane gives:

$$\oint_C (x^2 + y^2)dx + 3xy^2 dy = \iint_D (3y^2 - 2y)dxdy,$$

where D is the interior of the circle $x^2 + y^2 = 4$. The integral on the right is best evaluated using plane polar co-ordinates as the domain inside the circle centre with the origin radius 2 will have constant limits. Using $x = r\cos\theta, y = r\sin\theta$ transforms the right-hand side plane surface integral into a repeated integral

$$\iint_D (3y^2 - 2y)dxdy = \int_0^{2\pi} \int_0^2 (3r^2\sin^2\theta - 2r\sin\theta)rdrd\theta$$

of the type evaluated in the last chapter. Here are the details of the computation:

$$\int_0^{2\pi} \int_0^2 (3r^3\sin^2\theta - 2r^2\sin\theta)drd\theta = \int_0^{2\pi} \left[\frac{3}{4}r^4\sin^2\theta - \frac{2}{3}r^3\sin\theta\right]_0^2 d\theta$$

$$= \int_0^{2\pi} \left[12\sin^2\theta - \frac{16}{3}\sin\theta\right]d\theta$$

$$= \int_0^{2\pi} \left(6 - 6\cos 2\theta - \frac{16}{3}\sin\theta\right)d\theta$$

$$= \left[6\theta - 3\sin 3\theta + \frac{16}{3}\cos\theta \right]_0^{2\pi}$$

$$= 12\pi.$$

This can be confirmed by direct integration of the closed path integral (see Exercise 10.1). Let us now leave Green's theorem in the plane; it will crop up again in the next chapter. □

10.3 Integration over a Curved Surface

Integration over a curved surface means returning briefly to differential geometry. At the end of Chapter 5, it was briefly stated how to represent a surface parametrically by using $x = x(u, v)$, $y = y(u, v)$ and $z = z(u, v)$, and the example of a sphere was given. It is also possible to represent a surface by a single Cartesian equation $z = f(x, y)$, and this is convenient for contour-like surfaces that describe terrain as in walking over hill and dale. In parametric form, this would be $x = u$, $y = v$ and $z = f(u, v)$ of course. In general if an arbitrary point on a surface is represented by the values u, v, then the position vector of this point would be $\mathbf{r} = \mathbf{r}(u, v)$. Now if u is held constant $u = u_0$ say then this will represent a curve in the surface, similarly if v were held constant $v = v_0$ would also represent a curve in the surface. It was shown in Chapter 5 that $d\mathbf{r}(t)/dt$ is tangent to the curve $\mathbf{r} = \mathbf{r}(t)$ therefore using similar reasoning, the two derivatives

$$\frac{\partial \mathbf{r}}{\partial u} \quad \text{written} \quad \mathbf{r}_u \quad \text{and} \quad \frac{\partial \mathbf{r}}{\partial v} \quad \text{written} \quad \mathbf{r}_v$$

are both tangent to the surface $\mathbf{r} = \mathbf{r}(u, v)$. The normal to the surface $\mathbf{r} = \mathbf{r}(u, v)$ will thus be $\mathbf{r}_u \times \mathbf{r}_v$ and the unit normal $(\mathbf{r}_u \times \mathbf{r}_v)/|\mathbf{r}_u \times \mathbf{r}_v|$. It is now possible to compute a surface integral that might be written

$$\iint_S \mathbf{F} \cdot d\mathbf{S}.$$

The surface can be open with a border, like a sheet of rubber, or a hemisphere. It can also be closed like a sphere or an ellipsoid. The infinitesimal quantity $d\mathbf{S}$ may be represented as a very small locally plane patch on the surface (see figure 10.4). Therefore, the surface integral can be converted into a double (or repeated) integral, using

$$d\mathbf{S} = \hat{\mathbf{n}}dS = \frac{\mathbf{r}_u \times \mathbf{r}_v}{|\mathbf{r}_u \times \mathbf{r}_v|} h_1 h_2 dudv,$$

where h_1 and h_2 are the scale factors (see Chapter 7 on curvilinear coordinates). Hence,

$$\iint_S \mathbf{F} \cdot d\mathbf{S} = \int_{v_1}^{v_2} \int_{u_1}^{u_2} \mathbf{F} \cdot \frac{\mathbf{r}_u \times \mathbf{r}_v}{|\mathbf{r}_u \times \mathbf{r}_v|} h_1 h_2 dudv,$$

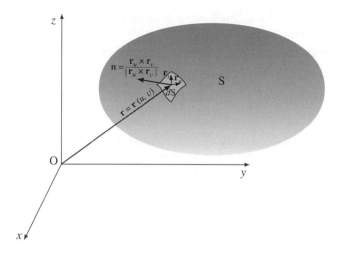

Figure 10.4 The surface S with the general position vector $\mathbf{r}$ marked, the normal n and tangent vectors $\mathbf{r}_u$ and $\mathbf{r}_v$ are also shown on the element $d\mathbf{S}$.

which is a repeated integral with constant limits. Further simplification is possible provided that the two sets of curves $u = $ constant and $v = $ constant are orthogonal. This means that with the direction of the normal, there is a local set of orthogonal curvilinear co-ordinates. As drawn in Figure 10.4, the tangents in the surface are $\mathbf{r}_u$ and $\mathbf{r}_v$, so from Chapter 7 $h_1 = |\mathbf{r}_u|$ and $h_2 = |\mathbf{r}_v|$. Moreover, since $\mathbf{r}_u$ and $\mathbf{r}_v$ are orthogonal,

$$|\mathbf{r}_u \times \mathbf{r}_v| = |\mathbf{r}_u||\mathbf{r}_v| = h_1 h_2,$$

the sine of the angle between them being unity. Hence,

$$d\mathbf{S} = \hat{\mathbf{n}} dS = \frac{\mathbf{r}_u \times \mathbf{r}_v}{|\mathbf{r}_u \times \mathbf{r}_v|} h_1 h_2 du dv = \mathbf{r}_u \times \mathbf{r}_v du dv \tag{10.1}$$

and evaluation becomes easier. Here is an example.

Example 10.3 Find the surface area of a sphere radius a from first principles.

Solution: Use spherical polar co-ordinates (r, ϕ, θ), see Figure 9.25, where the surface co-ordinates are θ and ϕ, the scale factors have been derived in Chapter 7 and are $h_1 = 1, h_2 = r$ and $h_3 = r \sin \phi$. (It is θ and ϕ that correspond to u and v, so it is h_2 and h_3 that are the scale functions for the surface co-ordinates). Thus, the surface area of the sphere is given by

$$\iint\limits_{S} dS = \int_0^{2\pi} \int_0^{\pi} h_2 h_3 d\phi d\theta = a^2 \int_0^{2\pi} \int_0^{\pi} \sin \phi d\phi d\theta$$

as on the surface of the sphere $r = a$. The limits are chosen so dS sweeps from north to south pole (ϕ limits, latitude), then this semi-circular arc completes a whole circle and sweeps out the surface of the sphere as it rotates about the 'North–South axis' (θ limits, longitudinal sweep of 2π), thus

$$a^2 \int_0^{2\pi} \int_0^{\pi} \sin\phi \, d\phi \, d\theta = a^2 \int_0^{2\pi} [-\cos\phi]_0^{\pi} d\theta = 2a^2 \int_0^{2\pi} d\theta = 4\pi a^2.$$

$\square$

The beauty of this method is its generality. For example at the stage where $r = a$ was stated for a sphere, another expression $r = r(\theta, \phi)$ could be inserted for, say, an ellipsoid or other surface that completely encloses the origin, the evaluation would proceed in the same way, though with possibly more difficult integration and after checking that the limits were still right. Integrals such as

$$\iint_S \mathbf{F} \times d\mathbf{S}; \quad \iint_S F dS \quad \text{or} \quad \iint_S \psi d\mathbf{S}$$

all follow as double integrals over ranges of the parameters u, v that define the surface S and specialist books on vector calculus deal with them. Here is a result for calculating a surface area when the curvilinear co-ordinates that generate the area are known. Let us state it in the form of a theorem:

Theorem 10.2   If $A(S)$ denotes the surface area of the curved surface S, then

$$A(S) = \iint_S \left[\left(\frac{\partial(y,z)}{\partial(u,v)} \right)^2 + \left(\frac{\partial(z,x)}{\partial(u,v)} \right)^2 + \left(\frac{\partial(x,y)}{\partial(u,v)} \right)^2 \right]^{1/2} du dv,$$

where $u = u(x, y), v = v(x, y)$ are the curves that generate the surface S.

Proof: From earlier, taking the modulus of Equation 10.1 gives

$$dS = |\mathbf{r}_u \times \mathbf{r}_v| du dv.$$

The way forward now is to relate the parameters to Cartesian co-ordinates. The position vector $\mathbf{r}$ has been introduced above, so as

$$\mathbf{r} = x\mathbf{i} + y\mathbf{j} + z\mathbf{k},$$

use the dependence of x, y, z on the parameters u, v to write

$$\mathbf{r}_u = x_u\mathbf{i} + y_u\mathbf{j} + z_u\mathbf{k} \quad \text{and} \quad \mathbf{r}_v = x_v\mathbf{i} + y_v\mathbf{j} + z_v\mathbf{k}.$$

Hence,

$$\mathbf{r}_u \times \mathbf{r}_v = \begin{vmatrix} \mathbf{i} & \mathbf{j} & \mathbf{k} \\ x_u & y_u & z_u \\ x_v & y_v & z_v \end{vmatrix}$$

$$= \mathbf{i}(y_u z_v - z_u y_v) + \mathbf{j}(x_v z_u - z_v x_u) + \mathbf{k}(x_u y_v - y_u x_v).$$

The right-hand side of this equation can be written in terms of Jacobians as follows

$$\mathbf{i}\left(\frac{\partial(y,z)}{\partial(u,v)}\right) + \mathbf{j}\left(\frac{\partial(x,z)}{\partial(u,v)}\right) + \mathbf{k}\left(\frac{\partial(x,y)}{\partial(u,v)}\right).$$

Thus, $|\mathbf{r}_u \times \mathbf{r}_v|$ can now be written as

$$|\mathbf{r}_u \times \mathbf{r}_v| = \left[\left(\frac{\partial(y,z)}{\partial(u,v)}\right)^2 + \left(\frac{\partial(x,z)}{\partial(u,v)}\right)^2 + \left(\frac{\partial(x,y)}{\partial(u,v)}\right)^2\right]^{1/2},$$

and once this is inserted into the expression

$$A(S) = \iint_S |\mathbf{r}_u \times \mathbf{r}_v| \, du dv$$

the theorem is proved. □

Here is an example of the use of this formula.

Example 10.4   Calculate the following surface areas:

a) the cone $x = u \cos v, y = u \sin v, z = u, 0 \le v < 2\pi, 0 \le u < 1$,
b) the helicoid $x = u \cos v, y = u \sin v, z = v, 0 \le v < 2\pi, 0 \le u < 1$.

Solution: One has a choice of using the Jacobian formula of the theorem, or using

$$A(S) = \iint_S |\mathbf{r}_u \times \mathbf{r}_v| \, du dv$$

directly. The latter is actually quicker. Therefore, starting with the cone:

a) Calculate $\mathbf{r}_u = (x_u, y_u, z_u) = (\cos v, \sin v, 1)$ and $\mathbf{r}_v = (x_v, y_v, z_v) = (-u \sin v, u \cos v, 0)$. This gives

$$\mathbf{r}_u \times \mathbf{r}_v = \begin{vmatrix} \mathbf{i} & \mathbf{j} & \mathbf{k} \\ x_u & y_u & z_u \\ x_v & y_v & z_v \end{vmatrix} = \begin{vmatrix} \mathbf{i} & \mathbf{j} & \mathbf{k} \\ \cos v & \sin v & 1 \\ -u \sin v & u \cos v & 0 \end{vmatrix}$$
$$= (-u \cos v, u \sin v, u).$$

Hence, $|\mathbf{r}_u \times \mathbf{r}_v| = \sqrt{u^2 \cos^2 v + u^2 \sin^2 v + u^2}) = u\sqrt{2}$ and the area if the cone is given by

$$\iint_S |\mathbf{r}_u \times \mathbf{r}_v| \, dS = \int_0^1 \int_0^{2\pi} u\sqrt{2} \, dv du = 2\pi\sqrt{2}\left[\frac{1}{2}u^2\right]_0^1 = \pi\sqrt{2}.$$

b) The second calculation proceeds in the same way:

$$\mathbf{r}_u = (x_u, y_u, z_u) = (\cos v, \sin v, 0)$$

and

$$\mathbf{r}_v = (x_v, y_v, z_v) = (-u \sin v, u \cos v, 1)$$

and this time the cross product is

$$\mathbf{r}_u \times \mathbf{r}_v = \begin{vmatrix} \mathbf{i} & \mathbf{j} & \mathbf{k} \\ x_u & y_u & z_u \\ x_v & y_v & z_v \end{vmatrix} = \begin{vmatrix} \mathbf{i} & \mathbf{j} & \mathbf{k} \\ \cos v & \sin v & 0 \\ -u \sin v & u \cos v & 1 \end{vmatrix} = (\sin v, -\cos v, u).$$

Hence, $|\mathbf{r}_u \times \mathbf{r}_v| = \sqrt{\sin^2 v + \cos^2 v + u^2} = \sqrt{1 + u^2}$ and the area if the helicoid is given by

$$\iint_S |\mathbf{r}_u \times \mathbf{r}_v| dS = \int_0^1 \int_0^{2\pi} \sqrt{1 + u^2} \, dv du = 2\pi \int_0^1 \sqrt{1 + u^2} du,$$

which is a bit more challenging, particularly since my well-known computer algebra package appears to give the wrong answer. It doesn't, and the sensible answer is restored once it is realised that $\ln(\sqrt{2} - 1) = -\ln(\sqrt{2} + 1)$. Because of this possible confusion, here is the integration in detail. Put

$$I = \int_0^1 \sqrt{u^2 + 1} du$$

and integrate by parts to give:

$$I = [u\sqrt{1 + u^2}]_0^1 - \int_0^1 u \cdot \frac{1}{2}(1 + u^2)^{-1/2} 2u du = \sqrt{2} - \int_0^1 \frac{u^2 du}{\sqrt{1 + u^2}}.$$

Writing the numerator $u^2 = 1 + u^2 - 1$ thus gives

$$I = \sqrt{2} - I + \int_0^1 \frac{du}{\sqrt{1 + u^2}},$$

and the last integral integrates to $\sinh^{-1} u$. Thus,

$$2I = \sqrt{2} + [\sinh^{-1}(u)]_0^1 = \sqrt{2} + \sinh^{-1}(1) = \sqrt{2} + \ln(1 + \sqrt{2})$$

using that $\sinh^{-1}(x) = \ln(x + \sqrt{1 + x^2})$ with $x = 1$, an expression easily derivable from the definition of $\sinh x$ by solving a quadratic. Hence, the surface integral

$$\iint_S |\mathbf{r}_u \times \mathbf{r}_v| dS = \pi[\sqrt{2} + \ln(1 + \sqrt{2})]$$

gives the area of the helicoid. □

Some books evaluate an integral over a surface by projection. This works well as long as the projection of $d\mathbf{S}$ on to one of the co-ordinate planes $x, y; y, z$ or z, x is unique. The surface has to be an open one with a bordering curve C and $d\mathbf{S}$ is replaced by $\mathbf{k}dxdy/|\hat{\mathbf{n}} \cdot \mathbf{k}|$ in the case of projecting on to the x, y plane. The shadow of the surface S then becomes the integrated region using the Cartesian co-ordinates x, y. Of course, z, x or y, z planes could also be used to project on to. This method is a good one when it works as it leads to the unfamiliar surface integral being converted to a repeated integral in Cartesian co-ordinates; sadly it doesn't work for spheres or cylinders. Let us do one example using this projection method.

Example 10.5   Use a projection method to evaluate the integral

$$\iint_S A dS,$$

where

$$A = \frac{z}{\sqrt{4x^2 + 4y^2 + 1}}$$

and S is the surface of the paraboloid $z = 4 - x^2 - y^2, 0 \le z \le 4$.

Solution: The projection method will work here as long as the projection is in the z direction. This means we use $\mathbf{k}$ so

$$\iint_S A dS = \iint_D A \frac{dD}{|\hat{\mathbf{n}} \cdot \mathbf{k}|},$$

where D represents the domain that results from projecting the paraboloid on to the x, y plane (see figure 10.5). You can think of it as the shadow cast by the paraboloid on the x, y plane if the sun was shining infinitely far away at the end of the z axis. The shadow, and hence D, will be circular in shape. The unit normal to the paraboloid $\hat{\mathbf{n}}$ is found using the grad. function:

$$\nabla(z - 4 + x^2 + y^2) = 2x\mathbf{i} + 2y\mathbf{j} + \mathbf{k}.$$

Therefore,

$$\hat{\mathbf{n}} = \frac{\nabla(z - 4 + x^2 + y^2)}{|\nabla(z - 4 + x^2 + y^2)|} = \frac{1}{\sqrt{4x^2 + 4y^2 + 1}}(2x\mathbf{i} + 2y\mathbf{j} + \mathbf{k}).$$

This means that

$$|\hat{\mathbf{n}} \cdot \mathbf{k}| = \frac{1}{\sqrt{4x^2 + 4y^2 + 1}}$$

and so

$$\frac{A}{|\hat{\mathbf{n}} \cdot \mathbf{k}|} = \frac{z}{\sqrt{4x^2 + 4y^2 + 1}}(\sqrt{4x^2 + 4y^2 + 1}) = z = 4 - x^2 - y^2,$$

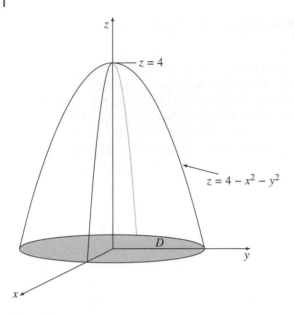

Figure 10.5 The paraboloid $z = 4 - x^2 - y^2$.

where the author admits selection of the scalar field to make the algebra easier. Thus,

$$\iint_S A \, dS = \iint_D A \frac{dD}{|\hat{\mathbf{n}} \cdot \mathbf{k}|} = \iint_D (4 - x^2 - y^2) dD$$

and the integral over the domain (circle) D is the same as those done in the last chapter. Here, it is convenient to use plane polar R, θ co-ordinates. The calculation proceeds as follows:

$$I = \iint_D (4 - x^2 - y^2) dD = \int_0^2 \int_0^{2\pi} (4 - R^2) R \, d\theta \, dR$$

and the right-hand side is quickly calculated as the limits are constant, so

$$I = \int_0^2 \int_0^{2\pi} (4R - R^3) d\theta \, dR = 2\pi \left[2R^2 - \frac{1}{4} R^4 \right]_0^2 = 8\pi.$$

$\square$

Given the self-confession that this last example had purposely simple algebra, coupled with the method failing for many surfaces means that this projection method is less robust than the parametric method given earlier. Therefore, let's leave it here and move on to applications.

10.4   Applications of Surface Integration

Surface integrals have many applications; in thermodynamics, fluid mechanics and elasticity, and the first example is about heat transfer. Most applications in these areas are similar and concern calculating the amount of a quantity entering or leaving a curved surface. More distinctive applications are in the field of electromagnetic theory where the variety of integration type often differs. Therefore, first let us calculate a flux of heat through a surface. In this chapter, direct calculation is appropriate, but it turns out that quantities such as heat flux are related to the gradient of a directly measurable variable. Heat flux $\mathbf{Q}$ is related to temperature T through $\mathbf{Q} = -\kappa \nabla T$ and in the next chapter use is made of this.

Example 10.6   The temperature field throughout space is given by $T = -z/a$ where a is a constant length. Calculate the heat flow given by $\mathbf{Q} = -\kappa \nabla T$ where κ is the thermal conductivity assumed constant. Hence, evaluate the heat flux out of the (closed) cone $x^2 + y^2 = z^2, 0 \leq z \leq a$.

Solution: The cone is shown in Figure 10.6. First of all, the vector field $\mathbf{Q}$ is calculated by finding $\kappa \nabla(-z/a)$ hence obtaining $\mathbf{Q} = (\kappa/a)\mathbf{k}$. Therefore, the

Figure 10.6 The cone $x^2 + y^2 = z^2$, the top is the circle $x^2 + y^2 = a^2$ at the level $z = a$.

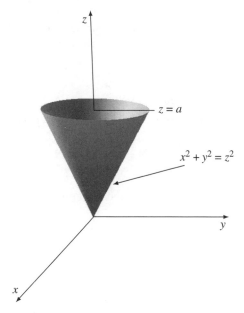

surface integral is calculated directly

$$\iint_S \mathbf{Q} \cdot d\mathbf{S} = \iint_{S_1} \mathbf{Q} \cdot d\mathbf{S} + \iint_{S_{z=a}} \mathbf{Q} \cdot d\mathbf{S},$$

where S_1 is the curved surface of the cone, and $S_{z=a}$ is the flat circular base of the inverted cone. Evaluating the integral over the curved surface using the parametrisation:

$$x = u \cos v, y = u \sin v, z = u$$

gives

$$\mathbf{r}_u = (\cos v, \sin v, 1) \quad \text{and} \quad \mathbf{r}_u = (-u \sin v, u \cos v, 0)$$

so that

$$\mathbf{r}_u \times \mathbf{r}_v = (-u \cos v, u \sin v, u),$$

and this gives the result

$$dS = |\mathbf{r}_u \times \mathbf{r}_v| du dv = u\sqrt{2} du dv.$$

The unit normal is

$$\hat{\mathbf{n}} = \frac{\mathbf{r}_u \times \mathbf{r}_v}{|\mathbf{r}_u \times \mathbf{r}_v|} = -\frac{1}{\sqrt{2}}(\cos v, -\sin v, 1)$$

choosing the sign that represents correctly the downwards $(-\mathbf{k})$ z component of the outward normal to the curved portion of the cone (see Figure 10.6), and this gives the result

$$\mathbf{Q} \cdot \hat{\mathbf{n}} = -\frac{\kappa}{a\sqrt{2}}.$$

Hence, the integral over the curved part of the cone is given by

$$\iint_{S_1} \mathbf{Q} \cdot d\mathbf{S} = \iint_{S_1} \mathbf{Q} \cdot \hat{\mathbf{n}} dS = \int_0^a \int_0^{2\pi} \left(-\frac{\kappa}{a\sqrt{2}}\right) u\sqrt{2} dv du.$$

Evaluating the repeated integral is straightforward:

$$\int_0^a \int_0^{2\pi} \left(-\frac{\kappa}{a\sqrt{2}}\right) u\sqrt{2} dv du = -\frac{\kappa}{a} \int_0^a \int_0^{2\pi} u dv du$$

$$= -\frac{\kappa}{a} \cdot 2\pi \cdot \left[\frac{1}{2}u^2\right]_0^a = -\pi\kappa a.$$

The integral over the flat part is

$$\iint_{S_{z=a}} \mathbf{Q} \cdot d\mathbf{S} = \frac{\kappa}{a} \iint_{S_{z=a}} dx dy = \frac{\kappa}{a}\pi a^2 = \pi\kappa a,$$

hence, the sum of the two is zero, leading to

$$\iint\limits_S \mathbf{Q} \cdot d\mathbf{S} = \iint\limits_{S_1} \mathbf{Q} \cdot d\mathbf{S} + \iint\limits_{S_{z=a}} \mathbf{Q} \cdot d\mathbf{S} = 0.$$

Thus, the net flux into or out of the cone is zero. In the next chapter, a result is derived that proves this result in much more generality. □

The fields associated with electromagnetic theory are electric or magnetic and typically an object in such a field is charged. If it is required to calculate the magnitude of the charge that has built up over an object, then is may be required to integrate over the shape of the object and this would be a surface integral. There are other ways surface integrals might occur. The calculation of surface integrals in fluid dynamics and thermodynamics does occur but is usually in the form of the last example. It is also true that the subject of the next chapter comes to our aid in solving many of these integrals, that is they can be converted to simpler volume integrals that are scalar rather than vectorial. Nevertheless here is an electromagnetic example.

Example 10.7 The vector fields **E** and **H** are the electric and magnetic fields, respectively, and are given by the expressions

$$\mathbf{E} = \hat{\theta}\frac{E_0}{r} \sin\theta \cos(\omega t - \beta r) \quad \text{and} \quad \mathbf{H} = \hat{\lambda}\frac{H_0}{r} \sin\theta \cos(\omega t - \beta r),$$

where E_0, H_0, ω and β are constants. Spherical polar co-ordinates r, θ, λ have been used. Calculate the Poynting vector $\mathbf{P} = \mathbf{E} \times \mathbf{H}$ and hence determine the instantaneous power

$$W = \iint\limits_S \mathbf{P}.d\mathbf{S}$$

over the sphere S of arbitrary radius a.

Solution: The electric and magnetic fields are both waves but they are directed at right angles to each other around the sphere $r = a$ on its surface. The vector $\mathbf{P} = \mathbf{E} \times \mathbf{H}$ and as spherical polar coordinates are right handed, this is

$$\mathbf{P} = \frac{E_0 H_0}{a^2} \sin^2\theta \cos^2(\omega t - \beta a)\hat{\mathbf{r}},$$

since $r = a$ on the sphere. [Simply write out the cross product $\mathbf{E} \times \mathbf{H}$ in spherical polars, and with two zeros in the each of rows two and three there is only one non-zero term. Then put $r = a$.] Using this version of the spherical polar co-ordinate system, the (scalar) element of surface area dS is given by

$$dS = a^2 \sin\theta d\theta d\lambda.$$

The direction of this element is $\hat{\mathbf{r}}$ being the direction of the normal to the sphere. Hence, $\mathbf{P}$ and $d\mathbf{S}$ are parallel and the integral becomes a scalar one:

$$W = \iint_S \mathbf{P} \cdot d\mathbf{S} = E_0 H_0 \int_0^{2\pi} \int_0^{\pi} \left(\frac{\sin^2\theta}{a^2} \right) \cos^2(\omega t - \beta a) a^2 \sin\theta d\theta d\lambda.$$

Evaluation proceeds as follows:

$$W = 2\pi E_0 H_0 \cos^2(\omega t - \beta a) \int_0^{\pi} \sin^3\theta d\theta$$

and the integration of $\sin^3\theta$ between 0 and π is 4/3; hence, finally

$$W = \frac{8}{3}\pi E_0 H_0 \cos^2(\omega t - \beta a),$$

a progressive wave twice the frequency of the parent fields with an average magnitude of $4\pi E_0 H_0/3$ for those who understand about such things. This average has no dependence on the radius of the sphere a. □

It is tempting to do further relevant examples here, but they are richer left until the end of the next chapter as they all benefit from knowledge of the various integral theorems stated and proved there.

Exercises

10.1 Confirm the result of Example 10.2 by direct evaluation of the integral:

$$\oint_C (x^2 + y^2)dx + 3xy^2 dy.$$

10.2 By converting from Cartesian x, y co-ordinates to plane polar R, θ co-ordinates, show that the area enclosed by a closed curve C can be written

$$\frac{1}{2}\oint_C R^2 d\theta,$$

and use this to calculate the area inside the cardioid $R = a(1 + \cos\theta)$.

10.3 Draw the graph then use polar co-ordinates to calculate the area enclosed by the leaf of the *Folium of Descartes* $x^3 + y^3 = 3axy$.

10.4 Use the formula

$$A(S) = \iint_S |\mathbf{r}_u \times \mathbf{r}_v| dudv$$

for surface area to calculate the area of the curved surface of the paraboloid: $x = u \sin v, y = u \cos v, z = u^2, 0 \le u, 1; 0 \le v < 2\pi$.

10.5 Use the parametrisation

$$\mathbf{r} = (a + b \cos \phi) \cos \theta \mathbf{i} + (a + b \cos \phi) \sin \theta \mathbf{j} + b \sin \phi \mathbf{k}, a > b > 0$$

to find the surface area of a torus. [Note: θ here is the cylindrical polar θ and ϕ is the angle measured vertically around the circular intersection of the torus with either the x, z or the y, z plane. There is no connection to spherical polar co-ordinates.]

10.6 Evaluate

$$\iint_S (x^2 + y^2 + z^2) dS,$$

where S is that part of the surface of the cone $x^2 + y^2 = 2z^2$ that lies between the planes $z = 1$ and $z = 3$.

10.7 If D is the projection of the surface $S : z = f(x, y)$ on to the x, y plane, show that the surface area of S is given by

$$\iint_D \sqrt{1 + \left(\frac{\partial z}{\partial x}\right)^2 + \left(\frac{\partial z}{\partial y}\right)^2} \, dx dy,$$

and deduce the form this expression takes if S is given instead by the equation $F(x, y, z) = 0$.

10.8 Evaluate by parametrisation the scalar surface integral

$$\iint_S (x + y + z) dS,$$

where S is that part of the unit sphere $x^2 + y^2 + z^2 = 1$ that lies in the positive octant $x, y, z \ge 0$. Repeat the evaluation using the projection method on to the x, y plane. Which is the better method in this instance?

10.9 Use projection methods to evaluate the surface integral

$$\int_S \mathbf{F} \times d\mathbf{S},$$

where $\mathbf{F} = z\mathbf{k}$ and S is that portion of a sphere of the unit sphere $x^2 + y^2 + z^2 = 1$ that is in the positive octant $x, y, z \ge 0$.

10.10   In the atmosphere, it can be assumed under some circumstances that the temperature decreases linearly with height above the ground, taken as $z = 0$. By direct integration, show that the flux of heat through the surface of a spherical weather balloon of arbitrary radius is zero.

10.11   The electric field $\mathbf{E}$ obeys an inverse square law of the form $(\mu/r^2)\hat{\mathbf{r}}$. Calculate the flux of this field through:
a) a sphere of radius a centred at the origin;
b) the cylinder $x^2 + y^2 = a^2, -a \le z \le a$.

11

Integral Theorems

11.1 Introduction

In this last chapter of the book, general relationships between integral types are explored. It transpires that these relationships are both mathematically interesting and of much practical value. In Section 10.2, Theorem 10.1 was stated and proved, and that is the starting point here. When expressions are derived between either surface and path integrals or surface and volume integrals, there will be integrals of either curl or divergence of vector fields; the loose analogy is with integration by parts where the integrand contains derivatives. This together with the geometry necessary for interpreting these expressions means that, to be completely rigorous and capture all possibilities, all proofs can be very long, difficult to follow and, to be frank, tedious. Therefore, a path has to be trodden between being rigorous, capturing all possibilities in a single theorem, and being merely descriptive. This chapter thus encapsulates the kind of decision that has had to be made throughout the book. To get specific, for example, a surface in space might be pierced by a line parallel with the z-axis just once. Restricting surfaces to be of this type together with assuming all functions smooth enough to be differentiable everywhere not only makes proving theorems much more straightforward, but also questions whether the theorems are valid for many common applications. To resolve this, a common-sense approach is used. The theorems are stated and proved in a way to show, it is hoped, what is going on. Once this understanding is established then there follows arguments that show how the theorems are extended. Finally, the derived relationships are applied to practical problems. Let us start with the relationship between surface and path integrals, called Green's theorem in the plane in Chapter 10 which is now generalised.

Two and Three Dimensional Calculus: With Applications in Science and Engineering, First Edition. Phil Dyke.
© 2018 John Wiley & Sons Ltd. Published 2018 by John Wiley & Sons Ltd.

11.2 Stokes' Theorem

The starting point for Stokes' theorem is to restate Green's theorem in the plane:

$$\oint_C (Pdx + Qdy) = \iint_D \left(\frac{\partial Q}{\partial x} - \frac{\partial P}{\partial y} \right) dx\, dy.$$

The first task is to write this in vectorial terms. Defining $\mathbf{F} = \mathbf{i}P + \mathbf{j}Q$, then gives

$$\mathbf{F} \cdot d\mathbf{r} = P\, dx + Q\, dy.$$

The integrand on the right-hand side can be written $\nabla \times \mathbf{F} \cdot \mathbf{k}\, dx\, dy$ and $\mathbf{k}\, dx\, dy$ is the element of the flat rectangular region in the x, y plane, so Green's theorem in the plane can be written:

$$\oint_C \mathbf{F} \cdot d\mathbf{r} = \iint_S \nabla \times \mathbf{F} \cdot d\mathbf{S}. \tag{11.1}$$

Changing this slightly to $\mathbf{F} = \hat{\mathbf{e}}_2 P + \hat{\mathbf{e}}_2 Q$ with $d\mathbf{S} = \hat{\mathbf{e}}_3 h_1 h_2\, du\, dv$ where now S is a curved surface, as if made of thin flexible plastic then hardened, and $\hat{\mathbf{e}}_1, \hat{\mathbf{e}}_2$ and $\hat{\mathbf{e}}_3$ are curvilinear co-ordinates with the first two unit vectors embedded in S. Equation 11.1 now reads exactly the same but is now more general and is termed *Stokes' theorem*. This is an example of a co-ordinate-free equation derived in a particular way that because it now is free of the co-ordinates, immediately generalises. It does however warrant a proof. Here it is as a formal theorem.

Theorem 11.1 (Stokes' Theorem) For an orientable surface S with a bounding curve C and vector field $\mathbf{F}$ continuous and differentiable on S, it is true that

$$\oint_C \mathbf{F} \cdot d\mathbf{r} = \iint_S \nabla \times \mathbf{F} \cdot d\mathbf{S}.$$

Proof: There are many ways to approach the proof of Stokes' theorem. Pure mathematicians couch the theorem in general terms and the proof then needs knowledge of such esoteric topological concepts as compact manifolds and de Rham cohomologies. This is unsuitable here, as is the slick, few-line proof using tensors. Some are quite content to believe the truth of Stokes' theorem based on the preliminary vectorisation of Green's theorem in the plane, Equation 11.1. More traditional approaches in textbooks aimed at engineers and scientists contain rather lengthy partial differentiation as mentioned in the introduction. Here, the decision is made to go along these more traditional lines and make use of Green's theorem, explaining restrictions (that turn out to be removable) at the end. It uses only results from Chapter 2 on partial differentiation and, of course, Green's theorem in the plane. Therefore, let's start with a Cartesian co-ordinate view of the surface integral in Stokes' theorem, but writing only a single component. Write $\mathbf{F} = F_1\mathbf{i} + F_2\mathbf{j} + F_3\mathbf{k}$, then

$$\nabla \times F_1\mathbf{i} = \begin{vmatrix} \mathbf{i} & \mathbf{j} & \mathbf{k} \\ \frac{\partial}{\partial x} & \frac{\partial}{\partial y} & \frac{\partial}{\partial z} \\ F_1 & 0 & 0 \end{vmatrix} = \frac{\partial F_1}{\partial z}\mathbf{j} - \frac{\partial F_1}{\partial y}\mathbf{k}.$$

Write the surface S as $z = f(x,y)$ and this straight away presumes that the surface S is an open one, that is the function $f(x,y)$ is single valued so that for each pair of values x, y there is only one z. It describes a landscape with no overhangs or vertical cliffs. The unit normal to this surface is written $\hat{\mathbf{n}}$ and any point on the surface will have position vector $\mathbf{r} = x\mathbf{i} + y\mathbf{j} + f(x,y)\mathbf{k}$. It was found in Chapter 10 that this normal is perpendicular to both $\mathbf{r}_x$ and $\mathbf{r}_y$, see Equation 10.1 and the preamble (in this part of Chapter 10 substitute $u = x$ and $v = y$, of course). From Chapter 10, the unit normal $\hat{\mathbf{n}}$ is perpendicular to both $\mathbf{r}_x$ and $\mathbf{r}_y$; hence,

$$\hat{\mathbf{n}} \cdot \mathbf{i} + \frac{\partial f}{\partial x}\hat{\mathbf{n}} \cdot \mathbf{k} = 0 \quad \text{and} \quad \hat{\mathbf{n}} \cdot \mathbf{j} + \frac{\partial f}{\partial y}\hat{\mathbf{n}} \cdot \mathbf{k} = 0.$$

The second of these results is used now. As the curl in Stokes' theorem is in scalar product with $d\mathbf{S}$ and this vector is in the direction of the normal let us form

$$(\nabla \times F_1\mathbf{i}) \cdot \hat{\mathbf{n}} = \frac{\partial F_1}{\partial z}(\hat{\mathbf{n}} \cdot \mathbf{j}) - \frac{\partial F_1}{\partial y}(\hat{\mathbf{n}} \cdot \mathbf{k})$$

and so

$$(\nabla \times F_1\mathbf{i}) \cdot \hat{\mathbf{n}} = (\hat{\mathbf{n}} \cdot \mathbf{k})\left[-\frac{\partial f}{\partial y}\frac{\partial F_1}{\partial z} - \frac{\partial F_1}{\partial y}\right]$$

upon eliminating $(\hat{\mathbf{n}} \cdot \mathbf{j})$. This result is laid aside while some partial derivative results from Chapter 2 are utilised. Now

$$F_1(x, y, z) = F_1(x, y, f(x,y)), \quad \text{on the surface } S = G(x,y) \text{ say.}$$

Using the chain rule from Chapter 2, thus gives

$$\frac{\partial G}{\partial y} = \frac{\partial F_1}{\partial y} + \frac{\partial F_1}{\partial f}\frac{\partial f}{\partial y} = \frac{\partial F_1}{\partial y} + \frac{\partial F_1}{\partial z}\frac{\partial f}{\partial y}$$

since $z = f(x,y)$ on the surface S. Therefore, picking up the expression for the curl of the first component of $\mathbf{F}$ derived above, this gives

$$(\nabla \times F_1\mathbf{i}) \cdot \hat{\mathbf{n}} = (\hat{\mathbf{n}} \cdot \mathbf{k})\left[-\frac{\partial f}{\partial y}\frac{\partial F_1}{\partial z} - \frac{\partial F_1}{\partial y}\right] = -\frac{\partial G}{\partial y}(\hat{\mathbf{n}} \cdot \mathbf{k}).$$

Multiplying this by dS noting that $\hat{\mathbf{n}} \cdot \mathbf{k} \, dS = dx \, dy$ as this is the projection of the element of area dS on to the x, y plane,

$$(\nabla \times F_1 \mathbf{i}) \cdot \hat{\mathbf{n}} \, dS = -\frac{\partial G}{\partial y} dx \, dy.$$

This last step is larger than it looks and gives the projected result:

$$\iint_S \nabla \times F_1 \mathbf{i} \cdot d\mathbf{S} = \iint_D -\frac{\partial G}{\partial y} dx \, dy$$

writing D for the projection of S on to the x, y plane, see Figure 11.1. In this figure, the orientation is also shown that ensures the normal points in the $\mathbf{k}$ direction and not the $-\mathbf{k}$ direction. Green's theorem in the plane is now used on the plane region D; here it is again from Chapter 10,

$$\oint_\Gamma (Pdx + Qdy) = \iint_D \left(\frac{\partial Q}{\partial x} - \frac{\partial P}{\partial y} \right) dx \, dy$$

and it is applied with $P = G$ and $Q = 0$; hence,

$$\oint_\Gamma G \, dx = \iint_D \left(-\frac{\partial G}{\partial y} \right) dx \, dy,$$

where the border of D is labelled Γ as C is reserved for the border of S itself. Hence,

$$\iint_S \nabla \times F_1 \mathbf{i} \cdot d\mathbf{S} = \oint_\Gamma G \, dx = \oint_C F_1 \, dx,$$

where the final integral around the border of S must be the same as the integral around the border of D as neither contains any z by construction. For now, let us assume that the projections on to the other two co-ordinate planes lead to similar relationships, namely

$$\iint_S \nabla \times F_2 \mathbf{j} \cdot d\mathbf{S} = \oint_C F_2 \, dy$$

and

$$\iint_S \nabla \times F_3 \mathbf{k} \cdot d\mathbf{S} = \oint_C F_3 \, dz.$$

Taking $\mathbf{i}$ times the first adding it to $\mathbf{j}$ times the second then to $\mathbf{k}$ times the third, remembering that

$$(\mathbf{a} \cdot \mathbf{i})\mathbf{i} + (\mathbf{a} \cdot \mathbf{j})\mathbf{j} + (\mathbf{a} \cdot \mathbf{k})\mathbf{k} = \mathbf{a}$$

results in Stokes' theorem:

$$\oint_C \mathbf{F} \cdot d\mathbf{r} = \iint_S \boldsymbol{\nabla} \times \mathbf{F} \cdot d\mathbf{S},$$

the unit vectors are constant and can be attached to the components of $\mathbf{F}$, but this is only justified if the projections can be done, and a glance at Figure 11.1 tells us that is certainly not always the case. In the figure, the projection on to the x, z plane would be invalid due to the 'fold' in S. The real problem is that the factor $\hat{\mathbf{n}} \cdot \mathbf{j}$ becomes zero when the vectors are at right angles to each other, and this happens in such folds. Algebraically, this singularity is due to a zero of $\hat{\mathbf{n}} \cdot \mathbf{j}$ or similar dot products of the normal with $\mathbf{i}$ or $\mathbf{k}$ in the denominator of the projected surface integral. In order to avoid this, an immediate strategy would be to treat S as a collection of surfaces, all of which obey the rules. This is fine as long as the isolated singularity at the junctions does not present a serious problem, and usually it does not. It does if the surface is a cylinder and the projection is parallel to the axis of the cylinder; the projection of any part of the curved area would then be just an arc, or maybe a circle, the area of which would be zero. However, re-orient the axes a little and this solves the problem. In fact, this is the key. As already mentioned in the beginning of this section, Stokes' theorem is co-ordinate independent, so any restriction on its validity that springs from the properties of the co-ordinate system one happens to be using must be able to be removed. The pure mathematical proofs will show this rigorously, but in this applied text, we will be content to leave it here. □

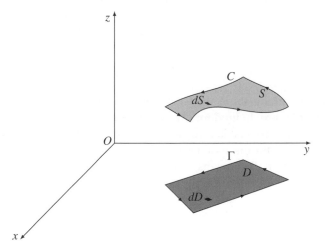

Figure 11.1 The curved surface S and its bounding closed curve C, together with its projection on to the x, y plane D and its bounding curve Γ, bounding curves are directed. Notice that as the boundaries are traversed, the domain is on the left; this is the correct orientation.

Stokes' theorem applies only if the surface S has a bounding curve, it has nothing to say about closed surfaces such as the surface of a sphere or a cylinder with the ends closed. Stokes' theorem does work for surfaces that have a bounding closed curve, that is, the curve C does not have breaks. It can have corners, like a square, as long as there are only a finite number of them (fractals are not allowed) and these corners are joined by curves that are smooth, that is, they have gradients that are continuous. These curves are called *piecewise smooth* curves. Likewise, the surface inside the curve must also be piecewise smooth; for surfaces, this means that almost everywhere there is a well-defined tangent plane, and the slope of the plane can twist and turn as the point of contact moves around, as long as the changes are not jerky. A cylinder will have a curved surface that is smooth, and if one end is closed, that too is a disc and is smooth. The rim of the disc where it joins the curved part is not smooth. However, the word 'piecewise' comes to our rescue and permits us to join two smooth surfaces together even if the join is not itself smooth. Again, as with curves, surfaces of a fractal nature are not allowed. Therefore, Stokes' theorem applies to all normal surfaces that have a bounding closed curve.

Fields occur all over the place in physics and engineering, fluid mechanics has flow, thermodynamics has heat transfer, and electromagnetism as the electric and magnetic fields. Mechanics itself has force, a field, and in elasticity theory the displacement vector is a field. In many of these applications, the equation $\nabla \times \mathbf{F} = \mathbf{0}$ will hold. In mechanics, this is a conservative field of force, in fluid mechanics, it is an irrotational flow and in electromagnetic theory, it can arise from Maxwell's equations. If a surface is such that for all closed curves C drawn in that surface $\nabla \times \mathbf{F} = \mathbf{0}$ inside C, then applying Stokes' theorem implies that

$$\oint_C \mathbf{F} \cdot d\mathbf{r} = 0.$$

In mechanics this says something about the work done traversing the closed curve, any closed curve. It is zero. In fluid mechanics, it defines the circulation around any closed curve; once more, it is zero. Section 6.4 fields with zero curl were discussed briefly, but now the discussion can be rounded off. Suppose that $\nabla \times \mathbf{F} = \mathbf{0}$ throughout a domain, this means that for *every* surface S drawn in the domain

$$\iint_S \nabla \times \mathbf{F} \cdot d\mathbf{S} = 0.$$

Applying Stokes' theorem to this implies that for *every* closed curve C

$$\oint_C \mathbf{F} \cdot d\mathbf{r} = 0$$

too. This implies that $\mathbf{F} \cdot d\mathbf{r} = d\phi$ is an exact differential. However, from Section 6.2 $d\phi = \nabla\phi \cdot d\mathbf{r}$, thus

$$\mathbf{F} \cdot d\mathbf{r} = d\phi = \nabla\phi \cdot d\mathbf{r}$$

so

$$(\mathbf{F} - \nabla\phi) \cdot d\mathbf{r} = 0$$

and since $d\mathbf{r}$ is arbitrary, it must be true that $\mathbf{F} = \nabla\phi$. In Chapter 6, it was established that $\nabla \times \nabla\phi = \mathbf{0}$, hence we have the following theorem.

Theorem 11.2   If $\mathbf{F}$ is a vector field, then $\nabla \times \mathbf{F} = \mathbf{0}$ if and only if $\mathbf{F} = \nabla\phi$.

This has now been established. In Chapter 6, it was shown that if $\mathbf{F} = \nabla\phi$, then $\nabla \times \mathbf{F} = \mathbf{0}$. To show that if $\nabla \times \mathbf{F} = \mathbf{0}$, then $\mathbf{F} = \nabla\phi$ is extremely difficult without Stokes' theorem, that is why this theorem has been postponed until now. A specific example is worth doing at this point. It is couched in general vector calculus terms, the applications to specific fields is left until later.

Example 11.1   Prove that

$$\oint_C \phi \, d\mathbf{r} = \iint_S d\mathbf{S} \times \nabla\phi.$$

Solution: The key here lies in choosing the correct form of the vector $\mathbf{F}$ in Stokes' theorem. The form $\mathbf{F} = \mathbf{a}\phi(x, y, z)$ where $\mathbf{a}$ is a constant but otherwise arbitrary vector works here. Stokes' theorem is

$$\oint_C \mathbf{F} \cdot d\mathbf{r} = \iint_S \nabla \times \mathbf{F} \cdot d\mathbf{S},$$

and writing $\mathbf{F} = \mathbf{a}\phi(x, y, z)$, this implies $\nabla \times (\mathbf{a}\phi) = \phi\nabla \times \mathbf{a} + \nabla\phi \times \mathbf{a}$. The first term vanishes as $\mathbf{a}$ is a constant vector. Thus, $\nabla \times (\mathbf{a}\phi) = \nabla\phi \times \mathbf{a}$. Stokes' theorem is thus

$$\oint_C \mathbf{a}\phi \cdot d\mathbf{r} = \iint_S \nabla\phi \times \mathbf{a} \, d\mathbf{S}.$$

The left-hand side is

$$\mathbf{a} \cdot \oint_C \phi \, d\mathbf{r}$$

since $\mathbf{a}$ is constant, and using the properties of the scalar triple product, it is possible to write

$$\nabla\phi \times \mathbf{a} \cdot d\mathbf{S} = d\mathbf{S} \times \nabla\phi \cdot \mathbf{a}$$

by cyclic permutation of the three vectors. Hence,

$$\iint_S \nabla\phi \times \mathbf{a} \cdot d\mathbf{S} = \iint_S d\mathbf{S} \times \nabla\phi \cdot \mathbf{a} = \mathbf{a} \cdot \iint_S d\mathbf{S} \times \nabla\phi$$

again as $\mathbf{a}$ is a constant vector. Thus,

$$\mathbf{a} \cdot \oint_C \phi \, d\mathbf{r} = \mathbf{a} \cdot \iint_S d\mathbf{S} \times \nabla\phi$$

or

$$\mathbf{a} \cdot \left[\oint_C \phi \, d\mathbf{r} - \iint_S d\mathbf{S} \times \nabla\phi \right] = 0$$

and since $\mathbf{a}$ is arbitrary, it must be the case that

$$\oint_C \phi \, d\mathbf{r} = \iint_S d\mathbf{S} \times \nabla\phi.$$

This completes the proof. The notation on the right is perhaps a little odd, and for those of a nervous disposition, it is better to write

$$\oint_C \phi \, d\mathbf{r} = \iint_S (d\mathbf{S} \times \nabla\phi).$$

□

Here is a more computational example where the power of Stokes' theorem is used.

Example 11.2 Use Stokes' theorem to evaluate the integral

$$\iint_S \nabla \times \mathbf{F} \cdot d\mathbf{S},$$

where $\mathbf{F} = (x^2 + y - 4)\mathbf{i} + 3xy\mathbf{j} + (2xz + z^2)\mathbf{k}$ and S is that part of the paraboloid $z = 4 - (x^2 + y^2)$ that lies above the x, y plane.

Solution: Stokes' theorem tells us that

$$\oint_C \mathbf{F} \cdot d\mathbf{r} = \iint_S \nabla \times \mathbf{F} \cdot d\mathbf{S}$$

so it appears easier to evaluate the integral on the left which will be a path integral over the border of the paraboloid, the circle $x^2 + y^2 = 4, z = 0$. In order to evaluate this integral, parameterise C by spotting that, on $C, z = 0, x = 2\cos\theta$ and $y = 2\sin\theta$, see Figure 11.2. Hence,

$$\mathbf{F} = (4\cos^2\theta + 2\cos\theta - 4)\mathbf{i} + 12\cos\theta\sin\theta\mathbf{j}$$

Figure 11.2 The paraboloid $z = 4 - x^2 - y^2$ and its bounding curve $x^2 + y^2 = 4$ on the x, y plane.

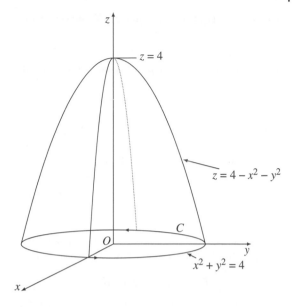

and the z component is zero because $z = 0$. In addition, on C,

$$d\mathbf{r} = dx\,\mathbf{i} + dy\,\mathbf{j} = -2\mathbf{i}\sin\theta\,d\theta + 2\mathbf{j}\cos\theta\,d\theta.$$

A couple of points to note here. It might be expected to abandon Cartesian co-ordinates as we are dealing with a circle with centre at the origin, tailor-made for plane polars. However, the vector $\mathbf{F}$ is in terms of Cartesians and to convert this into plane polar co-ordinates is more awkward than using Cartesian co-ordinates throughout. Secondly, note that on the curve C, $d\mathbf{r}$ has zero z component, so even if $\mathbf{F}$ did have a non-zero z component, it would play no part in the calculation of the path integral. Using the derived expressions we have, on C,

$$\mathbf{F} \cdot d\mathbf{r} = (-8\cos^2\theta\sin\theta - 4\sin^2\theta + 8\sin\theta + 24\cos^2\theta\sin\theta)d\theta$$

and the right-hand side is integrated between the limits 0 and 2π to traverse the circle C. The only non-zero contribution comes from the $-4\sin^2\theta$ term, and the integration is straightforward (use $2\sin^2\theta = 1 - \cos 2\theta$) giving the answer -4π. Any piecewise smooth surface having the border $x^2 + y^2 = 4, z = 0$ will have the same surface integral as a consequence of Stokes' theorem. Those who know about electric charge might find this less surprising, but it shows the power of Stokes' theorem. □

There are more applications of Stokes' theorem to come, but first there is another theorem to introduce, this time connecting surface and volume integrals.

11.3   Gauss' Divergence Theorem

In the introduction to Chapter 10, there was a section explaining how surface integrals occur naturally when calculating the flux of a quantity through a surface. If that surface is closed, then it is tempting to think that this flux must be zero, but that discounts the possibility that material is either being created or lost from inside this closed surface. Such a surface does not have contain something exotic like a black hole, it could be a two-dimensional fluid problem that contains a drain or a tap. Therefore, although the flux through a closed surface may not be zero, there must be a relationship between the amount of a material either being lost or being created inside the closed surface and the flux of the same material through the surface. This relationship is encapsulated in the following theorem that connects surface and volume integrals.

Theorem 11.3   If V, a simple solid region with bounding surface S and **F** is a vector field differentiable inside V and on S, then

$$\iiint_V \nabla \cdot \mathbf{F}\, dV = \oiint_S \mathbf{F} \cdot d\mathbf{S}.$$

Proof: First this will be proved for the case where V is a cuboid. The cuboid is shown in Figure 11.3. Consider the direction indicated by the arrows, and the volume integral

$$\iiint_V \frac{\partial F_2}{\partial y}\, dV = \iint_{S_2} \int_{y_1}^{y_2} \frac{\partial F_2}{\partial y}\, dy\, dS_2$$

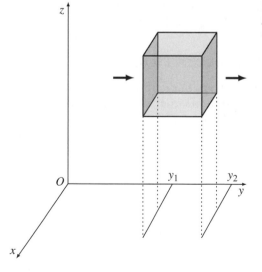

Figure 11.3 The cuboid V of sides $x_2 - x_1, y_2 - y_1, z_2 - z_1$. Only the y dimension is shown.

$$= \iint\limits_{S_2} [F_2(x, y_2, z) - F_2(x, y_1, z)]dS_2.$$

Here, S_2 denotes the two rectangles that form the left and right sides of the cuboid shown. On both faces $dS_2 = dx\,dz$ but on the left, $d\mathbf{S}_2 = -\mathbf{j}\,dx\,dz$ whereas on the right, $d\mathbf{S}_2 = \mathbf{j}\,dx\,dz$. Hence,

$$\iiint\limits_{V} \frac{\partial F_2}{\partial y}dV = \iint\limits_{S_2} [F_2(x, y_2, z) - F_2(x, y_1, z)]dS_2 =$$

$$= \iint\limits_{S_2} \mathbf{j}[F_2(x, y_2, z) + F_2(x, y_1, z)] \cdot d\mathbf{S}_2,$$

which can be written

$$\iiint\limits_{V} \frac{\partial F_2}{\partial y}dV = \iint\limits_{S_2} F_2\mathbf{j} \cdot d\mathbf{S}_2,$$

if it is recognised that the surface integral on the right indicates the two rectangles on the left and right of the cuboid in Figure 11.3. In a similar way, using the x-facing and z-facing sides of the same cuboid,

$$\iiint\limits_{V} \frac{\partial F_1}{\partial x}dV = \iint\limits_{S_1} F_1\mathbf{i} \cdot d\mathbf{S}_1$$

and

$$\iiint\limits_{V} \frac{\partial F_3}{\partial z}dV = \iint\limits_{S_3} F_3\mathbf{k} \cdot d\mathbf{S}_3.$$

Summing these three, and indicating by $\mathbf{S}$ the entire surface of the cuboid with an outward drawn normal on each of the six faces,

$$\iiint\limits_{V} \left\{ \frac{\partial F_1}{\partial x} + \frac{\partial F_2}{\partial y} + \frac{\partial F_3}{\partial z} \right\} dV = \iint\limits_{S} \mathbf{F} \cdot d\mathbf{S}$$

is obtained, and the left-hand side should be recognised as the divergence of $\mathbf{F} = F_1\mathbf{i} + F_2\mathbf{j} + F_3\mathbf{k}$. Hence, Gauss' flux theorem

$$\iiint\limits_{V} \mathbf{\nabla} \cdot \mathbf{F}\,dV = \oiint\limits_{S} \mathbf{F} \cdot d\mathbf{S}$$

is established for a cuboid. Now, suppose two cuboids are placed side by side. The contributions to the net surface integral from the two touching faces will exactly cancel, and Gauss' flux theorem will remain true for the combination. This argument will still hold for a whole agglomeration of cuboids as seen in

Figure 11.4 Three approximations to a sphere using progressively more cuboids.

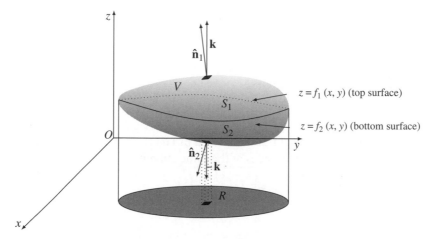

Figure 11.5 The solid V with upper surface S_1 and lower surface S_2 shown together with their normals $\hat{n}_1$ on dS_1 and $\hat{n}_2$ on dS_2 respectively. The normals also show the orientation of the infinitesimal areas dS_1 and dS_2, the normal to $dR = dx\, dy$ is of course vertical.

Figure 11.4, so in this way the proof of Gauss' flux theorem can be extended certainly to any simple solid and with thought to others too, provided as the number of cuboids increases, the resulting outer surface is such that all curves on it remain rectifiable. This is another way of saying that fractals are avoided. In case this is a problem, a more traditional proof is given below for completeness. Figure 11.5 shows the situation. A volume V is surrounded by a closed surface S. The surface is divided in two, an upper surface S_1 and a lower surface S_2 such that a typical vertical line cuts each just once. The surface S is thus assumed convex. The proof uses components rather along the lines of the proof of Stokes' theorem; so let $\mathbf{F} = \mathbf{i}F_1 + \mathbf{j}F_2 + \mathbf{k}F_3$ and consider the volume integral of the z derivative of the third component:

$$\iiint\limits_{V} \frac{\partial F_3}{\partial z} dV.$$

The choice, indicated in Figure 11.5 is to choose $dV = dz\, dx\, dy$, that is integrated with respect to z first. Therefore,

$$\iiint_V \frac{\partial F_3}{\partial z} dV = \iint_R \left\{ \int_{z=f_2(x,y)}^{z=f_1(x,y)} \frac{\partial F_3}{\partial z} dz \right\} dx\, dy$$

$$= \iint_R [F_3(x,y,f_1) - F_3(x,y,f_2)] dx\, dy.$$

Although the above integration looks straightforward enough, assuming $dV = dz\, dx\, dy$ means that the surfaces S_1 and S_2 have been projected on to the x, y plane and this is not useful for proving Gauss flux theorem. The top limit of the integral gives rise to the $F_3(x,y,f_1)dx\, dy$ term, and using Figure 11.5 for guidance, it can be seen that on $S_1, dx\, dy = \hat{\mathbf{n}}_1 \cdot \mathbf{k}\, dS_1$. The bottom limit of the integral gives rise to the $F_3(x,y,f_2)dx\, dy$ term, and similarly on $S_2, dx\, dy = \hat{\mathbf{n}}_2 \cdot (-\mathbf{k})dS_2$. On this lower surface S_2, the minus sign could use a little explanation. Several are available. Here, the outward drawn normal points downwards whereas $\mathbf{k}$ points upwards, thus the scalar product involves the product $|\hat{\mathbf{n}}||\mathbf{k}|$ times the cosine of the angle between these vectors. The angle is obtuse, so the cosine is negative and it is this that gives rise to the minus sign. The shortcut (using $-\mathbf{k}$) also works. Thus, the combination above gives

$$\iiint_V \frac{\partial F_3}{\partial z} dV = \iint_{S_1} F_3(x,y,f_1)\hat{\mathbf{n}}_1 \cdot \mathbf{k}\, dS_1 + \iint_{S_2} F_3(x,y,f_2)\hat{\mathbf{n}}_2 \cdot \mathbf{k}\, dS_2$$

and the right-hand pair amounts to the surface integral

$$\iint_S F_3\mathbf{k} \cdot \mathbf{n}\, dS = \iint_S F_3\mathbf{k} \cdot d\mathbf{S},$$

so

$$\iiint_V \frac{\partial F_3}{\partial z} dV = \iint_S F_3\mathbf{k} \cdot d\mathbf{S}.$$

Projecting on to the other co-ordinate planes gives rise to similar relationships:

$$\iiint_V \frac{\partial F_2}{\partial y} dV = \iint_S F_2\mathbf{j} \cdot d\mathbf{S} \quad \text{and} \quad \iiint_V \frac{\partial F_1}{\partial x} dV = \iint_S F_1\mathbf{i} \cdot d\mathbf{S}.$$

Adding these three together results in

$$\iiint_V \nabla \cdot \mathbf{F}\, dV = \iint_S \mathbf{F} \cdot d\mathbf{S},$$

which is Gauss' flux theorem. Generalising this to odd-shaped (non-convex), closed surfaces can be done by dividing such volumes into smaller ones.

As for Green's theorem in the plane and Stokes' theorem, the notation ∂V instead of the smooth surface S can be used so Gauss' flux theorem could be written:

$$\iiint_V \nabla \cdot \mathbf{F}\, dV = \oiint_{\partial V} \mathbf{F} \cdot d\mathbf{S}.$$

□

Here is a useful result couched as an example.

Example 11.3   If ϕ is a function continuously differentiable inside and on a closed surface S of a volume V, show that

$$\iiint_V \nabla \phi\, dV = \oiint_S \phi\, d\mathbf{S}.$$

Solution: This is a reasonably straightforward application of Gauss' flux theorem. It is here because of its importance in fluid mechanics. Put $\mathbf{F} = \phi \mathbf{a}$ where $\mathbf{a}$ is a constant but arbitrary vector and use the identity
$$\nabla \cdot (\phi \mathbf{a}) = \mathbf{a} \cdot \nabla \phi + \phi \nabla \cdot \mathbf{a} = \mathbf{a} \cdot \nabla \phi \text{ since } \mathbf{a} \text{ is a constant. Therefore,}$$

$$\iiint_V \nabla \cdot (\phi \mathbf{a}) dV = \iiint_V \nabla \phi \cdot \mathbf{a}\, dV = \mathbf{a} \cdot \iiint_V \nabla \phi\, dV.$$

The right-hand side of Gauss' flux theorem becomes

$$\oiint_{\partial V} (\mathbf{a}\phi) \cdot d\mathbf{S} = \mathbf{a} \cdot \oiint_{\partial V} \phi\, d\mathbf{S}.$$

Hence,

$$\mathbf{a} \cdot \iiint_V \nabla \phi\, dV = \mathbf{a} \cdot \oiint_{\partial V} \phi\, d\mathbf{S}$$

or

$$\mathbf{a} \cdot \left\{ \iiint_V \nabla \phi\, dV - \oiint_{\partial V} \phi\, d\mathbf{S} \right\} = 0.$$

Since the constant vector $\mathbf{a}$ is arbitrary, it must be true that

$$\iiint_V \nabla \phi\, dV = \oiint_{\partial V} \phi\, d\mathbf{S}.$$

With $\phi = p$ fluid pressure, this equation is used in deriving Euler's equation in fluid mechanics.

□

Here is a computational example.

Example 11.4   Use Gauss' flux theorem to calculate the values of the surface integral

$$\oiint_S \mathbf{F} \cdot d\mathbf{S}$$

with $\mathbf{F} = 2xy\mathbf{i} + yz^2\mathbf{j} + xz\mathbf{k}$ in the following two cases:

a) The surface of the region bounded by the planes: $x = 0, y = 0, z = 0$, $x = 2, y = 1$ and $z = 3$.
b) The surface of the cylindrical region bounded by $x^2 + y^2 = 1, z = 0$ and $z = 1$.

Solution: Using Gauss' flux theorem means that we need to calculate $\nabla \cdot \mathbf{F}$. Therefore, differentiating each component and adding gives, thus

$$\nabla \cdot \mathbf{F} = 2y + z^2 + x.$$

For part (a), Cartesian co-ordinates are used and the order z then y then x is chosen. All limits are constant here, so some of the tricky features explored in Chapter 9 do not occur.

$$\iiint_V \nabla \cdot \mathbf{F} \, dV = \int_0^2 \int_0^1 \int_0^3 (2y + z^2 + x) dz \, dy \, dx$$

$$= \int_0^2 \int_0^1 \left[2yz + \frac{1}{3}z^3 + xz \right]_0^3 dy \, dx$$

$$= \int_0^2 \int_0^1 6y + 9 + 3x \, dy \, dx$$

$$= \int_0^2 [3y^2 + 9y + 3xy]_0^1 dx = \int_0^2 12 + 3x \, dx$$

$$= \left[12x + \frac{3}{2}x^2 \right]_0^2 = 24 + 6 = 30.$$

Hence,

$$\oiint_S \mathbf{F} \cdot d\mathbf{S} = 30.$$

This is much easier than performing six integrals to cover the surface. For part (b), cylindrical polar co-ordinates are used. The calculation proceeds as before, but this time

$$\nabla \cdot \mathbf{F} = 2r \sin \theta + z^2 + r \cos \theta$$

so with the order z then θ then r, the limits are once again constant and the calculation proceeds as follows:

$$\iiint_V \nabla \cdot \mathbf{F}\, dV = \int_0^1 \int_0^{2\pi} \int_0^1 (2r \sin\theta + z^2 + r\cos\theta)dz\, d\theta\, dr$$

$$= \int_0^1 \int_0^{2\pi} \left[2rz \sin\theta + \frac{1}{3}z^3 + rz\cos\theta \right]_0^1 d\theta\, dr$$

$$= \int_0^1 \int_0^{2\pi} \left\{ 2r \sin\theta + \frac{1}{3} + r\cos\theta \right\} d\theta\, dr$$

and the integration with respect to θ eliminates the trigonometric terms and only

$$\int_0^1 \frac{2\pi}{3} dr$$

remains. The answer is thus

$$\frac{2\pi}{3}.$$

Once more, this is much more straightforward than adding the individual surface integrals. □

In Exercise 10.11 at the end of Chapter 10, the calculation of flux through a closed surface was addressed. This problem showed that the flux due to a source was the same for both a cylinder and a sphere. Let us generalise this result. Consider the integral

$$\iint_S \mathbf{E} \cdot d\mathbf{S},$$

where $E = (\mu \mathbf{r})/r^3$ and S is a simple closed surface. By 'simple', this means that any line that cuts this surface only does so in a maximum of two places. This is not strictly necessary but helps simplify the arguments. Generalisations can be made later. Notice that

$$\nabla \cdot \mathbf{E} = \frac{\mu}{r^2} \frac{\partial}{\partial r} \left(r^2 \cdot \frac{1}{r^2} \right) = 0$$

provided there are no zeros of r inside the closed surface S. Let there be one zero at the point P. Surround this by a small sphere of radius ϵ labelled S_ϵ, then

$$\iint_{S-S_\epsilon} \mathbf{E} \cdot d\mathbf{S} = \iint_S \mathbf{E} \cdot d\mathbf{S} - \iint_{S_\epsilon} \mathbf{E} \cdot d\mathbf{S} = 0,$$

where $S - S_\epsilon$ denotes the volume with the zero of r removed. Now,

$$\iint_{S_\epsilon} \mathbf{E} \cdot d\mathbf{S} = \mu \int_0^{2\pi} \int_0^\pi \frac{\epsilon^3 \sin \phi}{\epsilon^3} d\phi \, d\theta = 4\mu\pi,$$

hence in this circumstance

$$\iint_{S_\epsilon} \mathbf{E} \cdot d\mathbf{S} = 4\mu\pi.$$

We thus have the result that, for a simple closed surface S,

$$\iint_S \mathbf{E} \cdot d\mathbf{S} = \begin{cases} 0 & \text{if } r \neq 0 \text{ inside } S, \\ 4\mu\pi & \text{if } r = 0 \text{ once inside } S. \end{cases}$$

For those who are uncomfortable subtracting areas, translate them into volumes through Gauss' flux theorem and subtract them. This result itself is sometimes called Gauss' theorem. This result generalises straightforwardly for a surface that contains many zeros of r and for regions that are not simple. It is left as an exercise to find out what the result is if there is a zero of r actually on the surface (see Exercise 11.3).

Here is another useful general result that springs from Gauss' flux theorem.

11.3.1 Green's Second Identity

For his first identity, see the exercises at the end of this chapter. Green's second identity can be stated as a theorem as follows:

Theorem 11.4 If ϕ and ψ are scalar variables with continuous second order partial derivatives inside and on a volume V that is bordered by the closed surface S then

$$\iiint_V (\psi \nabla^2 \phi - \phi \nabla^2 \psi) dV = \oiint_S (\phi \nabla \psi - \psi \nabla \phi) \cdot d\mathbf{S}.$$

Proof: Consider the div. of a product expansion

$$\nabla \cdot (\phi \nabla \psi) = \nabla \phi \cdot \nabla \psi + \phi \nabla^2 \psi.$$

Reverse the roles of ϕ and ψ to obtain

$$\nabla \cdot (\psi \nabla \phi) = \nabla \psi \cdot \nabla \phi + \psi \nabla^2 \phi.$$

Subtract these to obtain

$$\nabla \cdot (\phi \nabla \psi) - \nabla \cdot (\psi \nabla \phi) = \phi \nabla^2 \psi - \psi \nabla^2 \phi$$

or

$$\nabla \cdot (\phi \nabla \psi - \psi \nabla \phi) = \phi \nabla^2 \psi - \psi \nabla^2 \phi.$$

Now, apply Gauss' flux theorem to this expression with $\mathbf{F} = \phi \nabla \psi - \psi \nabla \phi$ to deduce that

$$\iiint_V (\psi \nabla^2 \phi - \phi \nabla^2 \psi) dV = \iiint_V \nabla.(\phi \nabla \psi - \psi \nabla \phi) dV$$

$$= \oiint_S (\phi \nabla \psi - \psi \nabla \phi) \cdot d\mathbf{S},$$

which immediately establishes Green's second identity. $\qquad \square$

A very important consequence of Green's second identity now follows. It is a little outside mainstream vector calculus, but it is a uniqueness theorem. It says that if you have found one solution to the problem of solving Laplace's equation

$$\nabla^2 \phi = 0$$

inside and on a volume V, then that is the only one there is apart perhaps from a scale constant. Let's do it as an example.

Example 11.5 Show that the solution of the boundary value problem $\nabla^2 \phi = 0$ inside and on a volume V with bounding surface S is unique apart perhaps for a constant.

Solution: Green's second identity will be used here, so first suppose there are two solutions ϕ and ψ to the boundary value problem. Therefore, both $\nabla^2 \phi = 0$ and $\nabla^2 \psi = 0$ then Green's second identity gives

$$\iiint_V (\psi \nabla^2 \phi - \phi \nabla^2 \psi) dV = 0 = \oiint_S (\phi \nabla \psi - \psi \nabla \phi) \cdot d\mathbf{S}.$$

Let $\hat{\mathbf{n}}$ be a normal to the surface S then

$$\nabla \phi \cdot d\mathbf{S} = \nabla \phi \cdot \hat{\mathbf{n}} \, dS = \frac{\partial \phi}{\partial n} dS$$

using the definition of the directional derivative, see Chapter 2. The statement of Green's second identity thus gives

$$\oint_S \left(\phi \frac{\partial \psi}{\partial n} - \psi \frac{\partial \phi}{\partial n} \right) dS = 0.$$

As n can be in any direction, the normal to a closed surface has to sweep over all directions in fact, so it must be true that

$$\phi \frac{\partial \psi}{\partial n} - \psi \frac{\partial \phi}{\partial n} = 0.$$

Therefore, dividing by $\phi\psi$,

$$\frac{1}{\psi}\frac{\partial\psi}{\partial n} = \frac{1}{\phi}\frac{\partial\phi}{\partial n}$$

so $\quad \dfrac{\partial}{\partial n}(\ln\psi) = \dfrac{\partial}{\partial n}(\ln\phi).$

Integrating with respect to n gives

$$\ln\psi = \ln\phi + C,$$

where C has to be constant as $\mathbf{n}$, as said earlier can be in any direction. Writing $C = \ln K$ gives on taking exponentials

$$\psi = K\phi,$$

which is a statement that the two solutions are equal apart from a scaling constant. This is an important result. It means that by whatever means a solution has been found, fair means or foul, as long as it is a solution, there is no other. This is particularly useful when, on problems that are hard to solve, numerical means have been used. This result still applies. Of course, many problems are only approximately of this type and are in reality non-linear. This result is then no longer valid. Modellers retain a hope that since their problem is usually approximately linear, the result is still approximately valid. Such discussions belong in a different book. □

Before getting to more physical applications, it is helpful to state co-ordinate-free definitions of grad., div. and curl, and the theorems of Stokes and Gauss enable us to do this.

11.4 Co-ordinate-Free Definitions

For divergence, letting the volume tend to zero in Gauss' flux theorem immediately tells us that

$$\boldsymbol{\nabla} \cdot \mathbf{F} = \lim_{V \to 0} \frac{1}{V} \oiint_S \mathbf{F} \cdot d\mathbf{S}$$

and this can be used to define $\boldsymbol{\nabla} \cdot \mathbf{F}$ where V is any arbitrary volume surrounding the also arbitrary point P. This is essentially a mean value theorem: at some point inside V, there exists the point P where this limit must hold. Example 11.3 can be used to give a co-ordinate-free definition of grad. as follows. First of all, the result from this example

$$\iiint_V \boldsymbol{\nabla}\phi \, dV = \oiint_{\partial V} \phi \, d\mathbf{S} = \oiint_{\partial V} \hat{\mathbf{n}}\phi \, dS$$

and as with div. the volume is shrunk around an arbitrary point P to give the definition

$$\nabla\phi = \lim_{V\to 0} \frac{1}{V} \oiint_{\partial V} \hat{\mathbf{n}}\phi \, dS.$$

The normal $\hat{\mathbf{n}}$ will be the outward drawn unit normal to the surface S; however, the direction this takes depends on how the volume shrinks to zero, and if a practical definition is required, it is easiest to let $\hat{\mathbf{n}}$ equal, in turn, $\mathbf{i}, \mathbf{j}$ and $\mathbf{k}$ to generate the three components of $\nabla\phi$. As an example, here is the x direction scalar product:

$$\nabla\phi \cdot \mathbf{i} = \frac{\partial\phi}{\partial x} = \lim_{V\to 0} \frac{1}{V} \oiint_{\partial V} \phi \, dS$$

and this equates the x partial derivative to the limiting value of ϕ integrated over a surface whose normal is in the x direction, with the equation being exact in the limit of the surface becoming infinitesimally small. Therefore, as the normal to a surface is indeed in the same direction as $\nabla\phi$, this limit should come as no surprise. Although the very small area ∂V, think of it as a disc, is in the local y, z plane, it is so small that ϕ itself varies little over it, so the limit is that of the value of ϕ this disc multiplied by the ratio of the area of the disc divided by the original volume, both getting smaller and smaller, in the limit, the result is much like a mean value theorem, see Section 1.2.2. The other components of $\nabla\phi$ are defined the same way. It is also possible to couch this in terms of curvilinear co-ordinates, see Chapter 7, but this is not done here.

Using Stokes' theorem to define curl is a little more difficult, but it is the essentially the same story as was used for grad. Stokes' theorem is

$$\oint_C \mathbf{F} \cdot d\mathbf{r} = \iint_S \nabla\times\mathbf{F} \cdot d\mathbf{S}$$

and write this as

$$\oint_C \mathbf{F} \cdot d\mathbf{r} = \iint_S \mathbf{n} \cdot \nabla\times\mathbf{F} \, dS.$$

Now, the right-hand side can take the form $A\mathbf{n} \cdot \nabla\times\mathbf{F}$ by the mean value theorem, where A is the area of S and the curl is evaluated at a point P inside S. This means it is possible to define

$$\mathbf{n} \cdot \nabla\times\mathbf{F} = \lim_{A\to 0} \left\{ \frac{1}{A} \oint_C \mathbf{F} \cdot d\mathbf{r} \right\}$$

as the component of curl in the direction of $\mathbf{n}$ at the point P. As $\mathbf{n}$ can be chosen arbitrarily, it can be, in turn, the three co-ordinate directions to give the three components of curl hence specifying $\nabla\times\mathbf{F}$ at the point P completely. The definitions presented here are essentially a halfway house between the practical application of grad., div. and curl and pure mathematics. Pure mathematicians will proceed to unify and tighten up these definitions by defining exterior derivatives, but this would be out of place here.

11.5 Applications of Integral Theorems

In this section, the two theorems are used to derive results in applied areas such as electromagnetism, fluid mechanics, elasticity and thermodynamics. This will show the general power of these theorems. Of all the areas of application, by far the richest is electromagnetism, so that is where we start.

11.5.1 Electromagnetic Theory

The subject of electromagnetism is not old. Although the Ancient Greeks knew about static electricity, not until the eighteenth century were humans aware of electricity as a current that flows. Just a little later, magnetism was put on a similar scientific footing, the two combining under Faraday and then Maxwell in the nineteenth century. The whole story cannot be covered here, but the elegant use of vector calculus is demonstrated. The laws of electromagnetism were first formulated by James Clerk Maxwell (1831–1879) in 1861 and 1862, modified and neatened by others. Let us list them here, and then use vector calculus to put them in mathematical terms. They are experimental and deduced through observing physics experiments so they cannot be proved in a mathematical sense. The temptation to couch Maxwell's equations in tensor notation as preferred in relativity theory will be resisted.

11.5.1.1 Maxwell's Equations

1) *Gauss' Law for Electricity*: The flux of electric field leaving a closed volume is equal to the amount of electric charge inside it.
2) *Gauss' Law for Magnetism*: The flux of magnetic field leaving a closed volume is zero (there are no magnetic monopoles).
3) *Faraday's Law of Induction*: The voltage induced in a closed circuit is proportional to the rate of change of the magnetic flux it encloses.
4) *Amperè's Circuital Law*: The magnetic field induced around a closed loop is proportional to the electric current plus displacement current (rate of change of electric field) it encloses.

Let us now, one by one, examine these laws. Apologies in advance for all the different quantities that appear here; an attempt has been made to make them as simple as possible. In mathematical terms, the first equation is

$$\oiint_S \mathbf{D} \cdot d\mathbf{S} = 4\pi \iiint_V \rho \, dV,$$

where $\mathbf{D}$ is the electric flux density, defined as $\mathbf{D} = \epsilon\mathbf{E}$ and $\mathbf{E}$ is the more familiar electric field vector. The scalar, ϵ constant in an isotropic medium is the *permittivity*. ρ is the electric charge density. This lends itself directly to Gauss' flux theorem, from which

$$\iiint_V \nabla \cdot \mathbf{D} \, dV = 4\pi \iiint \rho \, dV$$

and since V is arbitrary this leads to the first of Maxwell's equations

$$\nabla . \mathbf{D} = 4\pi\rho. \tag{11.2}$$

The second is derived similarly, but this time there is no right-hand side. The magnetic field is denoted by $\mathbf{H}$ and the magnetic flux density is $\mathbf{B}$ where $\mathbf{B} = \mu\mathbf{H}$ and μ is the permeability, also usually constant in an isotropic medium. The second of Maxwell's equations is

$$\oiint_S \mathbf{B} \cdot d\mathbf{S} = 0$$

and Gauss' theorem is used as before to convert this to a volume integral and thus

$$\nabla \cdot \mathbf{B} = 0, \tag{11.3}$$

the second of Maxwell's equations. It is a statement of the observation that isolated magnetic poles are never observed, they always occur in opposing pairs.

The third of Maxwell's equations is more challenging. As originally stated, Faraday explained the phenomena of induced magnetism by lines of force being cut and inducement of magnetism, all done verbally without any mathematics. Maxwell tightened up the explanation and thus we have the statement above. The closed circuit around which electricity flows can be written

$$\oint_C \mathbf{E} \cdot d\mathbf{r}$$

and the (time) rate of change of the induced magnetic flux $\mathbf{B}$ is

$$\frac{\partial}{\partial t} \iint_{S_1} \mathbf{B} \cdot d\mathbf{S}_1,$$

where S_1 is the surface contained by the circuit C. Applying Stokes' theorem to the circuit thus gives

$$\oint_C \mathbf{E} \cdot d\mathbf{r} = \iint_{S_1} \nabla \times \mathbf{E} \cdot d\mathbf{S}_1.$$

Maxwell's third equation implies the relation

$$\oint_C \mathbf{E} \cdot d\mathbf{r} = -\frac{1}{c}\frac{\partial}{\partial t}\iint_{S_1} \mathbf{B} \cdot d\mathbf{S}_1,$$

where $1/c$ is a constant of proportion; what this is will be picked up later. Stokes' theorem applied to the line integral on the left implies

$$\iint_{S_1} \mathbf{\nabla} \times \mathbf{E} \cdot d\mathbf{S}_1 = -\frac{1}{c}\frac{\partial}{\partial t}\iint_{S_1} \mathbf{B} \cdot d\mathbf{S}_1$$

and the arbitrariness of S_1 (because the circuit C is also chosen arbitrarily) leads to

$$\mathbf{\nabla} \times \mathbf{E} = -\frac{1}{c}\frac{\partial \mathbf{B}}{\partial t}. \tag{11.4}$$

Although this is experimental, many books include some kind of proof, and they are littered across the internet. The fact is it is shown to be reasonable. The placement in a mathematical equation to replace the verbiage is neat but does not actually prove anything; no more than writing $\mathbf{F} = m\ddot{\mathbf{r}}$ proves Newton's second law of motion.

The final Maxwell equation is a statement of Amperè's circuital law. The magnetic field induced by a closed loop is

$$\oint_C \mathbf{B} \cdot d\mathbf{r}.$$

Writing $\mathbf{J}$ as the electric current density, the total current across the circuit is

$$\iint_{S_1} \mathbf{J} \cdot d\mathbf{S}_1$$

and the rate of change of electric field mirrors that of the magnetic field in the third equation and is

$$\frac{\partial}{\partial t}\iint_{S_1} \mathbf{E} \cdot d\mathbf{S}_1$$

so the fourth equation is

$$\oint_C \mathbf{B} \cdot d\mathbf{r} = \frac{1}{c}\left(\iint_{S_1} \mathbf{J} \cdot d\mathbf{S}_1 + \frac{\partial}{\partial t}\iint_{S_1} \mathbf{E} \cdot d\mathbf{S}_1\right)$$

and once Stokes' theorem is invoked, and the integrals eliminated, the fourth equation becomes

$$\mathbf{\nabla} \times \mathbf{B} = \frac{1}{c}\left(4\pi \mathbf{J} + \frac{\partial \mathbf{E}}{\partial t}\right). \tag{11.5}$$

Thus, Equations (11.2)–(11.5) are Maxwell's equations. The form they take here have enough constants to preserve dimensionality, but little else. The constant c is reasonably standard for the speed of light. In terms of the other constants in the *cgs* system, the permittivity $\epsilon = 1/(4\pi c)$ and the permeability $\mu = 4\pi/c$. Multiplying these together gives the speed $c = 1/\sqrt{\mu\epsilon}$, this is of use when applying Maxwell's equations in the theory of special relativity, a common application as electromagnetic waves travel at the speed of light. In fact, it is a good exercise in vector calculus to show that both $\mathbf{E}$ and $\mathbf{H}$ obey the (three-dimensional) wave equation

$$\nabla^2 \mathbf{A} = \frac{1}{c^2} \frac{\partial^2 \mathbf{A}}{\partial t^2}.$$

The use of the bold $\boldsymbol{\nabla}$ rather than ∇ is necessary as, strictly, the scalar Laplacian ∇^2 can only be calculated directly in Cartesian co-ordinates. Sometimes, the symbol $\square$ called the *d'Alembertian* is used:

$$\square \equiv \frac{1}{c^2} \frac{\partial^2}{\partial t^2} - \nabla^2$$

so the wave equation would be written

$$\square \mathbf{A} = \mathbf{0}$$

but not here. It tends to make an appearance in more advanced texts that go on to a tensor treatment. The wave equation for $\mathbf{E}$ can be derived by eliminating $\mathbf{B}$ from the equations

$$\boldsymbol{\nabla} \times \mathbf{E} = -\frac{1}{c} \frac{\partial \mathbf{B}}{\partial t} \quad \text{and} \quad \boldsymbol{\nabla} \times \mathbf{B} = \frac{1}{c} \frac{\partial \mathbf{E}}{\partial t},$$

these are Maxwell equations number three and four but with $\mathbf{J} = \mathbf{0}$, that is no electric current density. Simply take the curl of the first equation and use the identity

$$\boldsymbol{\nabla} \times \boldsymbol{\nabla} \times \mathbf{E} = \boldsymbol{\nabla}(\boldsymbol{\nabla} \cdot \mathbf{E}) - \nabla^2 \mathbf{E}.$$

As the first Maxwell equation can also be written in terms of $\mathbf{E}$ as $\boldsymbol{\nabla} \cdot \mathbf{E} = 4\pi\rho/\epsilon$ and assuming the right-hand side is constant, the grad. of this will be zero; hence,

$$\boldsymbol{\nabla} \times \boldsymbol{\nabla} \times \mathbf{E} = -\nabla^2 \mathbf{E}.$$

Thus,

$$\boldsymbol{\nabla} \times \boldsymbol{\nabla} \times \mathbf{E} = -\nabla^2 \mathbf{E} = -\frac{1}{c} \frac{\partial}{\partial t}(\boldsymbol{\nabla} \times \mathbf{B})$$

and using the equation for $\boldsymbol{\nabla} \times \mathbf{B}$ gives

$$\nabla^2 \mathbf{E} = \frac{1}{c} \frac{\partial}{\partial t}(\boldsymbol{\nabla} \times \mathbf{B}) = \frac{1}{c^2} \frac{\partial^2 \mathbf{E}}{\partial t^2},$$

which is the wave equation. Showing that $\mathbf{B}$ and hence $\mathbf{H}$ obeys the wave equation too follows along similar lines and is left as an exercise. Given that both $\mathbf{E}$ and $\mathbf{B}$ obey the wave equation, the most natural solutions are plane wave solutions of the form

$$\mathbf{E} = \mathbf{E_0}\, e^{i(kx+ly+mz-\omega t)} \quad \text{and} \quad \mathbf{B} = \mathbf{B_0}\, e^{i(kx+ly+mz-\omega t)},$$

where the vectors $\mathbf{E_0}$ and $\mathbf{B_0}$ are constant, the variability is all in the wave. However, in order to keep the algebra in proportion, let us assume the electric wave is propagating in the x direction. In addition, suppose there are no external sources, so zero electric current density, $\mathbf{J} = \mathbf{0}$ and zero charge density $\rho = 0$. Therefore, with the simplification

$$\mathbf{E} = a\mathbf{i}\, e^{i(kx+ly-\omega t)}$$

together with the pared-down Maxwell equations

$$\nabla \times \mathbf{B} = \frac{1}{c}\frac{\partial \mathbf{E}}{\partial t} \quad \text{and} \quad \nabla \times \mathbf{E} = -\frac{1}{c}\frac{\partial \mathbf{B}}{\partial t},$$

it is now possible to calculate the magnetic flux density $\mathbf{B}$ as

$$\mathbf{B} = -\frac{cl}{\omega}a\mathbf{k}\, e^{i(kx+ly-\omega t)}$$

together with $\omega = cl$, the wave dispersion relation. These show that these electric and magnetic fields are at right angles; they are plane polarised waves. Further generalisations are possible; they are left for the exercises.

11.5.2 Fluid Mechanics

Now, let us turn to applications in fluid mechanics. Take an arbitrary volume of fluid shown in Figure 11.6 and consider the amount of fluid entering and leaving the volume. The total amount of mass leaving this volume does so through the surface surrounding the volume. The rate of increase of this volume must be

$$\frac{\partial}{\partial t}\int_V \rho\, dV = \int_V \frac{\partial \rho}{\partial t}\, dV,$$

Figure 11.6 An arbitrary volume of fluid showing the unit normal $\hat{\mathbf{n}}$ to the infinitesimal surface element $d\mathbf{S} = \hat{\mathbf{n}}\, dS$, the local fluid velocity $\mathbf{u}$ and the pressure p that is everywhere normal to the surface of the volume V.

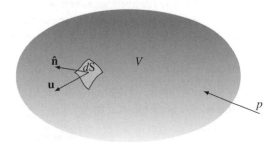

the time derivative can pass through the integration sign since the volume V although arbitrary does not depend on time. The flux of material into the surrounding surface S is

$$-\int_S \mathbf{u} \cdot d\mathbf{S},$$

where the direction of $d\mathbf{S}$ is the unit *outward* drawn normal, hence the negative sign. These must balance as the material entering V through its surface S has to be the same as the rate that V increases. Therefore, it must be the case that

$$\frac{\partial}{\partial t}\int_V \rho \, dV = -\int_S \mathbf{u} \cdot d\mathbf{S}.$$

Gauss' flux theorem is now used: for the fluid vector field $\mathbf{u}$, this takes the form

$$\int_S \mathbf{u} \cdot d\mathbf{S} = \int_V \boldsymbol{\nabla} \cdot \mathbf{u} \, dV.$$

Using this, we see that

$$\int_V \left\{ \frac{\partial \rho}{\partial t} + \boldsymbol{\nabla} \cdot \mathbf{u} \right\} dV = 0,$$

where V is our arbitrary volume. As it *is* arbitrary, this can only be true if

$$\frac{\partial \rho}{\partial t} + \boldsymbol{\nabla} \cdot \mathbf{u} = 0$$

throughout the domain of the fluid. This is the equation of conservation of mass, often called the continuity equation. If the density is constant, then the continuity equation takes the simple form:

$$\boldsymbol{\nabla} \cdot \mathbf{u} = 0.$$

Referring to the same volume in Figure 11.6, the acceleration of the current within this volume is

$$\frac{d\mathbf{u}}{dt}.$$

The volume itself moves around, so technically all variables are time dependent

$$\mathbf{u}(x, y, z, t) = \mathbf{u}(x(t), y(t), z(t), t).$$

This has the consequence that when we calculate the rate of change of $\mathbf{u}$ we really need to take account of the dependencies of x, y and z on t to do a proper job, and this is exactly what was covered by the chain rule introduced in Chapter 2:

$$\frac{d\mathbf{u}}{dt} = \frac{\partial \mathbf{u}}{\partial x}\frac{dx}{dt} + \frac{\partial \mathbf{u}}{\partial y}\frac{dy}{dt} + \frac{\partial \mathbf{u}}{\partial z}\frac{dz}{dt} + \frac{\partial \mathbf{u}}{\partial t}$$

or using vector notation and that

$$\frac{dx}{dt} = u, \quad \frac{dy}{dt} = v \quad \text{and} \quad \frac{dz}{dt} = w,$$

this can be written

$$\frac{d\mathbf{u}}{dt} = (\mathbf{u} \cdot \nabla)\mathbf{u} + \frac{\partial \mathbf{u}}{\partial t}$$

so this is another good application of vector calculus in fluid mechanics. The final application lies in applying Newton's second law to the volume. The left-hand side of the equation expressing Newton's second law is

$$\frac{\partial \mathbf{u}}{\partial t} + (\mathbf{u} \cdot \nabla)\mathbf{u}.$$

Let us turn to what is here the right-hand side of Newton's second law. This contains the forces that act on our small mass of fluid. In a fluid, there is always pressure. This is a force per unit area that acts on the surface of our small volume, but points inwards (in a direction $-\mathbf{n}$). Gravity also acts, but throughout the volume. Frictional forces, viscosity in laminar flow, but turbulence in general, are ignored. Therefore, we have

$$-\int_S p\mathbf{n} \, dS + \int_V \rho \mathbf{g} \, dV$$

as the total force. Before equating force to mass times rate of change of velocity in this volume, we need to convert the one surface integral into a volume integral. Example 11.3 gives the corollary to Gauss' flux theorem that is used here:

$$\int_S p\mathbf{n} \, dS = \int_V \nabla p \, dV.$$

Newton's second law is therefore

$$\int_V \rho \left(\frac{\partial \mathbf{u}}{\partial t} + (\mathbf{u} \cdot \nabla)\mathbf{u} \right) dV = -\int_V \nabla p \, dV + \int_V \rho \mathbf{g} \, dV,$$

and since the volume is arbitrary, and dividing by ρ, we have

$$\frac{\partial \mathbf{u}}{\partial t} + (\mathbf{u} \cdot \nabla)\mathbf{u} = -\frac{1}{\rho}\nabla p + \mathbf{g}. \tag{11.6}$$

This is Newton's second law for fluids and is called the Euler's equation for ideal fluid flow. Using the identity

$$\nabla(\mathbf{A} \cdot \mathbf{B}) = (\mathbf{B} \cdot \nabla)\mathbf{A} + (\mathbf{A} \cdot \nabla)\mathbf{B} + \mathbf{B} \times (\nabla \times \mathbf{A}) + \mathbf{A} \times (\nabla \times \mathbf{B})$$

with $\mathbf{A} = \mathbf{B} = \mathbf{u}$ gives

$$\nabla|\mathbf{u}|^2 = 2(\mathbf{u} \cdot \nabla)\mathbf{u} + 2\mathbf{u} \times (\nabla \times \mathbf{u})$$

so

$$(\mathbf{u} \cdot \nabla)\mathbf{u} = \nabla \left[\frac{1}{2}|\mathbf{u}|^2 \right] - \mathbf{u} \times (\nabla \times \mathbf{u}).$$

Euler's equation for fluids can thus be written

$$\frac{\partial \mathbf{u}}{\partial t} + \nabla \left[\frac{1}{2}|\mathbf{u}|^2 \right] - \mathbf{u} \times (\nabla \times \mathbf{u}) = -\frac{1}{\rho}\nabla p + \mathbf{g}. \tag{11.7}$$

Equation 11.7 can be integrated under some circumstances by using properties of vectors. Textbooks in fluid mechanics [1] are recommended for greater detail, so here let us be content to write

$$\zeta = \nabla \times \mathbf{u}$$

to define the *vorticity* of the fluid, and to assume $\zeta = \mathbf{0}$. A fluid with zero vorticity is called *irrotational* and vorticity in a fluid takes the same role as angular momentum does in mechanics. Vorticity is generated by friction, but in an inviscid flow is usually zero. Frictionless flows are sometimes called *ideal* flows. If $\nabla \times \mathbf{u} = \mathbf{0}$, then there exists a scalar function ϕ such that $\mathbf{u} = \nabla \phi$ and ϕ is called the *velocity potential*. Here, the concentration is on the application of vector calculus to fluid mechanics, so skipping over the implications to the physical world, assuming irrotational flow, Equation 11.7 can now be written:

$$\nabla \left[\frac{\partial \phi}{\partial t} \right] + \nabla \left[\frac{1}{2} |\mathbf{u}|^2 \right] = -\frac{1}{\rho} \nabla p + \mathbf{g}.$$

Spotting that gravity can also be written in terms of a potential as $\mathbf{g} = \nabla(-gz)$, this equation is in fact

$$\nabla \left[\frac{\partial \phi}{\partial t} + \frac{1}{2} |\mathbf{u}|^2 + \frac{p}{\rho} + gz \right] = 0.$$

If the scalar product of the left-hand side of this equation is taken with $d\mathbf{r}$, it becomes $\nabla \psi \cdot d\mathbf{r} = 0$ and this is $d\psi = 0$ from the definition of directional derivative. This integrates at once to $\psi = $ constant. Whence the equation

$$\frac{\partial \phi}{\partial t} + \frac{1}{2} |\mathbf{u}|^2 + \frac{p}{\rho} + gz = \text{constant}$$

has been derived. This is known as *Bernoulli's equation* after Daniel Bernoulli (1700–1782), one of the famous Swiss (originally Spanish Netherland Huguenots) Bernoulli family of eminent mathematicians, son of Johann and nephew of Jacob. It is an energy equation for fluid mechanics and has lots of interesting practical applications. The second term certainly looks like kinetic energy per unit mass, and fourth term potential energy per unit mass. The others need further explanation, found in more specialist texts. In many circumstances, only the middle two terms are important, so high flow means low pressure, and low flow means high pressure as the two sum to a constant (approximately). This can explain lift in aerofoils as the flow is greater over the top of a wing, hence reducing pressure compared to that below, promoting lift. It can also explain why newly emptied bins in a windy city centre delight in showing black bin liners flapping in the breeze; the breeze means low pressure compared to the pressure inside the bin and the net force pushes the bin-liner out. We leave fluid flow there and move on to elasticity.

11.5.3   Elasticity Theory

Elasticity and fluid mechanics are both part of continuum mechanics, so it is to be expected that many of the terms will be the same. The main difference is that elastic bodies retain their shape; a ball is an elastic solid and can be thrown through the air, and its motion has a lot in common with the behaviour of a rigid body in that, through the air it will rotate and travel in tune with Newton's laws. However, in addition, the ball will deform under a force but return to its original form once the force stops. The interesting part of elasticity theory is modelling how the elastic solid deforms; this is different. Rules have to be proposed (called *constitutive equations*). As in electromagnetic theory, there is much experimental science here, some of you will remember such quantities as Young's modulus and Poisson's ratio from school physics. These are measurable quantities and relations between the *stress*, a force per unit area (like pressure), and *strain* a dimensionless quantity proportional to the displacement due to the stress, remains a key area in the study of material science. In this section, applications of vector calculus are the focus, so in no way can all or even most of the essentials of elasticity theory be covered. However, it is important to devote some time to how both stress and strain are represented and to the relationship between them. Of all the applications of vector calculus, elasticity is the subject that can involve the most specialist (read advanced) mathematics. Having worked hard to avoid concepts such as manifolds and Lie groups, this near to the end of the textbook is not the place to suddenly face them, so those who need to know the full mathematical story are steered towards the book by Marsden and Hughes *The Mathematical Theory of Elasticity* published by Dover in 2003 [8]. Particularly tempting is to use tensors to their maximum extent, in particular introducing covariant and contravariant tensor notation; however, this too has been resisted as it is too much to face at this stage, so let us press on using only elementary mathematics.

It will be assumed that most readers will not be familiar with stress and strain, so starting with stress, Figure 11.7 has been drawn. It shows an infinitesimal cuboid of elastic material, and there are nine, yes nine, components of stress, three on each of the three orthogonal faces, the front face, top face and right face. The notation σ_{ij} has been adopted, the i refers to the direction of the normal to the surface and j refers to the actual direction of each force (stress is force per unit area). If i and j are the same, then the force is normal to the cuboid; this resembles pressure in a fluid which is always normal to a surface. Therefore, σ_{xx}, σ_{yy} and σ_{zz} are *normal* stresses. The other six forces $\sigma_{xy}, \sigma_{xz}, \sigma_{yx}, \sigma_{yz}, \sigma_{zx}$ and σ_{zy} are the *shear* stresses that resemble frictional forces. It is standard to refer to the nine components of stress in matrix form, and write

$$\mathcal{R} = \begin{pmatrix} \sigma_{xx} & \sigma_{xy} & \sigma_{xz} \\ \sigma_{yx} & \sigma_{yy} & \sigma_{yz} \\ \sigma_{zx} & \sigma_{zy} & \sigma_{zz} \end{pmatrix},$$

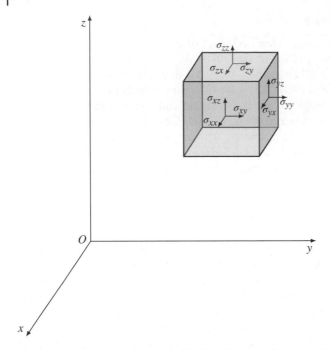

Figure 11.7 The infinitesimal cuboid of elastic material showing the nine components of the stress tensor.

where $\mathcal{R}$ is the *stress tensor*, a second-order tensor with nine components. Cartesian co-ordinates are used here for convenience, but the stress tensor as a second-order tensor is independent of co-ordinate system just as vectors or first-order tensors are. Whilst the Cartesian components are on display, the following facts are stated. The normal stress components σ_{xx}, σ_{yy} and σ_{zz} are the same as pressure and are indistinct from it. Moreover, since the infinitesimal cuboid is in equilibrium, in order that rotational motion is zero about each co-ordinate it has to be the case that $\sigma_{xy} = \sigma_{yx}, \sigma_{xz} = \sigma_{zx}$ and $\sigma_{yz} = \sigma_{zy}$, so the only unknowns are the pressure and three shear stresses. This is useful in the analysis of deformation of elastic material. The usual vector notation does not sit well with second-order tensors, so care is needed when interpreting both surface integrals and relevant integral theorems. Given this, time is now taken with understanding stress, and later strain too. Here is a vectorial equivalent description of the nine stress components shown in Figure 11.7:

$$\sigma_x = \sigma_{xx}\mathbf{i} + \sigma_{xy}\mathbf{j} + \sigma_{xz}\mathbf{k},$$
$$\sigma_y = \sigma_{yx}\mathbf{i} + \sigma_{yy}\mathbf{j} + \sigma_{yz}\mathbf{k},$$
$$\sigma_z = \sigma_{zx}\mathbf{i} + \sigma_{zy}\mathbf{j} + \sigma_{zz}\mathbf{k}.$$

This is useful, and using Cartesian co-ordinates turns out to be useful later. Co-ordinate-free approaches to stress are ideal and some will follow, but Cartesian co-ordinates are a useful for the introduction of strain, then the relationship between stress and strain. Strain is simply the name given to the deformation of a body caused by stress, compared to its unstressed position, hence it is a ratio of two distances and thus dimensionless. One of the similarities alluded to above is that both fluids and elastic bodies are continua. Thus, the total change in a quantity call it $\mathbf{X}(x, y, z)$ is, using the chain rule:

$$d\mathbf{X} = \frac{\partial \mathbf{X}}{\partial x} dx + \frac{\partial \mathbf{X}}{\partial y} dy + \frac{\partial \mathbf{X}}{\partial z} dz$$

usually written in vector form

$$d\mathbf{X} = (\nabla \mathbf{X}) \cdot d\mathbf{r}$$

only if you permit grad. of a vector (a second-order tensor). This resembles differentiation following the fluid or the Lagrangian view in fluid mechanics and the *material derivative* in elasticity. In Section 7.4.1 there is a brief section on Cartesian tensors, and in this notation

$$dX_i = X_{i,j} \, dx_j, \tag{11.8}$$

where there is summation over the j index. Suppose an arbitrarily shaped elastic solid V is subject to a body force $\mathbf{b}$ per unit mass as shown in Figure 11.8. In this figure, the infinitesimal surface element $d\mathbf{S}$ is also shown with its local normal $\hat{\mathbf{n}}$. For equilibrium of the elastic solid V, it has to be true that

$$\oiint_S \mathbf{t} \, dS + \iiint_V \rho \mathbf{b} \, dV = \mathbf{0}.$$

This, the equating of net force to zero, is a vector equation. It is not straightforward to convert the first term into a volume integral; it is best to write this equation in the tensor notation of Section 7.4.1 first. This is done by writing $\mathbf{t} = t_i$ and $\mathbf{b} = b_i$. Now, note that $t_i = \sigma_{ji} n_j$ so the surface integral converts to a volume integral and becomes, via Gauss' flux theorem,

$$\oiint_S \sigma_{ij} n_j \, dS = \iiint_V \sigma_{ij,j} \, dV$$

so the equilibrium condition is

$$\sigma_{ij,j} + \rho b_i = 0. \tag{11.9}$$

In vector terms, this has used that the divergence of a second-order tensor is a first-order tensor equation. This next bit is a good exercise in the manipulation in tensor notation. A volume in equilibrium does not rotate, so, again using tensor notation taking moments gives

$$\oiint_S \epsilon_{ijk} x_j \sigma_{kl} n_l \, dS + \iiint_V \rho \epsilon_{ijk} x_j b_k \, dV = 0.$$

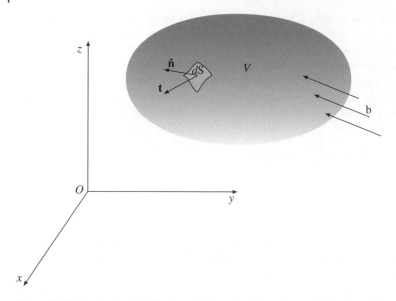

Figure 11.8 An elastic solid of volume V showing the unit normal $\hat{n}$ to the infinitesimal surface element $d\mathbf{S} = \hat{n}\, dS$ and the local direction of the stress $\mathbf{t}$. The body force $\mathbf{b}$ is also indicated.

Now, this looks a little daunting, but it is a first-order tensor or vector and i is the only free index, the others are summed over. This equation is $\mathbf{r}\times$ the equilibrium condition, and the alternating tensor ϵ_{ijk} is required to put vector products into tensor notation, $\mathbf{r} \times \mathbf{F} = \epsilon_{ijk} r_j F_k$, so it is not as bad as it first appears. The transformation of the first integral into a volume integral is achieved via

$$\oiint_S \epsilon_{ijk} x_j \sigma_{kl} n_l \, dS = \iiint_V \epsilon_{ijk}(x_j \sigma_{kl})_{,l}\, dV.$$

The above expression is thus

$$\iiint_V (\epsilon_{ijk}(x_j \sigma_{kl})_{,l} + \rho \epsilon_{ijk} x_j b_k) dV = 0$$

and as V can be any volume, it must be true that

$$\epsilon_{ijk}(x_j \sigma_{kl})_{,l} + \rho \epsilon_{ijk} x_j b_k = 0.$$

Using the product rule on the first term gives

$$\epsilon_{ijk}(x_{j,l}\sigma_{kl} + x_j \sigma_{kl,l}) + \rho \epsilon_{ijk} x_j b_k = 0.$$

Grouping the second and third terms extracting the common factor x_j leads to

$$\epsilon_{ijk} x_j (\sigma_{kl,l} + \rho b_k) + \epsilon_{ijk} x_{j,l}\sigma_{kl} = 0.$$

By Equation 11.9, the terms in parentheses add to zero. Hence,

$$\epsilon_{ijk} x_{j,l} \sigma_{kl} = 0$$

but

$$x_{j,l} = \frac{\partial x_j}{\partial x_l} = \delta_{jl}$$

so we have

$$\epsilon_{ijk} \delta_{jl} \sigma_{kl} = 0 \implies \epsilon_{ijk} \sigma_{kl} = 0.$$

Remembering the properties of the alternating tensor:

$$\epsilon_{ijk} = \begin{cases} 1 & \text{if } i, j, k \text{ are all different and permute cyclically,} \\ -1 & \text{if } i, j, k \text{ are all different and permute anti-cyclically,} \\ 0 & \text{in all other cases.} \end{cases}$$

So $\epsilon_{123} = \epsilon_{231} = \epsilon_{312} = 1, \epsilon_{132} = \epsilon_{321} = \epsilon_{213} = -1$ while every other ϵ_{ijk} for example ϵ_{112} or $\epsilon_{222} = 0$; this means that

$$\sigma_{12} = \sigma_{21}, \quad \sigma_{23} = \sigma_{32} \quad \text{and} \quad \sigma_{13} = \sigma_{31}$$

and the generalisation of the symmetric nature of the stress tensor σ_{ij} independent of co-ordinate system is established. Incidentally, this also demonstrates the power of tensor methods. However, because of the nature of the alternating tensor nothing can be said about the tensile stresses σ_{11}, σ_{22} and σ_{33} which is no surprise as the starting point was a moment and tensile stresses do not produce moments. Here in the form of an example is an introduction to calculating various elastic stresses.

Example 11.6 Determine how to calculate various stresses due to a stress given at a point by using the stress tensor

$$\mathcal{R} = \begin{pmatrix} 1 & 2 & 3 \\ 2 & 4 & 6 \\ 3 & 6 & 1 \end{pmatrix}.$$

Solution: First of all, let us calculate the force perpendicular to a particular plane. As an illustration, choose the y, z plane. This has unit normal **i** so the force will be

$$\mathcal{R} \cdot \mathbf{i} = \begin{pmatrix} 1 & 2 & 3 \\ 2 & 4 & 6 \\ 3 & 6 & 1 \end{pmatrix} \begin{pmatrix} 1 \\ 0 \\ 0 \end{pmatrix} = \begin{pmatrix} 1 \\ 2 \\ 3 \end{pmatrix}.$$

This is a force that is neither in the plane nor parallel to the normal to the plane, such is the nature of stress.

Here is another similar calculation: Suppose we want to find the force on the plane perpendicular to the direction $\mathbf{p} = \mathbf{i} + 2\mathbf{j} + 2\mathbf{k}$. The extra step is that you need to calculate the *unit* vector in the direction first. This is $\hat{\mathbf{p}} = (\mathbf{i} + 2\mathbf{j} + 2\mathbf{k})/3$. Then, the calculation proceeds as before:

$$\mathcal{R} \cdot \hat{\mathbf{p}} = \frac{1}{3} \begin{pmatrix} 1 & 2 & 3 \\ 2 & 4 & 6 \\ 3 & 6 & 1 \end{pmatrix} \begin{pmatrix} 1 \\ 2 \\ 2 \end{pmatrix} = \frac{1}{3} \begin{pmatrix} 11 \\ 22 \\ 17 \end{pmatrix}.$$

The answer is of course not a unit vector. It might be, sometimes. Here is a different calculation: Let us try and find the principal stress components at the point where the stress applies. This requires some theory first. Suppose the principal stress has the direction $\mathbf{p}$. Then, this will be the direction the stress (sometimes referred to as *traction*) has at this point so

$$\mathbf{p} \cdot \mathcal{R} = |\mathcal{R}|\mathbf{p}$$

or in tensors (vector notation does let us down here)

$$p_i R_{ij} = R p_i,$$

which is, written using the Kronecker delta δ_{ij}; that is, one if $i = j$ and zero otherwise,

$$(R_{ij} - R\delta_{ij})p_i = 0$$

and this in turn implies that the determinant

$$|R_{ij} - R\delta_{ij}| = 0$$

that is equivalent to determining the eigenvalues of the matrix whose components are the stress tensor. That is finding the roots of

$$\begin{vmatrix} 1-R & 2 & 3 \\ 2 & 4-R & 6 \\ 3 & 6 & 1-R \end{vmatrix} = 0,$$

which are 10, 0 and 2 with corresponding eigenvectors $3\mathbf{i} + 6\mathbf{j} + 5\mathbf{i}$, $-2\mathbf{i} + \mathbf{j}$ and $\mathbf{i} + 2\mathbf{j} - 3\mathbf{k}$. It is the first that gives the direction of the principal stress. Maybe you have not met eigenvalues and eigenvectors before; for details see a good text on linear algebra [7], but all we need here is the above determinant to get the eigenvalues, then the solution of a set of equations

$$\mathcal{R}\mathbf{v} = \lambda\mathbf{v}$$

to find the eigenvectors $\mathbf{v}$ where λ is one of the (three) eigenvalues. For example, the largest eigenvalue is 10 and so the directions of the principal stress

$\mathbf{p} = p_1\mathbf{i} + p_2\mathbf{j} + p_3\mathbf{k}$ is given by solving the set of three equations:

$$\begin{aligned}
-9p_1 + 2p_2 + 3p_3 &= 0, \\
2p_1 - 6p_2 + 6p_3 &= 0, \\
3p_1 + 6p_2 - 9p_3 &= 0.
\end{aligned}$$

Since by construction the determinant of these equations is zero, although it looks like $p_1 = p_2 = p_3 = 0$ is the solution, in fact non-zero solution to the ratios of the ps exist. Provided the eigenvalues are distinct, there is one eigenvalue for each eigenvector; special cases of coincident roots to the eigenvalue equation that can result in a whole plane of eigenvectors are dealt with in linear algebra texts [7], not here.

Now, let us turn to defining strain. Strain is another tensor, so the use of tensor notation is still very necessary. As stress is applied to a body, it undergoes a deformation that disappears as soon as the stress is removed. However, when it comes to representing this elastic distortion, it is simply a question of mapping where an arbitrary point in the volume V goes to when the stress is applied. Therefore, we are looking for a linear transformation. Linear transformations are represented using matrices, and matrices are also used to represent tensors of course, so all is consistent. The fact that the distances involved are infinitesimal is incidental and can be catered for. There may be some who are not that familiar with linear transformations, or have forgotten them, so here is a two-dimensional rotation in the x, y plane plus a translation

$$\begin{pmatrix} x \\ y \\ z \end{pmatrix} \rightarrow \begin{pmatrix} \cos\theta & \sin\theta & 0 \\ -\sin\theta & \cos\theta & 0 \\ 0 & 0 & 1 \end{pmatrix} \begin{pmatrix} x \\ y \\ z \end{pmatrix} + \begin{pmatrix} x_0 \\ y_0 \\ z_0 \end{pmatrix}.$$

Therefore, for example, if $\theta = \pi/3$ and $(x_0, y_0, z_0) = (1, 2, 3)$, then the point $(1, 0, 0)$ would transform to $(3/2, (4 - \sqrt{3})/2, 3)$. In complete generality, this kind of transformation would look like:

$$\begin{pmatrix} U_1 \\ U_2 \\ U_3 \end{pmatrix} = \begin{pmatrix} a_{11} & a_{12} & a_{13} \\ a_{21} & a_{22} & a_{23} \\ a_{31} & a_{32} & a_{33} \end{pmatrix} \begin{pmatrix} u_1 \\ u_2 \\ u_3 \end{pmatrix}$$

or using vectors and matrices

$$\mathbf{U} = A\mathbf{u}$$

or using Cartesian tensors

$$U_i = a_{ij}u_j.$$

Compare this with Equation 11.8. They are the same if u_j is the incremental change or deformation vector, a_{ij} is a tensor expressing the actual deformation caused by a stress and U_i is the deformed co-ordinate under this stress. In order to solve problems in elasticity, there has to be a relationship between the imposed stress and the deformation or strain in the elastic solid. The easiest relation, familiar to most is Hooke's law where stress and strain are in one dimension and the two are proportional, with the constant of proportionality being the modulus of elasticity, or Young's modulus, sometimes called the stiffness. However, at the other end of the complexity scale, the stress tensor σ_{ij} is related to the strain tensor usually written ϵ_{ij} via

$$\sigma_{ij} = C_{ijkl}\epsilon_{kl}$$

with the implied double summation. This is nine equations and the fourth-order tensor C_{ijkl} has $3^4 = 81$ terms and is a general *constitutive equation*. This is considerably more complex than a single Young's modulus, and yet this remains linear infinitesimal elasticity, no mention of finite (non-linear) elasticity at all. For a more detailed exposition, the interested reader is referred to specialist texts on elasticity or continuum mechanics. However, there are some relationships that are needed. First, some locally defined notation is required. The direct strain ϵ_{xx} is given by

$$\epsilon_{xx} = \frac{\partial u}{\partial x},$$

where the symbol u the displacement (*not* the speed) in the x direction. One can see this algebraically: the strain in the x direction on a surface that has normal in the same direction must be $\epsilon_{xx} = u(x + \delta x, y, z) - u(x, y, z)$ and the above expression is simply the first term in a Taylor expansion, all the rest of the expansion is ignored. Thus, we have assumed linearity and the infinitesimal nature of the strain has now been imposed. Similarly, the other two direct strains

$$\epsilon_{yy} = \frac{\partial v}{\partial y} \quad \text{and} \quad \epsilon_{zz} = \frac{\partial w}{\partial z},$$

where v and w are the displacements in the y and z directions respectively are also defined. It is useful to think of these displacements in the same way as stress, so that direct strain where the two suffices are the same has the normal and the direction of the displacement in the same direction. The *shearing strain* ϵ_{xy} is defined through rotation rather than stretch and is given by

$$\epsilon_{xy} = \frac{1}{2}\left[\frac{\partial u}{\partial y} + \frac{\partial v}{\partial x}\right].$$

Figure 11.9 shows one direct strain and one shear strain, others are not shown as these kind of diagrams can get very cluttered.

This is a little more difficult to see, but applying Taylor's theorem again might imply $\epsilon_{xy} = u(x, y + \delta y, z) - u(x, y, z)$; however, the expression ϵ_{yx} that has to be the same to prevent rotation would be $\epsilon_{yx} = v(x + \delta x, y, z) - v(x, y, z)$ so the

Figure 11.9 Elastic material showing two strains, the direct strain ϵ_{xx} and the shear strain ϵ_{xy}. The square A deforms into the square A'. No other strains are shown to avoid complexity.

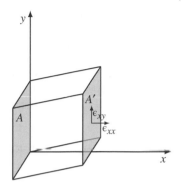

above expression which is half of each is thus the simplest that works. Similarly, the other two shear strains are

$$\epsilon_{zx} = \frac{1}{2}\left[\frac{\partial u}{\partial z} + \frac{\partial w}{\partial x}\right] \quad \text{and} \quad \epsilon_{yz} = \frac{1}{2}\left[\frac{\partial w}{\partial y} + \frac{\partial v}{\partial z}\right].$$

These definitions have been designed to ensure that $\epsilon_{xy} = \epsilon_{yx}, \epsilon_{zx} = \epsilon_{xz}$ and $\epsilon_{yz} = \epsilon_{zy}$, so the tensor ϵ_{ij} like σ_{ij} is also symmetric. Given the definitions of the nine components of strain, there are second-order relationships of the form

$$\frac{\partial^2 \epsilon_{xx}}{\partial y^2} + \frac{\partial^2 \epsilon_{yy}}{\partial x^2} = 2\frac{\partial^2 \epsilon_{xy}}{\partial x \partial y}$$

called compatibility equations. The notation $\mathbf{D} = u\mathbf{i} + v\mathbf{j} + w\mathbf{k}$ is used to denote the vector deformation. Therefore,

$$\nabla \cdot \mathbf{D} = \frac{\partial u}{\partial x} + \frac{\partial v}{\partial y} + \frac{\partial w}{\partial z} = \epsilon_{xx} + \epsilon_{yy} + \epsilon_{zz} = \delta,$$

which is called the *dilatation* and represents the overall change in the elastic volume.

Strain is what we need to find, and stress is directly measurable. The relation between them, theoretically, the 81 constants in the stress–strain relationship is in reality usually less complex. This is not an elasticity text, so only rudimentary relationships will be introduced here. A typical stress–strain relationship is

$$\sigma_{pp} = \lambda\delta + 2\mu\epsilon_{pp} = \lambda\nabla \cdot \mathbf{D} + 2\mu\epsilon_{pp}$$

for normal stress and

$$\sigma_{pq} = 2\mu\epsilon_{pq}, \quad p \neq q$$

for shear stress, where p and q are any of x, y or z. On the right is the dilatation δ, the strain ϵ_{pq} together with two constants λ and μ called elastic moduli or sometimes Lamé parameters (after Gabriel Lamé (1795–1870) a French mathematician who worked extensively on curvilinear co-ordinates and elasticity).

In terms of usual quantities, the Young's modulus E and Poisson's ratio v are

$$E = 2\mu(1 + v) \quad \text{and} \quad v = \frac{\lambda}{2(\lambda + \mu)}.$$

Equation 11.9 is the equilibrium condition for an arbitrary elastic solid. Therefore, we are now in a position to derive an equation for the vector displacement **D**. First, recall the vector expressions for the stresses

$$\sigma_x = \sigma_{xx}\mathbf{i} + \sigma_{xy}\mathbf{j} + \sigma_{xz}\mathbf{k},$$
$$\sigma_y = \sigma_{yx}\mathbf{i} + \sigma_{yy}\mathbf{j} + \sigma_{yz}\mathbf{k},$$
$$\sigma_z = \sigma_{zx}\mathbf{i} + \sigma_{zy}\mathbf{j} + \sigma_{zz}\mathbf{k}.$$

Using the constitutive equations, these become

$$\sigma_x = \lambda(\mathbf{\nabla} \cdot \mathbf{D})\mathbf{i} + \mu(u_y + v_x)\mathbf{j} + \mu(u_z + w_x)\mathbf{k},$$
$$\sigma_y = \mu(u_y + v_x)\mathbf{i} + \lambda(\mathbf{\nabla} \cdot \mathbf{D})\mathbf{j} + \mu(v_z + w_y)\mathbf{k},$$
$$\sigma_z = \mu(u_z + w_x)\mathbf{i} + \mu(v_z + w_y)\mathbf{j} + \lambda(\mathbf{\nabla} \cdot \mathbf{D})\mathbf{k}.$$

Written in terms of vectors, Equation 11.9 is

$$\frac{\partial \sigma_x}{\partial x} + \frac{\partial \sigma_y}{\partial y} + \frac{\partial \sigma_z}{\partial z} + \rho\mathbf{F} = \mathbf{0}$$

and substituting for the stresses this equation becomes

$$(\lambda + \mu)\mathbf{\nabla}(\mathbf{\nabla} \cdot \mathbf{D}) + \mu\nabla^2\mathbf{D} + \rho\mathbf{F} = \mathbf{0}.$$

Alternatively, using the identity

$$\nabla^2\mathbf{D} = \mathbf{\nabla}(\mathbf{\nabla} \cdot \mathbf{D}) - \mathbf{\nabla} \times (\mathbf{\nabla} \times \mathbf{D})$$

this is

$$(\lambda + \mu)\mathbf{\nabla} \times (\mathbf{\nabla} \times \mathbf{D}) + (\lambda + 2\mu)\nabla^2\mathbf{D} + \rho\mathbf{F} = \mathbf{0}.$$

This equation is for equilibrium, sometimes called the Navier equation (after another Frenchman, the engineer and physicist who specialised in mechanics, Claude Louis Navier (1785–1836)). It is not difficult to add a time-dependent term and the equation can be used for elastic waves of the type responsible for earthquakes, for example. Here is a reasonably simple example with practical applications.

Example 11.7   An elastic circular cylinder lies with its axis along the z-axis. The stresses are given by

$$\sigma_x = -\mu y\mathbf{k}, \quad \sigma_y = \mu x\mathbf{k} \quad \text{and} \quad \sigma_z = -\mu y\mathbf{i} + \mu x\mathbf{j},$$

where μ is an elastic constant. Determine the nature of the distortion given that there is no deformation in the z direction and that all direct strains are also zero.

Solution: Using the relationships

$$\sigma_{xz} = -\mu y = \mu(u_z + w_x) = \mu u_z$$
$$\sigma_{yz} = \mu x = \mu(v_z + w_x) = \mu v_z,$$

as $w = 0$, so upon integration with respect to z these give

$$u = -zy + f_1(x, y),$$
$$v = zx + f_2(x, y),$$

where $f_1(x, y)$ and $f_2(x, y)$ are arbitrary functions of integration. However, as all direct strains are zero $f_1 = f_1(y)$ only, and $f_2 = f_2(x)$ only. Therefore, the shear strain

$$\sigma_{xy} = \mu(u_y + v_x) = f_1' + f_2'$$

and if this is zero, then there is only rigid body rotation. Otherwise, there is a twisting motion called torsion. In the case of rigid body motion, both f_1 and f_2 are constant, and can be assumed zero as the addition of constants to u or v causes no strain. □

The last application of vector calculus is to heat transfer, and the next section concentrates on this.

11.5.4 Heat Transfer

The application of partial differentiation to thermodynamics is huge, see Section 2.9 for a taster. The application of vectors to heat transfer is less. In general, it is limited to conservation arguments and some that ride on fluid dynamics or more generally continuum mechanics. This section is thus more limited than those on electromagnetic theory or elasticity. First of all, let us recap the fluid continuity equation, this says what goes out through the surface S of an arbitrary volume V has to be the same as any loss of matter throughout the volume V.

$$\oiint_S \rho\mathbf{u} \cdot d\mathbf{S} = -\frac{d}{dt} \iiint_V \rho \, dV.$$

This integrates through use of Gauss' flux theorem to the continuity equation in fluid mechanics. Essentially, this can be applied to anything that flows, including heat. Heat is essentially a form of energy, and to measure the flow of heat means tracking the gradient of temperature, in fact putting $\mathbf{u} = -\kappa\boldsymbol{\nabla}T$ where κ is the thermal conductivity. Therefore, the flow of heat out of the closed surface S is

$$-\oiint_S \kappa\boldsymbol{\nabla}T \cdot d\mathbf{S},$$

where the negative sign is due to flow of heat out causing a drop in temperature. The change of heat inside the volume is the rate of change, not just of the density but of $c\rho T$ where c is the specific heat capacity. Both c and ρ are not time dependent, so the heat loss by the volume will be

$$-\iiint_V c\rho\frac{dT}{dt}\,dV.$$

Heat conservation thus means

$$-\oiint_S \kappa\boldsymbol{\nabla}T \cdot d\mathbf{S} = -\iiint_V c\rho\frac{dT}{dt}\,dV.$$

Transforming the left-hand side using Gauss' flux theorem and recognising that $\boldsymbol{\nabla}\cdot\kappa\boldsymbol{\nabla}T = \kappa\nabla^2 T$ thus

$$-\iint_V \kappa\nabla^2 T\,dV = -\iiint_V c\rho\frac{dT}{dt}\,dV.$$

Since V is arbitrary the equation

$$\kappa\nabla^2 T = c\rho\frac{dT}{dt}$$

must hold. This is the equation for heat conduction. The application of mathematics to heat flow now goes into how to solve differential equations, so we must leave it there.

Thus, three applications of vectors and tensors to physics have been briefly covered. Some of the later exercises below take things a little further, but they do need some specialist knowledge. Books such as [9], [11] and [12] have more applications.

Exercises

The final three exercises marked ** demand specialist knowledge and may not be suitable for every reader.

11.1 By writing $\mathbf{F} = F_1\mathbf{i} + F_2\mathbf{j}$, deduce Theorem 11.2 without using vectors for the two-dimensional case.

11.2 Given the vector field $\mathbf{F} = 2yz\mathbf{i} - (x + 3y - 2)\mathbf{j} + (x^2 + z)\mathbf{k}$, evaluate the surface integral

$$\iint_S \nabla \times \mathbf{F} \cdot d\mathbf{S},$$

where S is the intersection of the two cylinders $x^2 + y^2 = a^2$ and $x^2 + z^2 = a^2$ with $x, y, z \geq 0$.

11.3 Prove that

$$\oint_C \phi \, d\mathbf{r} = \iint_S d\mathbf{S} \times \nabla\phi,$$

where S is a surface with bounding curve C.

11.4 Use Gauss' flux theorem to evaluate

$$\oiint_S \mathbf{r} \cdot d\mathbf{S},$$

where S is any closed surface.

11.5 Given $\mathbf{B} = \nabla \times \mathbf{A}$ prove that

$$\iint_S \mathbf{B} \cdot d\mathbf{S} = 0$$

for any closed surface S.

11.6 Determine the value of the integral

$$\iint_S \mathbf{r} \times d\mathbf{S},$$

where S is any closed surface.

11.7 Let S be a smooth surface with a smooth bounding curve C and let $\mathbf{a}$ be a fixed vector. Show that

$$2 \iint_S \mathbf{a} \cdot d\mathbf{S} = \oint_C (\mathbf{a} \times \mathbf{r}) \cdot d\mathbf{r}.$$

11.8 Let S be a surface with smooth bounding curve C. Show that if ϕ and ψ are sufficiently differentiable scalar fields, then

$$\iint_S (\nabla\phi \times \nabla\psi) \cdot d\mathbf{S} = \oint_C (\phi\nabla\psi) \cdot d\mathbf{r}.$$

11.9 Establish Green's first identity

$$\oiint_S \phi \nabla \psi \cdot d\mathbf{S} = \iiint_V (\phi \nabla^2 \psi + \nabla \phi \cdot \nabla \psi) dV,$$

where ϕ and ψ are scalar fields with continuous partial derivatives inside and on the closed surface S that encloses the volume V.

11.10 Determine the value of the integral

$$\iint_S \mathbf{E} \cdot d\mathbf{S},$$

where $\mathbf{E} = (\mu \mathbf{r})/r^3$ and S is a closed surface with $r \neq 0$ inside S but $r = 0$ at a single location on S.

11.11 **The Poynting vector was defined in Chapter 10 as $\mathbf{P} = \mathbf{E} \times \mathbf{H}$, the vector product of the electric and magnetic fields. Use Maxwell's equations to show that the surface integral

$$\oiint_S \mathbf{P} \cdot d\mathbf{S}$$

over the closed surface S is the power or rate of energy flow across S.

11.12 **In a fluid, the quantity

$$\Gamma = \oint_C \mathbf{u} \cdot d\mathbf{r},$$

where C is a closed curve, is defined as the *circulation* of the fluid about C. The equation obeyed by the fluid velocity is

$$\frac{d\mathbf{u}}{dt} = -\frac{1}{\rho} \nabla p + \nabla \phi,$$

where p is the pressure, ρ is the density and ϕ a potential for the body force. Show that the Γ remains constant following the fluid. Show also that the vorticity flux through the area enclosed by the contour C must be zero. [*Note:* 'Following the fluid' means a derivative $\frac{d}{dt}$ where $\frac{d}{dt} \equiv \frac{\partial}{\partial t} + (\mathbf{u} \cdot \nabla).$]

11.13 **Interpret the equation

$$\frac{d}{dt} \iiint_V \rho \left(\frac{1}{2} v_i v_i + e \right) dV = \iiint_V b_i v_i \, dV + \oiint_S (\sigma_{ji} v_i - q_i) n_i \, dS$$

where e is the internal energy, σ_{ji} is the stress tensor and b_i is a body force vector.

12

Solutions and Answers to Exercises

12.1 Chapter 1

1.1 For this question, it is a matter of carefully evaluating the limits. Here they are.

a) The limit in this case is

$$\lim_{\delta x \to 0} \left\{ \frac{(x + \delta x)^3 - x^3}{\delta x} \right\}$$

$$= \lim_{\delta x \to 0} \left\{ \frac{x^3 + 3(\delta x)x^2 + O((\delta x)^2) - x^3}{\delta x} \right\}$$

so

$$\lim_{\delta x \to 0} \left\{ \frac{(x + \delta x)^3 - x^3}{\delta x} \right\} = \lim_{\delta x \to 0} [3x^2 + O(\delta x)] = 3x^2.$$

b)

$$\lim_{\delta x \to 0} \left\{ \frac{\sin(x + \delta x) - \sin x}{\delta x} \right\}$$

$$= \lim_{\delta x \to 0} \left\{ \frac{\sin x \cos \delta x + \cos x \sin \delta x - \sin x}{\delta x} \right\}$$

using that $\cos \delta x \approx 1$ and $\sin \delta x \approx \delta x$ gives the answer $\cos x$.

c)

$$\lim_{\delta x \to 0} \left\{ \frac{\ln(x + \delta x) - \ln x}{\delta x} \right\} = \lim_{\delta x \to 0} \left\{ \frac{1}{\delta x} \ln \left(1 + \frac{\delta x}{x} \right) \right\}$$

now we use the Taylor expansion for $\ln(1 + t)$ with $t = \delta x / x$

$$\ln \left(1 + \frac{\delta x}{x} \right) = \frac{\delta x}{x} + O((\delta x)^2)$$

so the limit gives $1/x$.

Two and Three Dimensional Calculus: With Applications in Science and Engineering, First Edition. Phil Dyke.
© 2018 John Wiley & Sons Ltd. Published 2018 by John Wiley & Sons Ltd.

1.2 For these Taylor polynomials, only a brief solution is given.
a) Use $\cos x = \cos(x - \pi/3 + \pi/3)$

$$= \cos(x - \pi/3)\cos \pi/3 - \sin(x - \pi/3)\sin \pi/3$$

and Maclaurin series for cosine and sine to obtain the polynomial

$$\cos x \approx \frac{1}{2} - \frac{\sqrt{3}}{2}\left(x - \frac{\pi}{3}\right) - \frac{1}{4}\left(x - \frac{\pi}{3}\right)^2 + \frac{\sqrt{3}}{12}\left(x - \frac{\pi}{3}\right)^3.$$

b) Use $\ln x = \ln(x - 1 + 1)$ to give

$$\ln x \approx (x - 1) - \frac{1}{2}(x - 1)^2 + \frac{1}{3}(x - 1)^3 - \frac{1}{4}(x - 1)^4.$$

c) This time with no approximation $x^4 = (x - 3 + 3)^4$

$$= 81 + 108(x - 3) + 54(x - 3)^2 + 12(x - 3)^3 + (x - 3)^4,$$

except of course the question asks for the last term to be omitted.

1.3 Rolle's Theorem is Theorem 1.2. What follows is an outline proof and not one that will satisfy pure mathematicians. If the function $f(x)$ is such that $f'(x) = 0$ throughout the interval $a \le x \le b$, then there is nothing to prove. If this is not the case, then as $f(a) = f(b)$ and $f(x)$ is differentiable throughout $a < x < b$ then speaking loosely, f rises through $x = a$ but has to fall again by the point $x = b$ to enable $f(a)$ to be the same as $f(b)$. In order to do this, it has to 'turn around' so to speak, reaching at least one peak at which $f'(x) = 0$ the differentiability of $f(x)$ ensures that the curve is smooth. Or if f falls at $x = a$, a similar argument ensures that there's at least one minimum in $f(x)$. In this way, Rolle's theorem is established.

1.4 The extended mean value theorem is proved by considering the function

$$F(x) = [f(b) - f(a)]g(x) - [g(b) - g(a)]f(x).$$

Now, $F(a) = F(b)$ so the conditions for Rolle's theorem apply. Hence, there is a $c \in (a, b)$ such that $F'(c) = 0$, thus

$$[f(b) - f(a)]g'(c) - [g(b) - g(a)]f'(c) = 0$$

or, rearranging

$$\frac{f'(c)}{g'(c)} = \frac{f(b) - f(a)}{g(b) - g(a)}$$

as required.

1.5 Set $h(x) = f(x) - f(a)$ and $k(x) = g(x) - g(a)$ so that $h(a) = 0$ and $k(a) = 0$. In addition, differentiating yields $k' = g'$ and $h' = f'$ so the extended mean value theorem implies

$$\frac{h'(c)}{k'(c)} = \frac{h(b)}{k(b)},$$

where $c \in (a, b)$. Now the interval (a, b) is assigned by us and we are free to choose it. The result follows if we let a get larger and larger, this forces a, b and c all to be infinite. Another way of proving the result is by utilising the transformation $x = 1/t$ differentiating and putting $t = 0$ to correspond to $x \to \infty$.

1.6 These are all reasonably straightforward.

a) This first limit is solved either by using L'Hôpital's rule:

$$\lim_{x \to 3} \left(\frac{x^3 - 2x^2 - x - 6}{x - 3} \right) = \lim_{x \to 3} \left(\frac{3x^2 - 4x - 1}{1} \right) = 14$$

or by spotting that $x^3 - 2x^2 - x - 6 = (x - 3)(x^2 + x + 2)$ and cancelling the factor $(x - 3)$, then putting $x = 3$ to get the same answer 14.

b) This limit is best solved by using Maclaurin series for cosine, so

$$\frac{1 - \cos x}{x^2} = \frac{1 - 1 + x^2/2 + O(x^4)}{x^2} = \frac{1}{2} + O(x^2)$$

and this gives the limit $1/2$.

c) This limit is found by rationalisation as follows:

$$\frac{\sqrt{x + 1} - 1}{x} = \frac{\sqrt{x + 1} - 1}{x} \frac{\sqrt{x + 1} + 1}{\sqrt{x + 1} + 1},$$

which equals

$$\frac{x + 1 - 1}{x(\sqrt{x + 1} + 1)} = \frac{1}{\sqrt{x + 1} + 1}$$

and as $x \to 0$ this tends to $1/2$.

d) This one is done from first principles. First as $x \to 0$ from the right $|x| = x$, so the ratio

$$\lim_{x \to 0} \frac{|x|}{x} = \lim_{x \to 0} \frac{x}{x} = 1.$$

If $x \to 0$ from the left $|x| = -x$, so the ratio

$$\lim_{x \to 0} \frac{|x|}{x} = \lim_{x \to 0} \frac{-x}{x} = -1.$$

As the two are different, the limit is not unique and so does not exist.

e) This last one is the most difficult. One step of L'Hôpital's rule gives

$$\lim_{x \to 0} \frac{\sin x - x}{x - \tan x} = \lim_{x \to 0} \frac{\cos x - 1}{1 - \sec^2 x},$$

which is still of the form $0/0$ but differentiating again gives

$$\lim_{x \to 0} \frac{\cos x - 1}{1 - \sec^2 x} = \lim_{x \to 0} \frac{-\sin x}{-2 \tan x \sec^2 x},$$

and the use of a little trigonometry gives

$$\lim_{x \to 0} \frac{-\sin x}{-2 \tan x \sec^2 x} = \lim_{x \to 0} \frac{\sin x}{2 \frac{\sin x}{\cos^3 x}} = \lim_{x \to 0} \frac{1}{2} \cos^3 x,$$

which is $1/2$.

1.7 These are trickier limits.

a) This limit is solved by taking logs. Put $f(x) = (\ln x)^{1/x}$, so

$$\ln f(x) = \frac{\ln(\ln x)}{x}$$

$$\lim_{x \to \infty} \left\{ \frac{\ln(\ln x)}{x} \right\} = \lim_{x \to \infty} \left\{ \frac{1/(x \ln x)}{1} \right\} = 0.$$

Therefore, the required limit is $e^0 = 1$

b) Once again, this is solved by taking logs. Put

$$f(x) = (3^x + 5^x)^{1/x},$$

so

$$\ln f(x) = \frac{1}{x} \ln(3^x + 5^x) = \frac{1}{x} \ln \left[5^x \left(1 + \left(\frac{3}{5} \right)^x \right) \right]$$

that becomes

$$\ln f(x) = \frac{1}{x} \ln 5^x + \frac{1}{x} \ln \left[1 + \left(\frac{3}{5} \right)^x \right]$$

and so

$$\ln f(x) \approx \frac{1}{x} x \ln 5 = \ln 5$$

as $x \to \infty$, since the second more complex log term will tend to zero as $x \to \infty$. The required limit is thus the exponential of this, which is 5.

c) This is solved by finding the two Maclaurin series:

$$\sin(\tan x) = x + \frac{1}{6}x^3 - \frac{1}{40}x^5 - \frac{55}{1008}x^7 + \cdots$$

and

$$\tan(\sin x) = x + \frac{1}{6}x^3 - \frac{1}{40}x^5 - \frac{107}{5040}x^7 + \cdots,$$

and these involve long calculation. It is better to use computer algebra. The first three terms are identical, so subtracting these gives the leading term

$$\sin(\tan x) - \tan(\sin x) = -\frac{1}{30}x^7 + \cdots ,$$

and so dividing by x^7 and letting $x \to 0$ lead to the value of the limit $=$ $-1/30$.

1.8 This is procedural but there is some algebra

a) $y = x^4 - 12x^3 + 48x^2 - 64x$. Differentiating gives $y' = 4x^3 - 36x^2 + 96x - 64 = 0$ for extrema. This factorises that means $y' = 4(x - 1)$ $(x - 4)^2 = 0$ so $x = 1$ or $x = 4$. As $y'' = 12x^2 - 72x + 96$ with $x = 1, y'' = 36 > 0$ so $x = 1$ is a minimum. With $x = 4, y'' = 0$ so $x = 4$ is a point of inflection.

b) With $y = 2x(x^2 + 2)^{-1/2}, y' = 4(x^2 + 2)^{-3/2} > 0$ for all x, so there are no extrema.

c) With $y = 2x^{5/3} - 5x^{4/3}$ so $y' = (10/3)x^{2/3} - (20/3)x^{1/3}$, and this is only zero if $x = 8$, which is a minimum.

1.9 With $f(x) = x^2 - a, f'(x) = 2x$, so the Newton–Raphson formula gives

$$x_{n+1} = x_n - \frac{f(x_n)}{f'(x_n)} = x_n - \frac{x_n^2 - a}{2x_n} = \frac{1}{2}\left(x_n + \frac{a}{x_n}\right)$$

as required. With $a = 5$, only three iterations of the method are required for the answer 2.236068; it converges rapidly. I used EXCEL.

1.10 Here is some reasonably routine integration,

a) Integrating by parts gives:

$$\int x \ln x \, dx = \left[\frac{1}{2}x^2 \ln x\right] - \int \frac{1}{2}x^2 \cdot \frac{1}{x} dx.$$

This gives the answer

$$\frac{1}{2}x^2 \ln x + \frac{1}{4}x^2 + C,$$

where C is an arbitrary constant.

b) Integrating by parts gives

$$\int \frac{xe^x}{(x+1)^2} dx = -\frac{x}{x+1} - \int -\frac{1}{x+1}(1+x)e^x \, dx,$$

where in the integration by parts formula $du = dx/(x+1)^2$ and $v = xe^x$ so that $u = -1/(x+1)$ and $dv = e^x \, dx + xe^x \, dx = (1+x)e^x \, dx$.

The integration is now straightforward and gives the answer

$$\frac{1}{1+x}e^x + C,$$

where C is an arbitrary constant.

c) Again this is an exercise in integration by parts:

$$I_n = \int \sin^{n-1}x \sin x \, dx = -\cos x \sin^{n-1}x$$

$$+ \int (n-1)\sin^{n-2}x \cos^2 x \, dx;$$

the second integral is, using $\cos^2 x = 1 - \sin^2 x$ simply, $(n-1)I_{n-2} - (n-1)I_n$ from which the result follows. Using this formula,

$$4I_4 = -\cos x \sin^3 x + 3I_2,$$

$$2I_2 = -\cos x \sin x + I_0 = -\cos x \sin x + x,$$

whence

$$I_4 = -\frac{1}{4}\cos x \sin^3 x + \frac{3}{8}(x - \cos x \sin x) + C.$$

Similarly,

$$5I_5 = 4I_3 - \cos x \sin^4 x$$

$$3I_3 = 2I_1 - \cos x \sin^2 x = -2\cos x - \cos x \sin^2 x$$

so

$$I_5 = \frac{4}{15}(-2\cos x - \cos x \sin^2 x) - \frac{1}{5}\cos x \sin^4 x + C',$$

where C and C' are arbitrary constants.

d) Writing $3 + 2x - x^2 = 4 - (1 - x)^2$, then

$$\int \frac{1}{\sqrt{3+2x-x^2}}dx = \int \frac{dx}{\sqrt{2^2 - (1-x)^2}}$$

$$= -\sin^{-1}\left\{\frac{1}{2}(1-x)\right\} + C$$

or

$$\sin^{-1}\left\{\frac{1}{2}(x-1)\right\} + C.$$

e) Using the hint given, the following partial fractions emerge from a rather long calculation:

$$\frac{1}{x^6+1} = \frac{1}{3}\left[\frac{1}{x^2+1} + \frac{1-x\sqrt{3}/2}{x^2 - x\sqrt{3}+1} + \frac{1+x\sqrt{3}/2}{x^2 + x\sqrt{3}+1}\right].$$

Using standard methods of integration (the logarithm form and completing the square that leads to inverse tangents), the answer is

$$\int \frac{dx}{1+x^6} = \frac{1}{3}\tan^{-1}x - \frac{\sqrt{3}}{12}\ln(x^2 - x\sqrt{3} + 1)$$
$$+ \frac{\sqrt{3}}{12}\ln(x^2 + x\sqrt{3} + 1) + \frac{1}{6}\tan^{-1}(2x - \sqrt{3})$$
$$+ \frac{1}{6}\tan^{-1}(2x + \sqrt{3}).$$

1.11 Here are some salient steps:

$$\frac{\cos x}{\alpha \cos x + \sin x} = \frac{1}{\alpha + \tan x}$$

makes things a bit easier though it is not strictly necessary. The integral is solved by putting $t = \tan\frac{1}{2}x$ so that $dt = \frac{1}{2}\sec^2\frac{1}{2}x\,dx$. The easiest thing to do next is to draw two right-angled triangles: the first with sides $t, 1$ and hypotenuse $\sqrt{1+t^2}$, this has angles $x/2$, $\pi/2 - x/2$ and $\pi/2$ (see Figure 12.1). The second triangle has sides $1 - t^2$, $2t$ and hypotenuse $1 + t^2$, this has angles x, $\pi/2 - x$ and $\pi/2$. The upshot is that you can read off any trigonometric expression in x in terms of t straight away. Doing this the integral given can be done using (rather awkward) partial fractions:

$$\int_0^{\pi/2} \frac{\cos x}{\alpha \cos x + \sin x}dx = \int_0^1 \frac{2(1 - t^2)}{(\alpha - \alpha t^2 + 2t)(1 + t^2)}dt$$

and the partial fractions give

$$\frac{2(1 - t^2)}{(\alpha - \alpha t^2 + 2t)(1 + t^2)} = \frac{2}{1 + \alpha^2}\left\{\frac{-t + \alpha}{1 + t^2} + \frac{\alpha^2 - \alpha t}{\alpha - \alpha t^2 + 2t}\right\},$$

so the integral is

$$\int_0^1 \frac{2(1 - t^2)}{(\alpha - \alpha t^2 + 2t)(1 + t^2)}dt$$

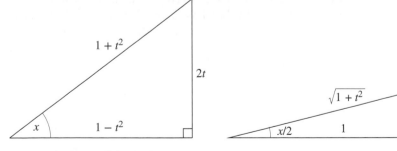

Figure 12.1 Two useful triangles.

$$= \frac{2}{1 + \alpha^2} \left[\alpha \tan^{-1} t - \frac{1}{2} \ln(1 + t^2) - \frac{1}{2} \ln(\alpha - \alpha t^2 - 2t) \right]_0^1,$$

and this gives the result in the question.

12.2 Chapter 2

2.1

$$f_x = 4x^3 y^2 \cos z, \quad f_y = 2x^4 y \cos z, \quad f_z = -x^4 y^2 \sin z.$$

2.2 The derivatives of $h(x, y)$ are

$$h_x = 3x^2 - 3y^2 \quad \text{and} \quad h_y = 3y^2 - 6xy$$

and at $(1, 1)$ these are $h_x = 0, h_y = -3$.

$$z_x = 4x^3 \quad \text{and} \quad z_y = -2y.$$

2.3 Using the chain rule,

$$\frac{\partial z}{\partial \xi} = \frac{\partial z}{\partial x} \frac{\partial x}{\partial \xi} + \frac{\partial z}{\partial y} \frac{\partial y}{\partial \xi},$$

where (ξ, η) are the (x, y) axes rotated anti-clockwise through an angle θ. Thus

$$x = \xi \cos \theta - \eta \sin \theta \quad \text{and} \quad y = \xi \sin \theta + \eta \cos \theta,$$

so $z_\xi = 4x^3 \cos \theta - 2y \sin \theta$ and at the point $(1, 1)$ this has the value $z_\xi = 4 \cos \theta - 2 \sin \theta$. Hence, $z_\xi = 0$ where $\tan \theta = 2$ or

$$\theta = \tan^{-1}(2).$$

2.4 The volume of the box is $V = abc$ independent of whether it is either open or closed at the top. The absolute error would be in terms of cubic centimetres. With no error, the box has volume $V = 90$ cm^3. With maximum error (take them all positive or all negative), this volume is $V = (5.2) \times (6.3) \times (3.1) = 101.56$ or $V = (4.8) \times (5.7) \times (2.9) = 79.34$ giving the range as $79.34 \le V \le 101.56$. This is precise.

The (approximate) relative error is, by logarithmic differentiation,

$$\frac{\Delta V}{V} \approx \frac{\Delta a}{a} + \frac{\Delta b}{b} + \frac{\Delta c}{c}.$$

Therefore,

$$\frac{\Delta V}{V} \approx \pm \frac{0.2}{5} \pm \frac{0.3}{6} \pm \frac{0.1}{3} = \pm 0.04 \pm 0.05 \pm 0.033 = \pm 0.123.$$

Therefore, according to this approximation the volume will be in the range 90 ± 11.07 or $78.93 \leq V \leq 101.07$. The difference between this and the previous range is due to linearisation.

2.5 a) Logarithmic differentiation gives

$$\frac{\Delta y}{y} \approx \frac{\Delta w}{w} + 3\frac{\Delta l}{l} - 4\frac{\Delta d}{d}.$$

Inserting the error values given (taking account of sign) gives

$$\frac{\Delta y}{y} \approx 2 - 3 - 4 \times 3 = -13\%.$$

b) The maximum error is

$$\frac{\Delta y}{y} \approx 2 + 3 + 4 \times 3 = 17\%.$$

2.6 In triangle ABC (see Figure 12.2), drop a perpendicular from A to the opposite edge c. This divides this side c into c_1 and c_2 such that $c = c_1 + c_2$ (see Figure 12.2). Therefore, $\cos A = c_1/b$ and $\cos B = c_2/a$ which means that $c = c_1 + c_2 = b\cos A + c\cos B$. The cosine rule gives $a^2 = b^2 + c^2 - 2bc\cos A$, hence

$$2a\delta a = 2b\delta b + 2c\delta c - (2\delta bc + 2\delta cb)\cos A.$$

Dividing by $2a$ gives

$$\delta a = \frac{\delta b}{a}(b - c\cos A) + \frac{\delta c}{a}(c - b\cos A),$$

Figure 12.2 The triangle.

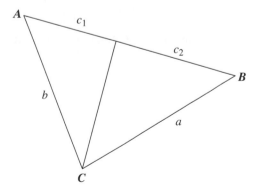

and using $(c - b \cos A) = a \cos B$ and $(b - c \cos A) = a \cos C$, this gives

$$\delta a = \delta b \cos C + \delta c \cos B.$$

2.7 This exercise follows Example 2.9 closely. The answer is

$$\frac{dy}{dx} = \frac{y^2 - xy \ln y}{x^2 - xy \ln x}.$$

2.8 Here, we use Leibniz rule

a) In the given integral, put $u = xa$ to obtain

$$\int_0^\infty \frac{dx}{a^4 + x^4} = \frac{\pi}{2a^3 \sqrt{2}}$$

then differentiate with respect to a and differentiate again to get the two answers:

$$\int_0^\infty \frac{dx}{(a^4 + x^4)^2} = \frac{3\pi}{8a^7 \sqrt{2}} \quad \text{and} \quad \int_0^\infty \frac{dx}{(a^4 + x^4)^3} = \frac{21\pi}{64a^8 \sqrt{2}}$$

from which, putting $a = 1$, we get the answers

$$\int_0^\infty \frac{dx}{(1 + x^4)^2} = \frac{3\pi}{8\sqrt{2}} \quad \text{and} \quad \int_0^\infty \frac{dx}{(1 + x^4)^3} = \frac{21\pi}{64\sqrt{2}}.$$

b) This is done by differentiating the given integral with respect to a, twice, then putting $a = 2$ then $a = 3$.

$$\int_{-\pi}^{\pi} \frac{dx}{(2 + \sin x)^2} = \frac{4\pi}{9} \sqrt{3} \quad \text{and} \quad \int_{-\pi}^{\pi} \frac{dx}{(3 + \sin x)^3} = \frac{19\pi}{256} \sqrt{2}.$$

c) Starting with

$$\int_0^1 x^p \, dx = \frac{1}{p + 1}$$

and differentiating both sides with respect to p gives

$$\int_0^1 \ln x x^p \, dx = -\frac{1}{(p + 1)^2},$$

doing this m times gives

$$\int_0^1 (\ln x)^m x^p \, dx = \frac{(-1)^m m!}{(p + 1)^{m+1}},$$

which is the required answer.

2.9 Formally, the method here should be to use Taylor's series in two variables: therefore, calculate the first and second derivatives f_x, f_y, f_{xx}, f_{xy} and

f_{yy} and insert into the formula

$$f \approx f + xf_x + yf_y + \frac{1}{2}(x^2 f_{xx} + 2xy f_{xy} + y^2 f_{yy})$$

however, the use of 'known' series can be used to speed up the calculation.

a) $\sin(x + y) = (x + y) + \frac{1}{3!}(x + y)^3$, using the Maclaurin series for sine the formal answer is $(x + y)$ as all second-order terms are zero.

b) In general,

$$x \ln y = h \ln(1 + k) = h \left(k - \frac{1}{2}k^2 + \cdots \right),$$

so the approximation required is hk or xy, and the next term is third order.

c) This part has to use Taylor's series in two dimensions, it is tedious but routine. The expansion $(x + h)/[(x + h) + (y + k)]$ is

$$\approx \frac{x}{x + y} + \left[h \frac{y}{(x + y)^2} - k \frac{x}{(x + y)^2} \right]$$

$$+ \left[-h^2 \frac{y}{(x + y)^3} + hk \frac{x - y}{(x + y)^3} + k^2 \frac{x}{(x + y)^3} \right]$$

to second order in h and k.

2.10 On the curves $\phi(x, y) = a_n$, we have

$$d\phi = \frac{\partial \phi}{\partial x} dx + \frac{\partial \phi}{\partial y} dy = 0,$$

suppose the slope of the curve is m_1 then on these curves

$$m_1 = \frac{dy}{dx} = -\frac{\phi_x}{\phi_y}.$$

Similarly, on the curves $\psi(x, y) = b_n$, we have

$$d\psi = \frac{\partial \psi}{\partial x} dx + \frac{\partial \psi}{\partial y} dy = 0,$$

suppose the slope of the curve is m_2, then on these curves

$$m_2 = \frac{dy}{dx} = -\frac{\psi_x}{\psi_y}.$$

However, we are told that the functions ϕ and ψ are conjugate, thus the Cauchy–Riemann equations hold and thus

$$\phi_x = \psi_y \quad \text{and} \quad \phi_y = -\psi_x.$$

This gives

$$m_1 m_2 = \frac{\phi_x \psi_x}{\phi_y \psi_y} = -\frac{\phi_x \phi_y}{\phi_y \phi_x} = -1$$

and as the product of the slopes is -1, the curves are at right angles to each other.

2.11 As this is a 'verify' question, one could simply perform the somewhat grisly differentiation. Spotting that the functions are, for part (a), the real and imaginary parts of $\frac{1}{1+z}$; and for part (b), those of $\frac{1}{1+z^2}$ immediately answers the question.

2.12 With

$$c_V \, dT = \frac{RT}{V} dV \quad \text{rearranging gives} \quad c_V \frac{dT}{T} = R \frac{dV}{V}.$$

Integrating,

$$c_V \ln T = R \ln V + A \quad \text{gives} \quad T^{c_V} = KV^{-R},$$

where C and K are constants of integration. Using the ideal gas law $pV = RT$,

$$\left(\frac{pV}{R}\right)^{c_V} = KV^{-R} \quad \text{or} \quad pV = kV^{-R/c_V},$$

where k is another constant. Lumping all the Vs together yields

$$pV^{1+R/c_V} = k.$$

It is true that $c_p = R + c_V$, so $1 + R/c_V = c_p/c_V = \gamma$, which gives the desired result.

2.13 From the relationship given by changing the subject of the formula, it is easily derived that:

$$\frac{\partial p}{\partial V} = -\frac{mRT}{V^2}, \quad \frac{\partial T}{\partial p} = \frac{V}{mR} \quad \text{and} \quad \frac{\partial V}{\partial T} = \frac{mR}{p}$$

from which upon multiplication the result follows.

12.3   Chapter 3

3.1 These are problems about finding extrema.
a) $f(x, y) = x^2 + 2y^2 - x^2 y$ so that

$$f_x = 2x - 2xy \quad f_y = 4y - x^2,$$

so that for an extremum,

$$f_x = 0 \Rightarrow x(1 - y) = 0 \quad \text{so that} \quad x = 0 \quad \text{or} \quad y = 1,$$
$$f_y = 0 \Rightarrow 4y - x^2 = 0,$$

which gives the above values for x and y means

$$(x, y) = (0, 0) \quad \text{or} \quad y = 1, x^2 = 4, \quad \text{giving} \quad (-2, 1), (2, 1).$$

Differentiating again gives

$$f_{xx} = 2 - 2y, \quad f_{xy} = -2x, \quad f_{yy} = 4,$$

whence

$$(0, 0) \Rightarrow f_{xx} = 2, f_{xy} = 0, f_{yy} = 4,$$

giving $D = f_{xx}f_{yy} - f_{xy}^2 > 0, f_{xx} > 0$ minimum;

$$(2, 1) \Rightarrow f_{xx} = 0, f_{xy} = -4, f_{yy} = 4,$$

giving $D = f_{xx}f_{yy} - f_{xy}^2 < 0$, saddle;

$$(-2, 1) \Rightarrow f_{xx} = 0, f_{xy} = 4, f_{yy} = 4,$$

giving $D = f_{xx}f_{yy} - f_{xy}^2 < 0$, saddle.

b) $f(x, y) = x^3 + y^3 - 3x - 3y$ so that

$$f_x = 3x^2 - 3 = 0, f_y = 3y^2 - 3 = 0,$$

giving the points $(1, 1), (-1, 1), (1, -1), (-1, -1)$.

$$f_{xx} = 9x, f_{xy} = 0, f_{yy} = 9y,$$

$$(1, 1) \Rightarrow f_{xx} = 9, f_{xy} = 0, f_{yy} = 9,$$

giving $D = f_{xx}f_{yy} - f_{xy}^2 > 0, f_{xx} > 0$ minimum;

$$(-1, 1) \Rightarrow f_{xx} = -9, f_{xy} = 0, f_{yy} = 9,$$

giving $D = f_{xx}f_{yy} - f_{xy}^2 < 0$, saddle;

$$(1, -1) \Rightarrow f_{xx} = 9, f_{xy} = 0, f_{yy} = -9,$$

giving $D = f_{xx}f_{yy} - f_{xy}^2 < 0$, saddle;

$$(-1, -1) \Rightarrow f_{xx} = -9, f_{xy} = 0, f_{yy} = -9$$

giving $D = f_{xx}f_{yy} - f_{xy}^2 > 0, f_{xx} < 0$ maximum.

c) $f(x, y) = 3x^2 - 2xy + y^2 - 8y$ so that

$$f_x = 6x - 2y = 0, \quad f_y = -2x + 2y - 8 = 0$$

giving the only root $(2, 6)$.

$$(2, 6) \Rightarrow f_{xx} = 6, \quad f_{xy} = -2, \quad f_{yy} = 2,$$

giving $D = f_{xx}f_{yy} - f_{xy}^2 > 0, \quad f_{xx} > 0$ minimum.

d) $f(x, y) = 4xy - x^4 - y^4$ so that

$$f_x = 4y - 4x^3 = 0, f_y = 4x - 4y^3 = 0$$

giving the roots $(0, 0), (1, 1)$ and $(-1, -1)$. Differentiating again gives

$$f_{xx} = -12x^2, f_{xy} = 4, \quad f_{yy} = -12y^2.$$

Hence

$$(0, 0) \Rightarrow f_{xx} = 0, \quad f_{xy} = 4, \quad f_{yy} = 0 \text{ hence } D < 0 \text{ and}$$

it is a classic saddle point.

$$(1, 1) \Rightarrow f_{xx} = -12, f_{xy} = 0, f_{yy} = -12,$$

giving $D = f_{xx}f_{yy} - f_{xy}^2 > 0, f_{xx} < 0$ maximum;

$$(-1, -1) \Rightarrow f_{xx} = -12, f_{xy} = 0, f_{yy} = -12$$

giving $D = f_{xx}f_{yy} - f_{xy}^2 > 0, f_{xx} < 0$ maximum too.

e) $f(x, y) = -xye^{(x^2+y^2)/2}$ so that

$$f_x = -ye^{(x^2+y^2)/2} - x^2ye^{(x^2+y^2)/2} = 0$$

$$\text{thus} - y - x^2y = 0 \text{ or } y(1 + x^2) = 0,$$

$$f_y = -xe^{(x^2+y^2)/2} - y^2xe^{(x^2+y^2)/2} = 0$$

$$\text{thus} - x - y^2x = 0 \text{ or } x(1 + y^2) = 0,$$

so the only solution to both $f_x = 0$ and $f_y = 0$ is $(x, y) = (0, 0)$. It is easier to do the rest of the problem from first principles as the second
derivatives are long to calculate. Near the origin, we write $x = \epsilon \cos \theta$
and $y = \epsilon \sin \theta$ so that

$$f(x, y) = -\epsilon^2 \cos \theta \sin \theta e^{\epsilon^2/2}$$

near the origin. At the origin itself $f = 0$ and the value varies positive
and negative as θ varies from 0 to 2π, hence the origin is a saddle.

f) $f = e^x \cos y$ so that $f_x = e^x \cos y$ and $f_y = -e^x \sin y$ both of which cannot
be zero together. Hence, there are no extrema.

g)

$$f(x, y) = \frac{x^2 + 2y^2}{(x + y)^2},$$

$$f_x = 0 \Rightarrow y(x - 2y) = 0$$

and

$$f_y = 0 \Rightarrow x(x - 2y) = 0$$

after use of the quotient differentiation rule. Hence $(0, 0)$ and all points on the line $x = 2y$ are extreme values. The latter also passes through the origin, and therefore all extreme values are not isolated and must (technically) be saddle points. To deduce more, use first principles as follows. Any point on the line is of the form $(2t, t)$, so consider $f(2t + \epsilon \cos\theta, t + \epsilon \sin\theta) - f(2t, t)$ and first note that

$$f(2t, t) = \frac{2}{3}$$

a constant for all values of t. Expand $f(2t + \epsilon \cos\theta, t + \epsilon \sin\theta)$ as a power series in ϵ either manually or by using MAPLE to get:

$$f(2t + \epsilon \cos\theta, t + \epsilon \sin\theta) = \frac{2}{3}$$
$$+ \frac{\epsilon^2}{9t^2} \left\{ \cos^2\theta + 2\sin^2\theta + 2(\sin\theta + \cos\theta)^2 \right.$$
$$\left. - \frac{8}{3}(\sin\theta + \cos\theta)^2 \right\} + O(\epsilon^3).$$

Therefore,

$$f(2t + \epsilon \cos\theta, t + \epsilon \sin\theta) - f(2t, t)$$
$$= \frac{\epsilon^2}{27t^2}(\cos\theta - 2\sin\theta)^2 \geq 0$$

after a bit of algebra, to lowest order in ϵ. The right-hand side is positive except where $\tan\theta = 1/2$, which represents the direction of the line $x = 2y$. Thus the 'saddle point' is a trough along this straight line. The origin is a point at which $f(x, y)$ itself is not well defined, therefore to state it has an extremum there is silly.

3.2 With $f(x, y) = \sin x \cos y$, we have $f_x = \cos x \cos y$ and $f_y = -\sin x \sin y$. Putting both of these zero gives the following infinity of turning points:

$$\left((2n + 1)\frac{\pi}{2}, l\pi \right); \left(k\pi, (2m + 1)\frac{\pi}{2} \right), \quad k, l, m, n \text{ integers or zero.}$$

The second derivatives are $f_{xx} = f_{yy} = -f = -\sin x \cos y$, and this is equal to $-(-1)^n \cdot (-1)^l$ at the extrema $\left((2n + 1)\frac{\pi}{2}, l\pi \right)$. $f_{xy} = -\cos x \sin y = 0$, so

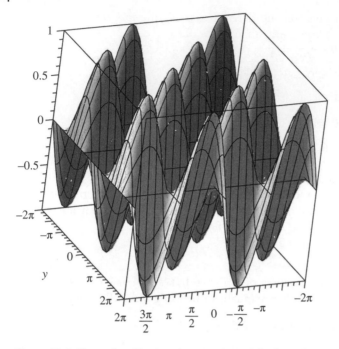

Figure 12.3 The surface $f(x, y) = \sin x \cos y$ is partially drawn here in the range $-2\pi \leq x, y \leq 2\pi$.

$D = f_{xx}f_{yy} - f_{xy}^2 = f^2 - 0 = 1 > 0$. Hence we have an array of maxima and minima. At the points $\left(k\pi, (2m+1)\frac{\pi}{2} \right), f = 0$ so are the two derivatives f_{xx} and f_{yy}. The mixed derivative f_{xy} is either 1 or -1, so $D = -f_{xy}^2 = -1 < 0$; hence these are all saddle points (see Figure 12.3).

3.3 The easiest way to solve this is by logarithmic differentiation

$$\ln f = \ln x + \ln y - \ln(a+x) - \ln(x+y) - \ln(b+y)$$

so

$$\frac{f_x}{f} = \frac{1}{x} - \frac{1}{a+x} - \frac{1}{x+y} = 0$$

and

$$\frac{f_y}{f} = \frac{1}{y} - \frac{1}{x+y} - \frac{1}{b+y} = 0.$$

The algebra is not too cumbersome; the first equation yields $x^2 = ay$, and the second $y^2 = bx$ so either $x = y = 0$ or

$$x = (ba^2)^{1/3}; \quad y = (ab^2)^{1/3}.$$

Avoid having to test second derivatives using the following arguments. Remember $x, y > 0$; for very small values of x, y; $a + x \approx a$ and $b + y \approx b$ so

$$f \approx \frac{xy}{ab(x+y)} = \frac{1}{ab(1/x + 1/y)} \to 0 \quad \text{as } x, y \text{ both approach } 0.$$

On the other hand for very large values of x, y, $a + x \approx x$ and $b + y \approx y$, so

$$f \approx \frac{xy}{xy(x+y)} = \frac{1}{x+y} \to 0 \quad \text{as } x, y \text{ both approach } \infty.$$

Hence, the value $x = (ba^2)^{1/3}$; $y = (ab^2)^{1/3}$ is a maximum. [The other solution $(0, 0)$ seems to be a minimum, but as $x, y > 0$ these are outside our domain.]

3.4 The equation $2x^2 + 3y^2 + z^2 = 1$ is an ellipsoid and the cuboid will have vertices $(\pm x, \pm y, \pm z)$ where all combinations of $\pm$ give the eight corners. Hence, the function

$$F(x, y, z, \lambda) = 8xyz + \lambda(2x^2 + 3y^2 + z^2 - 1)$$

partially differentiate with respect to x, y, z and λ and equate to zero to give

$$F_x = 8yz + 4x\lambda = 0, \tag{12.1}$$
$$F_y = 8xz + 6y\lambda = 0, \tag{12.2}$$
$$F_z = 8xy + 2z\lambda = 0, \tag{12.3}$$
$$F_\lambda = 2x^2 + 3y^2 + z^2 - 1 = 0. \tag{12.4}$$

Equation 12.1 $\times x$ − Equation 12.2 $\times y$ gives $\lambda(4x^2 - 6y^2) = 0$, so either $\lambda = 0$ or $x^2 = 3y^2/2$. Equation 12.2 $\times y$ − Equation 12.3 $\times z$ gives $\lambda(6y^2 - 2z^2) = 0$, so either $\lambda = 0$ or $z^2 = 3y^2$. $\lambda = 0$ is rejected as this leads to $V = 8xyz = 0$. Hence, $2x^2 + 3y^2 + z^2 = 1$ gives $3y^2 + 3y^2 + 3y^2 = 1$, so $y = \pm 1/3$. Therefore, $x = \pm\sqrt{6}/6$ and $z = \pm 1/\sqrt{3}$ so

$$V = 8xyz = 8 \frac{1}{\sqrt{3}} \frac{1}{3} \frac{\sqrt{6}}{6} = \frac{4}{9}\sqrt{2}.$$

3.5 Let the open end of the box be in the (y, z) plane, then the surface area of the box will be two sets of sides plus the bottom: $2xz + 2xy + yz$ and the volume $xyz = 100$. Form the function

$$F(x, y, z, \lambda) = 2xz + 2xy + yz + \lambda(xyz - 100)$$

and differentiate putting each derivative equal to zero:

$$F_x = 2z + 2y + \lambda yz = 0, \tag{12.5}$$
$$F_y = 2x + z + \lambda xz = 0, \tag{12.6}$$

$$F_z = 2x + y + \lambda xy = 0, \tag{12.7}$$
$$F_\lambda = xyz - 100 = 0. \tag{12.8}$$

If any of x, y or z is zero then Equation 12.8 is false, therefore Equations 12.6 and 12.7 subtracted gives $y = z$ and similarly Equations 12.5 and 12.6 or 12.5 and 12.7 give $2x = y = z$. Equation 12.8 then gives $4x^3 = 100$ thus $x = (25)^{1/3}$ and $y = z = 2(25)^{1/3}$. Thus, the surface area in this case is

$$12(25)^{2/3} \approx 102.6 \text{ cm}^2.$$

3.6 The basic set up is shown in Figure 12.4. In limiting the position of just being able to slide around the corner, the trolley is as depicted in

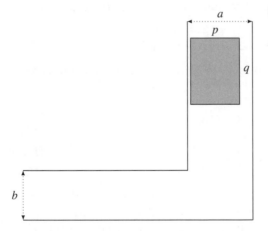

Figure 12.4 The general layout.

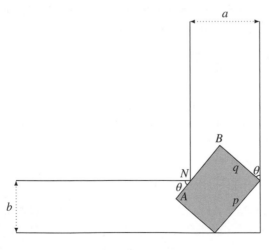

Figure 12.5 The limiting position.

Figure 12.5. From this figure,

$$AN = (b - q\cos\theta)\text{cosec}\theta \quad \text{and} \quad NB = (a - q\sin\theta)\sec\theta$$

therefore,

$$p = AB = (b - q\cos\theta)\text{cosec}\theta + (a - q\sin\theta)\sec\theta, \tag{12.9}$$

and the area of the trolley is

$$A(q,\theta) = pq = q(b\,\text{cosec}\theta - q\cot\theta + a\sec\theta - q\tan\theta).$$

For an extreme value,

$$\frac{\partial A}{\partial q} = b\,\text{cosec}\theta - 2q\cot\theta + a\sec\theta - 2q\tan\theta = 0,$$

$$\frac{\partial A}{\partial \theta} = -bq\,\text{cosec}\theta\cot\theta + q^2\text{cosec}^2\theta + aq\sec\theta\tan\theta - q^2\sec^2\theta = 0.$$

The first of these equations is used to isolate q. Multiplying by $\cos\theta\sin\theta$ gives

$$2q = b\cos\theta + a\sin\theta$$

so equation 12.9 is

$$p = -q(\cot\theta + \tan\theta) + \frac{b}{\sin\theta} + \frac{a}{\cos\theta},$$

which after substitution for q and simplifying is

$$p = \frac{1}{2}\left(\frac{b\cos\theta + a\sin\theta}{\cos\theta\sin\theta}\right) = \frac{q}{\sin\theta\cos\theta}$$

so $q = p\cos\theta\sin\theta$. The partial derivative of A with respect to θ equation gives, when simplified

$$b\cos^3\theta - a\sin^3\theta = q\cos 2\theta,$$

where we have rejected the solution $q = 0$ as an obvious minimum (the trolley has no width). Substituting for q gives the following trigonometric algebra:

$$b\cos^3\theta - a\sin^3\theta = q\cos 2\theta = \frac{1}{2}(b\cos\theta + a\sin\theta)\cos 2\theta$$

$$= \frac{1}{2}[2\cos^2\theta - 1]b\cos\theta + \frac{1}{2}[1 - 2\sin^2\theta]a\sin\theta$$

$$= b\cos^3\theta - \frac{1}{2}b\cos\theta + \frac{1}{2}a\sin\theta - a\sin^3\theta.$$

(Note the use of the two forms of $\cos 2\theta$ to generate $b\cos^3\theta - a\sin^3\theta$ on both sides of the equation. I'd be interested to know if MAPLE does this). Cancelling the factor $b\cos^3\theta - a\sin^3\theta$ from both sides gives

$$\tan\theta = \frac{b}{a} \quad \text{so} \quad \cos\theta = \frac{a}{\sqrt{a^2 + b^2}} \quad \text{and} \quad \sin\theta = \frac{b}{\sqrt{a^2 + b^2}}.$$

Hence,

$$q = \frac{1}{2}\left(b \cdot \frac{a}{\sqrt{a^2 + b^2}} + a \cdot \frac{b}{\sqrt{a^2 + b^2}} \right) = \frac{ab}{\sqrt{a^2 + b^2}}$$

and

$$p = \frac{\frac{ab}{\sqrt{a^2+b^2}}}{\frac{a}{\sqrt{a^2+b^2}}\frac{b}{\sqrt{a^2+b^2}}} = \sqrt{a^2 + b^2}$$

as required. Therefore, the maximum cross sectional area of the trolley is ab the product of the corridor widths.

3.7 The cuboid will have volume $2x \times 2y \times z = 4xyz$, so the function to be considered is

$$F(x, y, z, \lambda) = 4xyz + \lambda(z + \sqrt{x^2 + y^2} - 9),$$

whence the equations to solve are

$$F_x = 4yz + \lambda \frac{x}{\sqrt{x^2 + y^2}} = 0,$$

$$F_y = 4xz + \lambda \frac{y}{\sqrt{x^2 + y^2}} = 0,$$

$$F_z = 4xy + \lambda = 0,$$

$$F_\lambda = z + \sqrt{x^2 + y^2} - 9 = 0.$$

Rejecting zero values as not giving maxima the first two of these equations gives $x = y$ as both have to be positive. In addition, either of these now gives $z = x/\sqrt{2}$. $x = y$ in the last equation implies $z + x\sqrt{2} = 9$ from which

$$x = \frac{9}{\sqrt{2} + \frac{1}{\sqrt{2}}} = 3\sqrt{2}$$

and $y = 3\sqrt{2}$ and $z = 3$ giving the volume 216.

3.8 This problem is solved by putting the angles in terms of lengths. The time taken for light to travel through both media to and from the interface is given by:

$$\frac{\sqrt{d_1^2 + x^2}}{v_1} + \frac{\sqrt{d_2^2 + y^2}}{v_2},$$

where d_1 and d_2 are the perpendicular distances of source and destination of the light, respectively. The function to be differentiated is thus

$$F(x, y, \lambda) = \frac{\sqrt{d_1^2 + x^2}}{v_1} + \frac{\sqrt{d_2^2 + y^2}}{v_2} + \lambda(x + y - a)$$

differentiating and equating to zero gives

$$F_x = \frac{1}{2\sqrt{d_1^2 + x^2}} \times \frac{2x}{v_1} + \lambda = 0,$$

$$F_y = \frac{1}{2\sqrt{d_2^2 + y^2}} \times \frac{2y}{v_2} + \lambda = 0,$$

$$F_\lambda = x + y - a = 0.$$

Subtracting the first two gives

$$\frac{x}{v_1\sqrt{d_1^2 + x^2}} = \frac{y}{v_2\sqrt{d_2^2 + y^2}} \quad \text{or} \quad \frac{\sin\theta_1}{v_1} = \frac{\sin\theta_2}{v_2},$$

which is Snell's Law.

3.9 The solution is found using only two iterations as $x = 0.4, y = 1$ and $z = -0.6$. The Newton–Raphson method is exact for quadratic functions so the method gives the exact result. This question is done using software; it can be done manually but takes time.

12.4 Chapter 4

Some of the questions here are reasonably straightforward; where this is the case, only the answer is given.

4.1 This is $\mathbf{b} - \mathbf{a} = 2\mathbf{i} - 2\mathbf{k}$.

4.2 Straightforwardly

$$\frac{1}{2}(\mathbf{a} + \mathbf{b}) = 2\mathbf{i} + 2\mathbf{j} + 2\mathbf{k}.$$

A is the point $(1, 2, 3)$ and B is the point $(3, 2, 1)$, so the mid point is $(2, 2, 2)$ the same as $(1/2)(\mathbf{a} + \mathbf{b})$.

4.3 The lengths are $\sqrt{14}$, $\sqrt{14}$ and $2\sqrt{2}$.

4.4 As

$$\cos \alpha = \frac{\mathbf{a} \cdot \mathbf{i}}{|\mathbf{a}|}, \quad \cos \beta = \frac{\mathbf{a} \cdot \mathbf{j}}{|\mathbf{a}|}, \quad \text{and} \quad \cos \gamma = \frac{\mathbf{a} \cdot \mathbf{k}}{|\mathbf{a}|}$$

at once,

$$\cos^2\alpha + \cos^2\beta + \cos^2\gamma = \frac{1}{|\mathbf{a}|^2}((\mathbf{a} \cdot \mathbf{i})^2 + (\mathbf{a} \cdot \mathbf{j})^2 + (\mathbf{a} \cdot \mathbf{k})^2)$$

and since $(\mathbf{a} \cdot \mathbf{i}) = a_1$, and so on the result follows.

4.5 The triangle is isosceles and using the scalar product,

$$\mathbf{a} \cdot \mathbf{b} = 10 = |\mathbf{a}||\mathbf{b}| \cos(A\hat{O}B) = 14\cos(A\hat{O}B)$$

giving $\cos(A\hat{O}B) = 5/7$ so $A\hat{O}B = 44.4°$. Similarly, $O\hat{A}B = O\hat{B}A = 67.8°$. Note that when calculating angles in a triangle by this method, sometimes the cosine is negative implying the angle is obtuse. This minus sign can safely be ignored as the method caters not only for calculating the acute angle between two vectors, θ but also for calculating the obtuse angle $\pi - \theta$.

4.6 Using the standard notation for a triangle, the vertices given are the vectors joining the origin to each vertex, so

$$\mathbf{c} = \overrightarrow{OB} - \overrightarrow{OA} = (-3, 0, -3),$$
$$\mathbf{b} = \overrightarrow{OC} - \overrightarrow{OA} = (1, 2, -7),$$
$$\mathbf{a} = \overrightarrow{OB} - \overrightarrow{OC} = (-4, -2, 4),$$

whence $|\mathbf{a}| = 6$, $|\mathbf{b}| = 3\sqrt{6}$ and $|\mathbf{c}| = 3\sqrt{2}$ and using the scalar product as in the last question

$$\cos A = \frac{\mathbf{b} \cdot \mathbf{c}}{|\mathbf{b}||\mathbf{c}|} = \frac{18}{3\sqrt{6} \cdot 3\sqrt{2}} = \frac{\sqrt{3}}{3},$$
$$\cos B = \frac{\mathbf{a} \cdot \mathbf{c}}{|\mathbf{a}||\mathbf{c}|} = 0,$$
$$\cos C = \frac{\mathbf{a} \cdot \mathbf{b}}{|\mathbf{a}||\mathbf{b}|} = \frac{-36}{6 \cdot 3\sqrt{6}} = \frac{\sqrt{6}}{3},$$

ignoring the minus sign as explained in the previous question. $\cos(B) = 0$ means that B is a right angle.

4.7 This is reasonably procedural: $\langle \mathbf{a}, \mathbf{a} \rangle = \mathbf{a} \cdot \mathbf{a} = a_1^2 + a_2^2 + a_3^2$ that is always positive. If $\langle \mathbf{a}, \mathbf{a} \rangle = 0$ then $a_1^2 + a_2^2 + a_3^2 = 0$ and the only possibility is that all components are zero, so $\mathbf{a} = \mathbf{0}$. The last two laws follow from the properties of real numbers once the scalar products are put in terms of components.

4.8 Use the same method as the solution to question 4.5. The three angles are $78.46°$ and two equal angles of $50.77°$ as the triangle is isosceles.

4.9 Using the usual notation for a triangle the cosine rule is $c^2 = a^2 + b^2 - 2ab \cos C$. Now $\mathbf{c} = \mathbf{a} - \mathbf{b}$ using vector addition around the triangle. However, we can also write $c^2 = \mathbf{c} \cdot \mathbf{c} = (\mathbf{a} - \mathbf{b}) \cdot (\mathbf{a} - \mathbf{b})$. Expanding this gives $c^2 = a^2 + b^2 - 2\mathbf{a} \cdot \mathbf{b}$ and writing in the definition of scalar product gives the result.

4.10 Noticing that

$$|\mathbf{a} + \mathbf{b}|^2 = (\mathbf{a} + \mathbf{b}) \cdot (\mathbf{a} + \mathbf{b}) = a^2 + 2\mathbf{a} \cdot \mathbf{b} + b^2$$

and

$$|\mathbf{a} - \mathbf{b}|^2 = (\mathbf{a} - \mathbf{b}) \cdot (\mathbf{a} - \mathbf{b}) = a^2 - 2\mathbf{a} \cdot \mathbf{b} + b^2$$

subtracting these and dividing by 4 gives the proof.

4.11 The definition of $\mathbf{a} \times \mathbf{b}$ is $\mathbf{a} \times \mathbf{b} = |\mathbf{a}||\mathbf{b}| \sin \theta \hat{\mathbf{n}}$ where $\hat{\mathbf{n}}$ is the unit vector perpendicular to both $\mathbf{a}$ and $\mathbf{b}$ such that $\mathbf{a}, \mathbf{b}$ and $\hat{\mathbf{n}}$ form a right-handed system. If the vectors $\mathbf{a}$ and $\mathbf{b}$ were written in the reverse order, then in order for the right handedness to be preserved the triad would be $\mathbf{b}, \mathbf{a}$ and $-\hat{\mathbf{n}}$, so

$$\mathbf{b} \times \mathbf{a} = |\mathbf{b}||\mathbf{a}| \sin \theta(-\hat{\mathbf{n}}) = -|\mathbf{a}||\mathbf{b}| \sin \theta \hat{\mathbf{n}} = -\mathbf{a} \times \mathbf{b}$$

as required. The determinant definition of cross product can also be used. Try it.

4.12 Writing down the three expansions of the triple vector product:

$$\mathbf{a} \times (\mathbf{b} \times \mathbf{c}) = \mathbf{b}(\mathbf{a} \cdot \mathbf{c}) - \mathbf{c}(\mathbf{a} \cdot \mathbf{b}),$$
$$\mathbf{b} \times (\mathbf{c} \times \mathbf{b}) = \mathbf{c}(\mathbf{b} \cdot \mathbf{a}) - \mathbf{a}(\mathbf{b} \cdot \mathbf{c}),$$
$$\mathbf{c} \times (\mathbf{a} \times \mathbf{b}) = \mathbf{a}(\mathbf{c} \cdot \mathbf{b}) - \mathbf{b}(\mathbf{c} \cdot \mathbf{a}).$$

Adding them, and the commutativity of the scalar product, gives the result.

4.13 The plane $x + 2y - 6z = 0$ passes through the origin. Hence it is quite easy to find two vectors that lie entirely in this plane. Choose $z = 1, y = 1$ then $x = 4$ in order to lie on the plane. Therefore, point $(4, 1, 1)$ lies on the plane. Another point might be $x = 2, y = 2$ so $z = 1$ giving the point $(2, 2, 1)$ the vectors $4\mathbf{i} + \mathbf{j} + \mathbf{k}$ and $2\mathbf{i} + 2\mathbf{j} + \mathbf{k}$ both lie entirely on the plane, so their cross product will be perpendicular to this plane:

$$\begin{vmatrix} \mathbf{i} & \mathbf{j} & \mathbf{k} \\ 2 & 2 & 1 \\ 4 & 1 & 1 \end{vmatrix} = \mathbf{i} + 2\mathbf{j} - 6\mathbf{k},$$

which is the normal vector $\mathbf{n}$ in the question. Any scalar multiple of this will also be normal.

4.14 Generalising this is not difficult, but the choices of points on the plane $ax + by + cz = 0$ has to be made carefully to have easy algebra. The following work well:

$$\mathbf{r}_1 = \frac{1}{a}\mathbf{i} + \frac{1}{b}\mathbf{j} - \frac{2}{c}\mathbf{k}$$

and

$$\mathbf{r}_2 = \frac{2}{b}\mathbf{j} - \frac{2}{c}\mathbf{k}$$

then

$$\mathbf{r}_1 \times \mathbf{r}_2 = \begin{vmatrix} \mathbf{i} & \mathbf{j} & \mathbf{k} \\ 1/a & 1/b & -2/c \\ 0 & 2/b & -2/c \end{vmatrix} = \frac{2}{bc}\mathbf{i} + \frac{2}{ac}\mathbf{j} + \frac{2}{ab}\mathbf{k}$$

is normal to the plane $ax + by + cz = 0$, as are any scalar multiples of course. Multiplying by the scalar $abc/2$ gives the normal vector $a\mathbf{i} + b\mathbf{j} + c\mathbf{k}$ an answer one might expect from the previous special case.

4.15 This example is a good one for displaying that vectors are representative of length and direction and are not equations of lines. The paraboloid is $R^2 = 4z$ in cylindrical polar co-ordinates (R, θ, z), see Chapter 7. (If this is new, simply write $R^2 = x^2 + y^2$ as a convenient shorthand; its the only bit we need here. The problem is two dimensional on any plane through the z axis, see Figure 12.6.) The unit normal is

$$\hat{\mathbf{n}} = \frac{-x\mathbf{i} - y\mathbf{j} + 2\mathbf{k}}{\sqrt{R^2 + 4}}.$$

As reflection is occurring at the paraboloidal mirror, the angle between the normal and the vertical $\mathbf{k}$ is the same as the angle between the normal and the reflected ray, labelled $\mathbf{p}$ in Figure 12.6. Given $\mathbf{n} = -x\mathbf{i} - y\mathbf{j} + 2\mathbf{k}$, the angle between $\mathbf{n}$ and $\mathbf{k}$ shown as ϕ in Figure 12.6 is given by

$$\mathbf{n} \cdot \mathbf{k} = |\mathbf{n}||\mathbf{k}| \cos \phi$$

so

$$\cos \phi = \frac{2}{\sqrt{R^2 + 4}}.$$

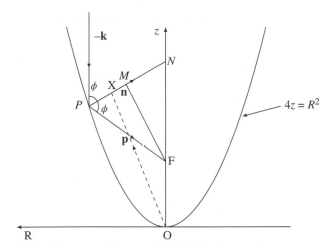

Figure 12.6 A cross-section through the paraboloid $4z = R^2$ showing the vectors $\mathbf{n}, \mathbf{p}$ and the position X of a point on the normal $\mathbf{n}$. M is the mid point of PN.

Now, P is the point $(R, z) = (R, \frac{1}{4}R^2)$, since it is on the paraboloid $4z = R^2$. Any point on the line $\overrightarrow{PN}$ labelled X in Figure 12.6 has position vector $\overrightarrow{OX}$ shown dashed in this figure. It takes the general form

$$\overrightarrow{OX} = \overrightarrow{OP} + \overrightarrow{PX} = \left(R, \frac{1}{4}R^2\right) + \lambda(-R, 2) = \left(R - \lambda R, \frac{1}{4}R^2 + 2\lambda\right).$$

As the parameter λ varies the line PN is generated. The value $\lambda = 1$ gives a zero for the R co-ordinate, hence corresponds to the point N. Thus, N is the point $(0, \frac{1}{4}R^2 + 2)$. The triangle PNF is isosceles as angles $F\hat{N}P$ and $F\hat{P}N$ are both equal to ϕ. Thus, the sides PF and NF are also equal. However, the length of PN is readily found as we know the co-ordinates of both ends:

$$|PN| = \sqrt{(0-R)^2 + \left(\frac{1}{4}R^2 + 2 - \frac{1}{4}R^2\right)^2} = \sqrt{R^2 + 4}.$$

Now from the properties of isosceles triangles, the mid point of PN, M in Figure 12.6 is at the foot of the perpendicular from the vertex F. Thus PM has length $\frac{1}{2}PN = \frac{1}{2}\sqrt{R^2 + 4}$ and

$$|\overrightarrow{PF}| = PM \sec\phi = \left[\frac{1}{2}\sqrt{R^2 + 4}\right]\left[\frac{1}{2}\sqrt{R^2 + 4}\right] = \frac{1}{4}(R^2 + 4).$$

Hence $NF(= PF)$ has length $\frac{1}{4}(R^2 + 4) = \frac{1}{4}R^2 + 1$, and since N has z co-ordinate $\frac{1}{4}R^2 + 2$ this means that F must have co-ordinate $(0, 1)$, which is independent of R. This proves that all reflected rays pass through F the focus of the paraboloid, and it is this property that is used by satellite dish manufacturers.

4.16 Consider the triangle ABC. The cross product $\mathbf{a} \times \mathbf{b} = |\mathbf{a}||\mathbf{b}| \sin C \hat{\mathbf{n}}$, and the magnitude of this will be the area of the triangle. The magnitudes of the other cross products $\mathbf{c} \times \mathbf{a}$ and $\mathbf{b} \times \mathbf{c}$ will also be equal to the area of the triangle, hence

$$|\mathbf{a}||\mathbf{b}| \sin C = |\mathbf{c}||\mathbf{a}| \sin B = |\mathbf{b}||\mathbf{c}| \sin A.$$

Dividing this by the product abc gives the sine rule.

4.17 The orthogonality follows straight from the definition of the primed vectors: $\mathbf{a} \cdot \mathbf{a}'$ gives the same triple product in numerator and denominator, hence unity, whereas $\mathbf{a} \cdot \mathbf{b}'$ gives a triple product with two vectors the same in the numerator, hence zero. With $V = \mathbf{a} \cdot \mathbf{b} \times \mathbf{c}$, we consider the awkward looking product

$$\frac{(\mathbf{b} \times \mathbf{c}) \cdot (\mathbf{c} \times \mathbf{a}) \times (\mathbf{a} \times \mathbf{b})}{V^3}$$

and we use $\mathbf{a} \times (\mathbf{b} \times \mathbf{c}) = \mathbf{b}(\mathbf{a} \cdot \mathbf{c}) - \mathbf{c}(\mathbf{a} \cdot \mathbf{b})$ to deduce that

$$(\mathbf{c} \times \mathbf{a}) \times (\mathbf{a} \times \mathbf{b}) = \mathbf{a}[(\mathbf{c} \times \mathbf{a}) \cdot \mathbf{b}] - \mathbf{b}[(\mathbf{c} \times \mathbf{a}) \cdot \mathbf{a}] = \mathbf{a}V$$

as the second term on the right is zero. Hence, multiplying on the left by $\mathbf{b} \times \mathbf{c}$ gives immediately

$$\mathbf{b} \times \mathbf{c} \cdot (\mathbf{c} \times \mathbf{a}) \times (\mathbf{a} \times \mathbf{b}) = \mathbf{b} \times \mathbf{c} \cdot \mathbf{a}V = V^2$$

and the result is proved. The result involving $\mathbf{r}$ is obtained by letting

$$\mathbf{r} = \alpha\mathbf{a} + \beta\mathbf{b} + \gamma\mathbf{c}$$

and successively taking the scalar product with $\mathbf{a}', \mathbf{b}'$ then $\mathbf{c}'$, taking advantage of $\mathbf{a}, \mathbf{b}$ and $\mathbf{c}$ not being coplanar, hence a basis for the three-dimensional vector space. Orthogonality of the reciprocal vectors with the undashed originals then gives the answer.

4.18 This problem is another involving triple vector products.
a) Start with the identity $\mathbf{x} \cdot (\mathbf{c} \times \mathbf{d}) = (\mathbf{x} \times \mathbf{c}) \cdot \mathbf{d}$ and put $\mathbf{x} = (\mathbf{a} \times \mathbf{b})$, this gives

$$(\mathbf{a} \times \mathbf{b}) \cdot (\mathbf{c} \times \mathbf{d}) = [(\mathbf{a} \times \mathbf{b}) \times \mathbf{c}] \cdot \mathbf{d},$$

expanding the triple cross product via

$$(\mathbf{a} \times \mathbf{b}) \times \mathbf{c} = \mathbf{b}(\mathbf{a} \cdot \mathbf{c}) - \mathbf{a}(\mathbf{b} \cdot \mathbf{c}),$$

gives the answer.

b) Using part (*a*) of this question three times:

$$(\mathbf{b} \times \mathbf{c}) \cdot (\mathbf{a} \times \mathbf{d}) = (\mathbf{c} \cdot \mathbf{d})(\mathbf{a} \cdot \mathbf{b}) - (\mathbf{b} \cdot \mathbf{d})(\mathbf{a} \cdot \mathbf{c}),$$
$$(\mathbf{c} \times \mathbf{a}) \cdot (\mathbf{b} \times \mathbf{d}) = (\mathbf{b} \cdot \mathbf{c})(\mathbf{a} \cdot \mathbf{d}) - (\mathbf{c} \cdot \mathbf{d})(\mathbf{a} \cdot \mathbf{b}),$$
$$(\mathbf{a} \times \mathbf{b}) \cdot (\mathbf{c} \times \mathbf{d}) = (\mathbf{a} \cdot \mathbf{c})(\mathbf{b} \cdot \mathbf{d}) - (\mathbf{a} \cdot \mathbf{d})(\mathbf{b} \cdot \mathbf{c}).$$

Adding these gives a zero right-hand side, hence establishing the result.

12.5 Chapter 5

5.1 Most of these are straightforward.

a) Writing in terms of components this is

$$\frac{d}{dt}(\mathbf{A} + \mathbf{B}) = \frac{d}{dt}[(A_1 + B_1)\mathbf{i} + (A_2 + B_2)\mathbf{j} + (A_3 + B_3)\mathbf{k}],$$

and as all the components are functions of time, the right-hand side of this is

$$= \frac{dA_1}{dt}\mathbf{i} + \frac{dA_2}{dt}\mathbf{j} + \frac{dA_3}{dt}\mathbf{k} + \frac{dB_1}{dt}\mathbf{i} + \frac{dB_2}{dt}\mathbf{j} + \frac{dB_3}{dt}\mathbf{k} = \frac{d\mathbf{A}}{dt} + \frac{d\mathbf{B}}{dt}$$

the required answer.

b) This follows from using the product rule on the result:

$$\frac{d}{dt}(\mathbf{A} \cdot \mathbf{B}) = \frac{d}{dt}(A_1B_1 + A_2B_2 + A_3B_3).$$

c) The cross product is

$$\frac{d}{dt}(\mathbf{A} \times \mathbf{B}) = \frac{d}{dt}\begin{vmatrix} \mathbf{i} & \mathbf{j} & \mathbf{k} \\ A_1 & A_2 & A_3 \\ B_1 & B_2 & B_3 \end{vmatrix}.$$

The first component of the right-hand side is

$$\frac{d}{dt}(A_2B_3 - A_3B_2)\mathbf{i} = (A_2'B_3 - A_3'B_2 + A_2B_3' - A_3B_2')\mathbf{i}$$
$$= (A_2'B_3 - A_3'B_2)\mathbf{i} + (A_2B_3' - A_3B_2')\mathbf{i},$$

using the product rule and a dash to represent time derivatives. Doing the same for the other two components and recombining into determinant form yield

$$\begin{vmatrix} \mathbf{i} & \mathbf{j} & \mathbf{k} \\ A_1' & A_2' & A_3' \\ B_1 & B_2 & B_3 \end{vmatrix} + \begin{vmatrix} \mathbf{i} & \mathbf{j} & \mathbf{k} \\ A_1 & A_2 & A_3 \\ B_1' & B_2' & B_3' \end{vmatrix},$$

which is

$$\frac{d\mathbf{A}}{dt} \times \mathbf{B} + \mathbf{A} \times \frac{d\mathbf{B}}{dt}$$

as required. Note how it is important to preserve the order here.
d) This last result follows directly from applying the product rule to the scalars in the expansion:

$$\frac{d}{dt}(\phi\mathbf{A}) = \frac{d}{dt}(\phi A_1 \mathbf{i} + \phi A_2 \mathbf{j} + \phi A_3 \mathbf{k})$$

and needs no more elaboration.

5.2 The scalar triple product is

$$\mathbf{A} \cdot \mathbf{B} \times \mathbf{C} = \begin{vmatrix} A_1 & A_2 & A_3 \\ B_1 & B_2 & B_3 \\ C_1 & C_2 & C_3 \end{vmatrix},$$

the expansion of this determinant is

$$A_1(B_2 C_3 - B_3 C_2) + A_2(B_3 C_1 - B_1 C_3) + A_3(B_1 C_2 - B_2 C_1)$$

and differentiating any of these with respect to t, indicating this differentiation with a dash leads to three terms to every one term, typically is

$$\frac{d}{dt}(A_2 B_3 C_1) = A_2' B_3 C_1 + A_2 B_3' C_1 + A_2 B_3 C_1'.$$

Hence

$$\frac{d}{dt} \begin{vmatrix} A_1 & A_2 & A_3 \\ B_1 & B_2 & B_3 \\ C_1 & C_2 & C_3 \end{vmatrix}$$

$$= \begin{vmatrix} A_1' & A_2' & A_3' \\ B_1 & B_2 & B_3 \\ C_1 & C_2 & C_3 \end{vmatrix} + \begin{vmatrix} A_1 & A_2 & A_3 \\ B_1' & B_2' & B_3' \\ C_1 & C_2 & C_3 \end{vmatrix} + \begin{vmatrix} A_1 & A_2 & A_3 \\ B_1 & B_2 & B_3 \\ C_1' & C_2' & C_3' \end{vmatrix},$$

which is

$$\frac{d}{dt}\mathbf{A} \cdot \mathbf{B} \times \mathbf{C} = \frac{d\mathbf{A}}{dt} \cdot \mathbf{B} \times \mathbf{C} + \mathbf{A} \cdot \frac{d\mathbf{B}}{dt} \times \mathbf{C} + \mathbf{A} \cdot \mathbf{B} \times \frac{d\mathbf{C}}{dt}.$$

5.3 Using the component form of $\mathbf{a}$ gives the scalar product:

$$\mathbf{a} \cdot \frac{d\mathbf{a}}{dt} = a_1 \frac{da_1}{dt} + a_2 \frac{da_2}{dt} + a_3 \frac{da_3}{dt},$$

hence, the right-hand side is

$$\frac{d}{dt}\left[\frac{1}{2}(a_1^2 + a_2^2 + a_3^2)\right] = \frac{d}{dt}\left[\frac{1}{2}a^2\right],$$

which gives the answer.

5.4 In the following, the notation $\dot{\mathbf{r}}$ for time derivative is used.

a) This first example of a vector differential equation is simply integrated twice to give

$$\mathbf{r} = \frac{1}{2}\mathbf{a}t^2 + \mathbf{A}t + \mathbf{B},$$

where $\mathbf{A}$ and $\mathbf{B}$ are arbitrary vector constants of integration.

b) This is more challenging. First take the vector product with $\mathbf{a}$ and expand $\mathbf{a} \times (\mathbf{a} \times \ddot{\mathbf{r}})$ using the dot to denote differentiation with respect to t. This gives

$$(\mathbf{a} \cdot \ddot{\mathbf{r}})\mathbf{a} - a^2\ddot{\mathbf{r}} = \mathbf{a} \times \mathbf{b}. \tag{12.10}$$

Advantage can now be taken that these are vector quantities. As $\mathbf{a}$ and $\mathbf{a} \times \mathbf{b}$ are not parallel (in fact they are at right angles), it can be stated that $\ddot{\mathbf{r}}$ has to be in the plane defined by $\mathbf{a}$ and $\mathbf{a} \times \mathbf{b}$:

$$\ddot{\mathbf{r}} = \lambda\mathbf{a} + \mu\mathbf{a} \times \mathbf{b},$$

where λ and μ are constants. For those troubled by this, it is the directions that matter; $\mathbf{a}$ is multiplied by $(\mathbf{a} \cdot \ddot{\mathbf{r}})$ in Equation 12.10 and this actually contains the factor $\ddot{\mathbf{r}}$, which we are trying to find, but it doesn't matter as it is inside a scalar product that renders its directional properties impotent. Comparing this with Equation 12.10 gives

$$\ddot{\mathbf{r}} = \lambda\mathbf{a} - \frac{\mathbf{a} \times \mathbf{b}}{a^2}$$

and λ is an unknown parameter that can depend upon t. Therefore, formally the solution is

$$\mathbf{r} = k(t)\mathbf{a} - \frac{\mathbf{a} \times \mathbf{b}t^2}{2a^2} + \mathbf{A}t + \mathbf{B},$$

where $k(t)$, $\mathbf{A}$ and $\mathbf{B}$ are unknown ($k(t)$ is λ integrated twice). They are determined by other conditions imposed on $\mathbf{r}$, for example initial conditions.

5.5 The parabola $y^2 = 4ax$ is parameterised $x = at^2, y = 2at$. Therefore, the position vector of any point on the parabola can be written

$$\mathbf{r} = at^2\mathbf{i} + 2at\mathbf{j}$$

and

$$\frac{d\mathbf{r}}{dt} = 2at\mathbf{i} + 2a\mathbf{j}, \quad \text{with its modulus} \quad \frac{ds}{dt} = 2a\sqrt{t^2 + 1}.$$

$\hat{\mathbf{T}}$ is found by finding the unit vector in the $\dfrac{d\mathbf{r}}{dt}$ direction, so

$$\hat{\mathbf{T}} = \frac{t\mathbf{i} + \mathbf{j}}{2a\sqrt{t^2 + 1}}.$$

Differentiating again gives

$$\frac{d\hat{\mathbf{T}}}{dt} = \mathbf{i}\frac{1}{(1+t^2)^{3/2}} - \mathbf{j}\frac{t}{(1+t^2)^{3/2}}$$

and dividing by ds/dt yields

$$\frac{d\hat{\mathbf{T}}}{ds} = \frac{\mathbf{i} - t\mathbf{j}}{2a(t^2+1)^2}.$$

The curvature κ is the modulus of this and $\kappa = 1/2a(t^2+1)^{3/2}$. Therefore,

$$\hat{\mathbf{N}} = \frac{1}{\kappa}\frac{d\hat{\mathbf{T}}}{ds} = 2a(t^2+1)^{3/2}\frac{\mathbf{i} - t\mathbf{j}}{2a(t^2+1)^2} = \frac{\mathbf{i} - t\mathbf{j}}{\sqrt{1+t^2}}.$$

$\hat{\mathbf{B}}$ is found by using $\hat{\mathbf{B}} = \hat{\mathbf{T}} \times \hat{\mathbf{N}}$

$$\hat{\mathbf{B}} = \frac{1}{1+t^2}\begin{vmatrix} \mathbf{i} & \mathbf{j} & \mathbf{k} \\ t & 1 & 0 \\ 1 & -t & 0 \end{vmatrix} = -\mathbf{k},$$

entirely in line with a curve in the x, y plane. Finally, the torsion $\tau = 0$ as with all plane curves. It is useful to check that $\hat{\mathbf{T}}$ and $\hat{\mathbf{N}}$ are unit vectors as there is a lot of calculation here.

5.6 Starting with $\mathbf{r} = (2\theta^3, 3\theta^2, \theta)$ differentiating gives

$$\frac{d\mathbf{r}}{d\theta} = (6\theta^2, 6\theta, 3) \quad \text{so} \quad \left|\frac{d\mathbf{r}}{d\theta}\right| = 3\sqrt{(4\theta^4 + 4\theta^2 + 1)} = 3(2\theta^2 + 1).$$

From this point, this follows Example 5.1 and the answers are

$$\hat{\mathbf{T}} = \frac{1}{2\theta^2 + 1}(2\theta^2, 2\theta, 1), \quad \kappa = \frac{2}{3(2\theta^2 + 1)^2},$$

$$\hat{\mathbf{N}} = \frac{1}{2\theta^2 + 1}(, 2\theta, 1 - 2\theta^2, -2\theta), \quad \tau = -\frac{2}{3(2\theta^2 + 1)^2},$$

$$\hat{\mathbf{B}} = \frac{1}{2\theta^2 + 1}(1, -2\theta, 2\theta^2),$$

from which $|\tau| = \kappa$.

5.7 With

$$\mathbf{r} = (2a\cos t, 2a\sin t, bt^2),$$

we have $\quad \dot{\mathbf{r}} = (-2a\sin t, 2a\cos t, 2bt)$

and $\quad \ddot{\mathbf{r}} = (-2a\cos t, -2a\sin t, 2b),$

from which $\dot{\mathbf{r}} \cdot \ddot{\mathbf{r}} = 4b^2 t$. Forming the cross product

$$\dot{\mathbf{r}} \times \ddot{\mathbf{r}} = \begin{vmatrix} \mathbf{i} & \mathbf{j} & \mathbf{k} \\ -2a\sin t & 2a\cos t & 2bt \\ -2a\cos t & -2a\sin t & 2b \end{vmatrix},$$

which is $(4ab(\cos t + \sin t), 4ab(\sin t - t \cos t), 4a^2)$. Finally,

$$\hat{\mathbf{T}} = \frac{\dot{\mathbf{r}}}{|\dot{\mathbf{r}}|} = \frac{1}{\sqrt{a^2 + b^2 t^2}} (-a \sin t, a \cos t, bt).$$

5.8 Taking the shapes in turn,

a) For a cylinder $x^2 + y^2 = R^2$ with parameters u, v, any point will have co-ordinates $\mathbf{r} = (R \cos u, R \sin u, v)$, so

$$\mathbf{r}_u = (-R \sin u, R \cos u, 0) \quad \text{and} \quad \mathbf{r}_v = (0, 0, 1)$$

therefore,

$$\hat{\mathbf{T}}_u = (-\sin u, \cos u, 0) \quad \text{and} \quad \hat{\mathbf{T}}_v = (0, 0, 1)$$

and $\hat{\mathbf{T}}_u \times \hat{\mathbf{T}}_v = (\cos u, \sin u, 0)$. Hence, $E = R, F = 0$ and $G = 1$ so $\mathbf{U} = (\cos u, \sin u, 0)$ and

$$\mathbf{U}_u = (-\sin u, \cos u, 0) \quad \text{and} \quad \mathbf{U}_v = (0, 0, 0)$$

giving

$$l = S(\mathbf{r}_u) \cdot \mathbf{r}_u = \frac{1}{R} \quad \text{and} \quad n = S(\mathbf{r}_v) \cdot \mathbf{r}_v = 0$$

with

$$S = \begin{pmatrix} \dfrac{1}{R} & 0 \\ 0 & 0 \end{pmatrix}.$$

Gaussian curvature is zero and mean is curvature $1/(2R)$.

b) For the sphere, following the same method as above, but omitting the details of calculation:

$$\mathbf{r} = (r \cos u \sin v, r \sin u \sin v, r \cos v),$$

$$\mathbf{r}_u = (-r \sin u \sin v, r \cos u \sin v, 0),$$

$$\mathbf{r}_v = (r \cos u \cos v, r \sin u \cos v, -r \sin v)$$

from which

$$\mathbf{U} = \frac{\mathbf{r}_u \times \mathbf{r}_v}{||\mathbf{r}_u \times \mathbf{r}_v||} = (-\cos u \sin v, -\sin u \sin v, -\cos v).$$

This is the normal to the sphere (pointing inwards, but no matter). The tangent vectors are

$$\mathbf{U}_u = (\sin u \sin v, -\cos u \sin v, 0),$$

$$\mathbf{U}_v = (-\cos u \cos v, -\sin u \cos v, \sin v).$$

These are the shape functions $S_p(\mathbf{u})$ and $S_p(\mathbf{v})$, and comparing these two results with the derivatives of $\mathbf{r}$ above yields

$$S_p(\mathbf{u}) = \frac{1}{r} \mathbf{r}_u \quad \text{and} \quad S_p(\mathbf{v}) = \frac{1}{r} \mathbf{r}_v$$

that confirms the two curvatures along the directions of the principal tangents are $1/r$ giving

$$S = \begin{pmatrix} 1/r & 0 \\ 0 & 1/r \end{pmatrix}.$$

The Gaussian curvature is $1/r^2$ and the mean curvature is $1/r$. In fact, as the sphere has curvature $1/r$ in all directions, for *all* points of the sphere $S_v = -cv$ with $c = 1/r$, r being the constant radius of the sphere. The form of S found laboriously by calculus could simply have been written down from the well-known property of the sphere.

5.9 First of all a little pure mathematics: if $\mathbf{v}$ and $\mathbf{w}$ are linearly independent vectors in the tangent space, then they form a basis for the set of all tangent vectors and in particular:

$$S(\mathbf{v}) = a\mathbf{v} + b\mathbf{w},$$

$$S(\mathbf{w}) = c\mathbf{v} + d\mathbf{w}$$

for some a, b, c and d where these are scalars. Therefore,

$$S = \begin{pmatrix} a & b \\ c & d \end{pmatrix}$$

is one representation of the shape operator S. However, taking the cross product gives

$$S(\mathbf{v}) \times S(\mathbf{w}) = (a\mathbf{v} + b\mathbf{w}) \times (c\mathbf{v} + d\mathbf{w}) = (ac - bd)\mathbf{v} \times \mathbf{w}$$

$$= K(\mathbf{p})\mathbf{v} \times \mathbf{w},$$

Since the determinant $(ac - bd)$ is the definition of the Gaussian curvature. In addition, the two cross products

$$S(\mathbf{v}) \times \mathbf{w} = (a\mathbf{v} + b\mathbf{w}) \times \mathbf{w} = a\mathbf{v} \times \mathbf{w}$$

and

$$\mathbf{v} \times S(\mathbf{w}) = \mathbf{v} \times (c\mathbf{v} + d\mathbf{w}) = d\mathbf{v} \times \mathbf{w}$$

when added together give

$$S(\mathbf{v}) \times \mathbf{w} + \mathbf{v} \times S(\mathbf{w}) = 2H(\mathbf{p})\mathbf{v} \times \mathbf{w}$$

as $H = \frac{1}{2}(a + d)$ half of the trace of the matrix of S. The next part uses Lagrange's identity (see Exercise 4.17).

$$(\mathbf{a} \times \mathbf{b}) \cdot (\mathbf{c} \times \mathbf{d}) = (\mathbf{a} \cdot \mathbf{c})(\mathbf{b} \cdot \mathbf{d}) - (\mathbf{a} \cdot \mathbf{d})(\mathbf{b} \cdot \mathbf{c}).$$

Temporarily set

$$\mathbf{a} = \mathbf{r}_u, \quad \mathbf{b} = \mathbf{r}_v, \; \mathbf{c} = S(\mathbf{r}_u), \quad \text{and} \quad \mathbf{d} = S(\mathbf{r}_v)$$

and Lagrange's Identity yields

$$(\mathbf{r}_u \times \mathbf{r}_v) \cdot (S(\mathbf{r}_u) \times S(\mathbf{r}_v))$$
$$= (\mathbf{r}_u \cdot S(\mathbf{r}_u))(\mathbf{r}_v \cdot S(\mathbf{r}_v)) - (\mathbf{r}_u \cdot S(\mathbf{r}_v))(\mathbf{r}_v \cdot S(\mathbf{r}_u)),$$

and the right-hand side is $ln - m^2$ as l, m and n are defined in the text. The above two equations for K and H are now scalar multiplied by $(\mathbf{v} \times \mathbf{w})$. Doing them in turn, first the equation for K becomes

$$(S(\mathbf{v}) \times S(\mathbf{w}))(\mathbf{v} \times \mathbf{w}) = K(\mathbf{p})||\mathbf{v} \times \mathbf{w}||^2.$$

Writing $\mathbf{v} = \mathbf{r}_u$ and $\mathbf{w} = \mathbf{r}_v$ and using Lagrange's Identity, together with:

$$||\mathbf{r}_u \times \mathbf{r}_v||^2 = EG - F^2$$

yields

$$K(\mathbf{p}) = \frac{ln - m^2}{EG - F^2}$$

as required. The equation for H treated in the same way gives

$$(S(\mathbf{r}_u) \times \mathbf{r}_v + \mathbf{r}_u \times S(\mathbf{r}_v))(\mathbf{r}_u \times \mathbf{r}_v) = 2H(\mathbf{p})||\mathbf{r}_u \times \mathbf{r}_v||^2.$$

Lagrange's Identity once more applied to the left-hand side gives

$$(S(\mathbf{r}_u) \times \mathbf{r}_v) \cdot (\mathbf{r}_u \times \mathbf{r}_v) = lG - mF$$

and

$$(\mathbf{r}_u \times S(\mathbf{r}_v)) \cdot (\mathbf{r}_u \times \mathbf{r}_v) = En - mF,$$

whence

$$H = \frac{En + lG - 2mF}{2(EG - F^2)}.$$

5.10 The vector $\mathbf{U}$ is normal to both $\mathbf{r}_u$ and $\mathbf{r}_v$ so

$$\mathbf{U} \cdot \mathbf{r}_u = 0 \quad \text{and} \quad \mathbf{U} \cdot \mathbf{r}_v = 0.$$

Differentiating these gives

$$\mathbf{U}_u \cdot \mathbf{r}_u + \mathbf{U} \cdot \mathbf{r}_{uu} = 0 \quad \text{and} \quad \mathbf{U}_v \cdot \mathbf{r}_u + \mathbf{U} \cdot \mathbf{r}_{uv} = 0$$

together with

$$\mathbf{U}_v \cdot \mathbf{r}_v + \mathbf{U} \cdot \mathbf{r}_{vv} = 0.$$

Hence, since

$$l = \mathbf{r}_u \cdot S(\mathbf{r}_u) = -\mathbf{r}_u \cdot \mathbf{U}_u, \quad m = \mathbf{r}_v \cdot S(\mathbf{r}_u) = -\mathbf{r}_u \mathbf{U}_u$$

and

$$n = \mathbf{r}_v \cdot S(\mathbf{r}_v) = -\mathbf{r}_v \mathbf{U}_v$$

give the three results.

5.11 Starting from

$$\mathbf{r} = (u \cos v, u \sin v, bv)$$

and differentiating gives

$$\mathbf{r}_u = (\cos v, \sin v, 0),$$
$$\mathbf{r}_v = (-u \sin v, u \cos v, b),$$
$$\mathbf{r}_{uu} = \mathbf{0},$$
$$\mathbf{r}_{uv} = (-\sin v, \cos v, 0),$$
$$\mathbf{r}_{vv} = (-u \cos v, -u \sin v, 0),$$

from which

$$\mathbf{U} = \frac{\mathbf{r}_u \times \mathbf{r}_v}{||\mathbf{r}_u \times \mathbf{r}_v||} = \left(\frac{b \sin v}{b^2 + u^2}, \frac{b \cos v}{b^2 + u^2}, \frac{u}{b^2 + u^2} \right).$$

Therefore,

$$l = \mathbf{r}_{uu} \cdot \mathbf{U} = 0,$$
$$m = \mathbf{r}_{uv} \cdot \mathbf{U} = (-\sin v, \cos v, 0),$$
$$n = \mathbf{r}_{vv} \cdot \mathbf{U},$$
$$= (-u \cos v, u \sin v, 0) \cdot \left(\frac{b \sin v}{b^2 + u^2}, \frac{b \cos v}{b^2 + u^2}, \frac{u}{b^2 + u^2} \right) = 0.$$

Thus, without showing the details (calculating E, F and G is not necessary, only $||\mathbf{r}_u \times \mathbf{r}_v||$), we get

$$K = \frac{ln - m^2}{EG - F^2} = -\frac{-b^2}{(b^2 + u^2)^2}$$

and $H = 0$, so the surface of a helicoid is a minimal surface.

5.12 As

$$\frac{d\mathbf{r}}{dt} = \dot{s}\hat{\mathbf{T}} \quad \text{and} \quad \frac{d^2\mathbf{r}}{dt^2} = \ddot{s}\hat{\mathbf{T}} + \kappa\dot{s}^2\hat{\mathbf{N}},$$

taking the cross product gives

$$\frac{d\mathbf{r}}{dt} \times \frac{d^2\mathbf{r}}{dt^2} = \kappa\dot{s}^3\hat{\mathbf{T}} \times \hat{\mathbf{N}} = \kappa\dot{s}^3\hat{\mathbf{B}}.$$

Taking the modulus gives

$$\kappa = \frac{1}{\dot{s}^3} \left| \frac{d\mathbf{r}}{dt} \times \frac{d^2\mathbf{r}}{dt^2} \right|.$$

Substituting back for κ gives the expression for $\hat{\mathbf{B}}$.

5.13 With $r = a(1 + \cos\theta)$, differentiating yields

$$\dot{\mathbf{r}} = -a\dot{\theta}\sin\theta\hat{\mathbf{r}} + a(1 + \cos\theta)\dot{\theta}\hat{\boldsymbol{\theta}}$$

hence,

$$v^2 = |\dot{\mathbf{r}}|^2 = a^2\dot{\theta}^2\sin^2\theta + a^2(1 + \cos\theta)^2\dot{\theta}^2$$

from which we get the first answer:

$$v = a\dot{\theta}\sqrt{2 + 2\cos\theta}.$$

We are told that v is constant, so differentiating with respect to t gives

$$0 = 4a^2\dot{\theta}\ddot{\theta}(1 + \cos\theta) - 2a^2\dot{\theta}^2\sin\theta$$

therefore,

$$\ddot{\theta} = \frac{\dot{\theta}^2\sin\theta}{2(1 + \cos\theta)} = \frac{v^2\sin\theta}{4a^2(1 + \cos\theta)^2}.$$

The radial component of acceleration is $\ddot{r} - r\dot{\theta}^2$;

$$\ddot{r} = \frac{d}{dt}(-a\dot{\theta}\sin\theta) = -a\ddot{\theta}\sin\theta - a\dot{\theta}^2\cos\theta.$$

Hence

$$\ddot{r} - r\dot{\theta}^2 = -a\ddot{\theta}\sin\theta - a\dot{\theta}^2\cos\theta - a(1 + \cos\theta)\dot{\theta}^2$$

substituting for both $\ddot{\theta}$ and $\dot{\theta}^2$ from above, the right-hand side becomes

$$= -\frac{v^2\sin^2\theta}{4a(1 + \cos\theta)^2} - \frac{v^2(1 + 2\cos\theta)}{2a(1 + \cos\theta)},$$

which is

$$= -\frac{v^2}{4a(1 + \cos\theta)^2}[1 - \cos^2\theta + 2 + 6\cos\theta + 4\cos^2\theta] = -\frac{3v^2}{4a},$$

a constant.

5.14 With $r = a(1 - \cos\theta)$, $\dot{r} = a\sin\theta\dot{\theta}$ and $\ddot{r} = a\ddot{\theta}\sin\theta + a\dot{\theta}^2\cos\theta$. With $\dot{\theta} = 3$, these become $\dot{r} = 3a\sin\theta$ and $\ddot{r} = a\ddot{\theta}\sin\theta + 9a\cos\theta$, so where $\theta = 0$, $\dot{r} = 0$ but $\ddot{r} = 9a$.

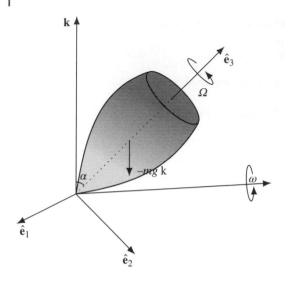

Figure 12.7 The spinning top showing both the rotation vectors ω and Ω, and that the centre of mass of the top is always in the $\hat{e}_2$, k plane.

5.15 The law that governs such a spinning top is the conservation of angular momentum. The total angular momentum of the spinning top, spinning with angular velocity Ω is that of the centre of mass plus that relative to the centre of mass (Figure 12.7):

$$\dot{L} + \Omega \times L,$$

and since the top itself is spinning, a notation is developed that separates this from any movement of the axis, so suppose that

$$\omega = \omega_1 \hat{e}_1 + \omega_2 \hat{e}_2 + \omega_3 \hat{e}_3,$$

where ω denotes the spin of the axes. In addition, the actual spin of the top is $\Omega = \Omega \hat{e}_3$. Therefore, if k is a fixed vertical axis, the angle between $\hat{e}_3$ and k is α. The torque of the top about the origin, the point of contact of the top with the solid ground is

$$a\hat{e}_3 \times (-mg k),$$

where a is the distance of the centre of mass of the top from the k axis. As we can write

$$k = (k \cdot \hat{e}_1)\hat{e}_1 + (k \cdot \hat{e}_2)\hat{e}_2 + (k \cdot \hat{e}_3)\hat{e}_3$$

so

$$k = \sin\alpha \hat{e}_2 + \cos\alpha \hat{e}_3.$$

Therefore, with these components the torque $a\hat{e}_3 \times (-mg k)$ is only in the $\hat{e}_1$ direction. The top is axisymmetric and so the moments of inertia about both $\hat{e}_1$ and $\hat{e}_2$ will be the same, say A and that about $\hat{e}_3$ will be C.

The equations of motion are due to Euler and are three conservation of angular momentum equations about each axis, but only one is required for this problem, the one about the $\hat{\mathbf{e}}_2$ axis. There is no applied torque, so the angular momentum of the centre of mass plus that relative to it sums to zero:

$$A\dot{\omega}_2 + (A - C)\omega_3\omega_1 + C\omega_1\Omega = 0.$$

Since the angle α is a constant, $\hat{\mathbf{e}}_2$ is in the plane of the fixed axis $\mathbf{k}$ and $\hat{\mathbf{e}}_3$ and so $\omega_1 = 0$. (The motion $\omega_1 \neq 0$ is called nutation assumed zero in this problem as α is constant.) Therefore, the above equation gives $\dot{\omega}_2 = 0$, or $\omega_2 = $ constant. Writing

$$\omega_2 = \frac{d\theta_2}{dt}$$

this means that θ_2 increases at a steady rate, which is the top precesses at a constant rate.

12.6   Chapter 6

6.1   Direct calculation gives $\nabla(2x^4y - xyz) = (8x^3y - yz, 2x^4 - xz, xy)$, so at the point $(1, 1, 2)$ this is $(6, 0, 1)$.

6.2   As

$$\nabla(\phi\psi) = \frac{\partial}{\partial x}(\phi\psi)\mathbf{i} + \frac{\partial}{\partial y}(\phi\psi)\mathbf{j} + \frac{\partial}{\partial z}(\phi\psi)\mathbf{k}$$

applying the product rule, the result follows at once.

6.3   With $\phi = (x^2 + y^2 + z^2)^{n/2}$,

$$\frac{\partial}{\partial x}((x^2 + y^2 + z^2)^{n/2}) = \frac{n}{2}(x^2 + y^2 + z^2)^{n/2-1}2x = \frac{nx}{r}r^{n-1}$$

so

$$\nabla(x^2 + y^2 + z^2)^{1/2} = n\mathbf{r}r^{n-2} = n\hat{\mathbf{r}}r^{n-1}.$$

6.4   Again using components, the answer is

$$\nabla(\ln r) = \frac{\mathbf{r}}{r^2}.$$

6.5   With $\phi = x^4 + y^4 + z^4 - 3$, $\nabla\phi = 4x^3\mathbf{i} + 4y^3\mathbf{j} + 4z^3\mathbf{k}$ which at $(1, 1, 1)$ gives the unit normal as

$$\frac{1}{\sqrt{3}}\{\mathbf{i} + \mathbf{j} + \mathbf{k}\}.$$

6.6 With $\phi = xy^2 + xyz$, then $\nabla\phi = (y^2 + yz)\mathbf{i} + (2xy + xz)\mathbf{j} + xy\mathbf{k}$. This is evaluated at the point $(1, 2, 3)$ and has the value $7\mathbf{i} + 7\mathbf{j} + 2\mathbf{k}$; and this has to be the direction of the maximum variation of ϕ as argued in the paragraph after Example 6.2, another name for the directional derivative. Its value will be the magnitude of this vector which is $\sqrt{102}$.

6.7 Given $F(x(t), y(t), z(t), t)$, then using the chain rule,

$$\frac{dF}{dt} = \frac{\partial F}{\partial x}\frac{dx}{dt} + \frac{\partial F}{\partial y}\frac{dy}{dt} + \frac{\partial F}{\partial z}\frac{dz}{dt} + \frac{\partial F}{\partial t}.$$

Writing the first three terms as a scalar product,

$$\frac{dF}{dt} = \frac{d\mathbf{r}}{dt} \cdot \nabla F + \frac{\partial F}{\partial t},$$

and if the last term is zero, this is the directional derivative of F in the direction of the vector $\dot{\mathbf{r}}$. In fluid mechanics, this the classic rate of change of motion following a fluid or Lagrangian derivative.

6.8 Writing

$$\nabla u = \frac{\partial u}{\partial x}\mathbf{i} + \frac{\partial u}{\partial y}\mathbf{j} + \frac{\partial u}{\partial z}\mathbf{k},$$

$$\nabla v = \frac{\partial v}{\partial x}\mathbf{i} + \frac{\partial v}{\partial y}\mathbf{j} + \frac{\partial v}{\partial z}\mathbf{k},$$

$$\nabla w = \frac{\partial w}{\partial x}\mathbf{i} + \frac{\partial w}{\partial y}\mathbf{j} + \frac{\partial w}{\partial z}\mathbf{k}.$$

Therefore,

$$\nabla u \cdot (\nabla v \times \nabla w) = \begin{vmatrix} u_x & u_y & u_z \\ v_x & v_y & v_z \\ w_x & w_y & w_z \end{vmatrix} = \frac{\partial(u, v, w)}{\partial(x, y, z)},$$

using the suffix derivative notation. The Jacobian was introduced in Chapter 2, and if it is zero, there is a functional relationship between u, v and w.

6.9 This question concerns the algebra of vectors under differentiation.
a) With $\mathbf{r} \cdot \mathbf{a} = a_1 x + a_2 y + a_3 z$, then as

$$\frac{\partial}{\partial x}(a_1 x + a_2 y + a_3 z) = a_1, \quad \text{and so on}$$

The result $\nabla(\mathbf{r} \cdot \mathbf{a}) = \mathbf{a}$ follows.
b) This is a little tricky, and the starting point, far from obvious, is to write $(\mathbf{a} \times \mathbf{r}) \cdot (\mathbf{a} \times \mathbf{r}) = |\mathbf{a} \times \mathbf{r}|^2$. The logic is that differentiating

the modulus of a function is always fraught with peril and using components is too messy. Thus,

$$\nabla |\mathbf{a} \times \mathbf{r}|^n = \nabla [(\mathbf{a} \times \mathbf{r}) \cdot (\mathbf{a} \times \mathbf{r})^{n/2}]$$

hence,

$$\nabla |\mathbf{a} \times \mathbf{r}|^n = \frac{n}{2} [(\mathbf{a} \times \mathbf{r}) \cdot (\mathbf{a} \times \mathbf{r})]^{(n/2)-1} (2(\mathbf{a} \times \mathbf{r}) \cdot \nabla)(\mathbf{a} \times \mathbf{r}).$$

With the ability to take ∇ through the constant vector the right-hand side is

$$= ((\mathbf{a} \times \mathbf{r}) \cdot \nabla)(\mathbf{a} \times \mathbf{r}) = (\mathbf{a} \times (\mathbf{a} \times \mathbf{r}) \cdot \nabla)\mathbf{r} = \mathbf{a} \times (\mathbf{a} \times \mathbf{r}),$$

from which the result follows.

c) With

$$\nabla \left(\frac{1}{r} \right) = -\frac{\mathbf{r}}{r^3},$$

the given expression then becomes

$$-\mathbf{a} \cdot \nabla \left(\frac{\mathbf{a} \cdot \mathbf{r}}{r^3} \right).$$

Using the product rule gives

$$\mathbf{a} \cdot \left[\frac{3(\mathbf{a} \cdot \mathbf{r})\mathbf{r}}{r^5} - \frac{\nabla(\mathbf{a} \cdot \mathbf{r})}{r^3} \right]$$

and using part (a), this is

$$\frac{3(\mathbf{a} \cdot \mathbf{r})^2 - a^2 r^2}{r^5}$$

as required.

6.10 By direct differentiation,

$$\nabla \left[-\frac{x\mathbf{i}}{\sqrt{x^2 + y^2}} - \frac{y\mathbf{j}}{\sqrt{x^2 + y^2}} \right]$$

$$= -\frac{\partial}{\partial x} \left(\frac{x}{\sqrt{x^2 + y^2}} \right) - \frac{\partial}{\partial y} \left(\frac{y}{\sqrt{x^2 + y^2}} \right)$$

$$= \frac{-x^2 - y^2}{(x^2 + y^2)^{3/2}} = -\frac{1}{\sqrt{x^2 + y^2}}.$$

This represents a sink of material at the origin.

6.11 With $\mathbf{F} = x^2 y z \mathbf{i} + x y^2 z \mathbf{j} + x y z^2 \mathbf{k}$, $\nabla \cdot \mathbf{F} = 6xyz$, so at $(1, 1, -1)$ its value is -6.

6.12 This follows directly from the identity proved in Example 6.7:

$$\nabla \cdot (\mathbf{A} \times \mathbf{B}) = \mathbf{B} \cdot \nabla \times \mathbf{A} - \mathbf{A} \cdot \nabla \times \mathbf{B}$$

with $\mathbf{A} = \nabla \phi$ and $\mathbf{B} = \nabla \psi$ the above identity yields

$$\nabla \cdot (\nabla \phi \times \nabla \psi) = \nabla \psi \cdot (\nabla \times \nabla \phi) - \nabla \phi \cdot (\nabla \times \nabla \psi)$$

the right-hand side of which is identically zero, hence $\nabla \phi \times \nabla \psi$ is solenoidal.

6.13 Computation gives $\nabla \times \mathbf{A} = xz^2\mathbf{i} + (xy^2 - yz^2)\mathbf{i} + (y^2 - x^2 z)\mathbf{k}$ and at $(1, 1, -1)$ this is $\mathbf{i} + 2\mathbf{k}$.

6.14 With $\mathbf{A} = (x^2 y, xy^2, xyz^2)$ and $\mathbf{B} = (yz, xz, xy)$ as $\nabla \times \mathbf{B} = 0$ so does $\mathbf{A} \times (\nabla \times \mathbf{B})$. To calculate $(\mathbf{A} \times \nabla) \times \mathbf{B}$ take a little more effort. The operator $(\mathbf{A} \times \nabla)$ is

$$\begin{vmatrix} \mathbf{i} & \mathbf{j} & \mathbf{k} \\ x^2 y & xy^2 & xyz^2 \\ \dfrac{\partial}{\partial x} & \dfrac{\partial}{\partial y} & \dfrac{\partial}{\partial z} \end{vmatrix}$$

$$= \mathbf{i} \left(xy^2 \frac{\partial}{\partial z} - xyz^2 \frac{\partial}{\partial y} \right) + \mathbf{j} \left(xyz^2 \frac{\partial}{\partial x} - x^2 y \frac{\partial}{\partial z} \right)$$

$$+ \mathbf{k} \left(x^2 y \frac{\partial}{\partial y} - xy^2 \frac{\partial}{\partial x} \right).$$

Operating on the vector $\mathbf{B}$ gives $(xy^2 z^2 - x^3 y - xy^2 z, x^2 yz + x^2 yz^2, x^3 y + x^2 yz)$, which is certainly not zero, hence the importance of the parentheses.

6.15 The easiest way to answer this exercise is to spot that the complex function $w = \ln(z)$ has real and imaginary parts

$$w = \ln|z| + i\arg(z) = \frac{1}{2}\ln(x^2 + y^2) + i\tan^{-1}(y/x),$$

and since the real and imaginary parts of a complex function are conjugate, QED. Of course, direct differentiation is also possible but longer.

12.7 Chapter 7

7.1 These are all reasonably straightforward, only demanding elementary algebra:

a) $y = x$ is a plane. In cylindrical and spherical polars, it is the same: $\theta = \pi/4$.

b) $z = x^2 + y^2$ is a paraboloid. In cylindrical polars, it is $z = R^2$. In spherical polars $\cos\phi = r\sin^2\phi$.

c) $z^2 = 4(x^2 + y^2)$ is a cone. In cylindrical polars, it is $z = 2R$ and in spherical polars, $\tan \phi = 1/2$.

d) $z = 0$ is the x, y plane, the same in cylindrical polars and $\phi = \pi/2$ in spherical polars.

e) $x^2 + y^2 + z^2 = 16$ is a sphere of radius 4 centred at the origin. In cylindrical polars, it is $R^2 + z^2 = 16$, and in spherical polars, it is $r = 4$.

7.2 These surfaces are given below:

a) $R = 3$ is a cylinder axis $x = y = 0$, the z-axis.

b) $\phi = \pi/3$ is a cone, axis $x = y = 0$ and vertex angle $2\pi/3$.

c) $R = \cos \theta$, this is a cylinder $(x - 1/2)^2 + y^2 = 1/4$ with z arbitrary. Axis $x = 1/2, y = 0$, radius $1/2$.

d) $R^2 = 9z$ is a paraboloid $x^2 + y^2 = 9z$.

e) $\theta = \pi/4$ in either cylindrical or spherical polar co-ordinates is the plane $x = y$ with z arbitrary.

7.3 The given vector is transformed in cylindrical polars through $x = R \cos \theta, y = R \sin \theta$ and

$$i = \hat{R} \cos \theta - \hat{\theta} \sin \theta,$$
$$j = \hat{R} \sin \theta + \hat{\theta} \cos \theta,$$

whence

$$F = (R^2 \cos^2 \theta + R \sin \theta)\hat{R} + (R \sin \theta - R^2 \cos^2 \theta \sin \theta)\hat{\theta} + z^3 k,$$

and in spherical polars,

$$i = \hat{r} \cos \theta \sin \phi - \hat{\phi} \cos \theta \cos \phi - \hat{\theta} \sin \theta,$$
$$j = \hat{r} \sin \theta \sin \phi - \hat{\phi} \sin \theta \cos \phi + \hat{\theta} \cos \theta,$$
$$k = \hat{r} \cos \phi - \hat{\phi} \sin \phi,$$

so substitution of $x = r \cos \theta \sin \phi, y = r \sin \theta \sin \phi, z = r \cos \phi$ together with the above unit vectors gives

$$F = \hat{r}(r^2 \cos^3 \theta \sin^3 \phi + r \sin^2 \theta \sin^2 \phi + r^3 \cos^4 \phi)$$
$$- \hat{\phi}(r^2 \cos^3 \theta \sin^2 \phi \cos \phi + r \sin^2 \theta \sin \phi \cos \phi + r^3 \cos^3 \phi \sin \phi)$$
$$+ \hat{\theta}(r^2 \cos^2 \theta \sin \theta \sin^2 \phi + r \sin \theta \cos \theta \sin \phi).$$

7.4 These are bipolar co-ordinates. Using the formulae

$$h_1 = \left| \frac{\partial r}{\partial u} \right|, \quad h_2 = \left| \frac{\partial r}{\partial v} \right|, \quad h_3 = \left| \frac{\partial r}{\partial z} \right|,$$

the answers are

$$h_1 = h_2 = \frac{a}{\cosh v - \cos u}, \quad h_3 = 1.$$

7.5 Use the arguments in the text before Example 7.1 that show the area element is $h_1 h_2 \, du \, dv$ and the volume element is $h_1 h_2 h_3 \, du \, dv \, dw$.

7.6 This result follows from the three results in the text:

$$\nabla u = \frac{\hat{e}_1}{h_1}, \quad \nabla v = \frac{\hat{e}_2}{h_2}, \quad \nabla w = \frac{\hat{e}_3}{h_3}$$

together with the three equations

$$\frac{\partial \mathbf{r}}{\partial u} = h_1 \hat{e}_1, \quad \frac{\partial \mathbf{r}}{\partial u} = h_2 \hat{e}_2, \quad \frac{\partial \mathbf{r}}{\partial w} = h_3 \hat{e}_3.$$

The triple scalar product of the first three is $1/(h_1 h_2 h_3)$ and that of the second three is $h_1 h_2 h_3$, which gives the result. This is an example of a reciprocal set of vectors.

7.7 With $x = uv \cos \phi, y = uv \sin \phi, z = \frac{1}{2}(u^2 - v^2)$ differentiating the position vector $\mathbf{r}$ yields

$$|\mathbf{r}_u| = h_1 = \sqrt{u^2 + v^2}, \quad |\mathbf{r}_v| = h_2 = \sqrt{u^2 + v^2}, \quad |\mathbf{r}_\phi| = h_3 = uv.$$

Substituting into the formulas for $\nabla \psi, \nabla \cdot \mathbf{F}$ and $\nabla \times \mathbf{F}$ is procedural and yields:

$$\nabla \psi = \frac{1}{\sqrt{u^2 + v^2}} \frac{\partial \psi}{\partial u} \hat{e}_1 + \frac{1}{\sqrt{u^2 + v^2}} \frac{\partial \psi}{\partial v} \hat{e}_2 + \frac{1}{uv} \frac{\partial \psi}{\partial \phi} \hat{e}_3,$$

$$\nabla \cdot \mathbf{F} = \frac{1}{uv(u^2 + v^2)} \left\{ \frac{\partial}{\partial u} [uv \sqrt{u^2 + v^2} F_1] + \frac{\partial}{\partial v} [uv \sqrt{u^2 + v^2} F_2] \right\}$$
$$+ \frac{\partial}{\partial \phi} [(u^2 + v^2) F_3]$$

and

$$\nabla \times \mathbf{F} = \frac{1}{uv(u^2 + v^2)} \begin{vmatrix} (u^2 + v^2)^{1/2} \hat{e}_1 & (u^2 + v^2)^{1/2} \hat{e}_2 & uv\hat{e}_3 \\ \dfrac{\partial}{\partial u} & \dfrac{\partial}{\partial v} & \dfrac{\partial}{\partial \phi} \\ (u^2 + v^2)^{1/2} F_1 & (u^2 + v^2)^{1/2} F_2 & uv F_3 \end{vmatrix}.$$

7.8 Write $a_{ij} = \frac{1}{2}(F_{ij} + F_{ji})$, then as $a_{ij} = a_{ji}$ this tensor is symmetric. Write $b_{ij} = \frac{1}{2}(F_{ij} - F_{ji})$, then as $b_{ij} = -b_{ji}$ this tensor is antisymmetric. Hence, $F_{ij} = a_{ij} + b_{ij}$ an arbitrary tensor is the sum of a symmetric and an anti-symmetric tensor.

7.9 The tensor expression to be considered is $\epsilon_{ijk}\epsilon_{rsk} = \delta_{ir}\delta_{js} - \delta_{is}\delta_{jr}$. Careful enumeration is the only sure way to establish this. The left-hand side is $\epsilon_{ij1}\epsilon_{rs1} + \epsilon_{ij2}\epsilon_{rs2} + \epsilon_{ij3}\epsilon_{rs3}$. With $i = j$, this is zero for all values of r and s, so take $i = 1, j = 2$ the last term is 1 if $r = 1, s = 2$ but -1 if $r = 2, s = 1$. If either r or s is 3, the whole is zero. Going through the same set of values for $\delta_{ir}\delta_{js} - \delta_{is}\delta_{jr}$ and the same results are found. This establishes the identity.

7.10 It is easily established that $\mathbf{a} \cdot \mathbf{b} = a_i b_i$ and that $\mathbf{a} \times \mathbf{b} = \epsilon_{ijk} a_j b_k$.

7.11 The transformations are

a) $\overline{A}^i_{jk} = \dfrac{\partial \overline{x}^i}{\partial x^r} \dfrac{\partial x^s}{\partial \overline{x}^j} \dfrac{\partial x^t}{\partial \overline{x}^k} A^r_{st}$,

b) $\overline{B}^n = \dfrac{\partial \overline{x}^n}{\partial x^m} B^m$,

c) $\overline{C}^{ijk}_{lmn} = \dfrac{\partial \overline{x}^i}{\partial x^r} \dfrac{\partial \overline{x}^j}{\partial x^s} \dfrac{\partial \overline{x}^k}{\partial x^t} \dfrac{\partial x^u}{\partial \overline{x}^l} \dfrac{\partial x^v}{\partial \overline{x}^m} \dfrac{\partial x^w}{\partial \overline{x}^n} C^{rst}_{uvw}$.

7.12 For δ^i_j, the transformation is

$$\delta^i_j = \dfrac{\partial \overline{x}^i}{\partial x^p} \dfrac{\partial x^r}{\partial \overline{x}^j} \delta^r_p.$$

However, as δ^r_p is zero unless $p = r$, putting $p = r$ on the right and using the chain rule gives

$$\delta^i_j = \dfrac{\partial \overline{x}^i}{\partial \overline{x}^j}.$$

This immediately tells us that δ^i_j is Kronecker's delta, since the partial derivative is zero unless $i = j$. Starting with either covariance δ_{ij} or contravariance δ^{ij} leads nowhere as the chain rule can no longer be applied. Thus, representing Kronecker's delta as a mixed tensor is correct, anything else is not.

7.13 The permutation tensor of any order is defined as:

$$\epsilon_{i_1 i_2 \ldots i_n} = \begin{cases} 1 & \text{if } i_1, i_2, \ldots, i_n \text{ permute cyclically,} \\ -1 & \text{if } i_1, i_2, \ldots, i_n \text{ permute anti-cyclically,} \\ 0 & \text{in all other cases.} \end{cases}$$

Therefore, even if all of the suffices are different, the permutation tensor is still zero if the order is neither cyclic nor anticyclic.

12.8   Chapter 8

8.1   Most of these are straightforward:

a) On $C, x = a \cos t, y = a \sin t$ and $ds = \sqrt{(dx)^2 + (dy)^2} = a\,dt$. The integral thus becomes

$$a^4 \int_0^\pi \cos^2 t \sin t \, dt = \frac{2}{3} a^4.$$

b) The parameterisation stays as in the previous part so the integral is

$$a^4 \int_0^\pi \cos t \sin^2 t \, dt = 0$$

a very different answer.

c) Here, the integrand is $\phi \, dx$ notice, not ds, so we have

$$\int_C (x^2 + y^4 + z^6) dx = \int_0^1 (2t^6 + t^8) 3t^2 \, dt = \frac{28}{33}.$$

d) Here $ds^2 = \sqrt{a^2 \sin^2\theta + b^2 \cos^2\theta} \, d\theta$ with the integrand also equalling

$$\sqrt{a^2 + b^2 - r^2} = \sqrt{a^2 \sin^2\theta + b^2 \cos^2\theta}.$$

Hence, the integral becomes

$$\int_0^{2\pi} [a^2 \sin^2\theta + b^2 \cos^2\theta] d\theta = \pi(a^2 + b^2).$$

8.2   For part (a), $\mathbf{F} = x^2 \mathbf{i}$ and $d\mathbf{r} = \mathbf{i} \, dx$ with $-1 \le x \le 1$ and $y = 0$. The integral is

$$\int_{-1}^1 x^2 \, dx = \frac{2}{3}.$$

On path (b), $x = \cos\theta, y = \sin\theta$ with $-\pi \le \theta \le 0$, so $\mathbf{F} = \cos^2\theta \mathbf{i} + \sin\theta \mathbf{j}$ and $d\mathbf{r} = -\sin\theta \, d\theta \mathbf{i} + \cos\theta \, d\theta \mathbf{j}$ giving the integral

$$\int_{-\pi}^0 \{-\cos^2\theta \sin\theta + \sin\theta \cos\theta\} d\theta = \left[\frac{1}{3}\cos^3\theta + \frac{1}{2}\sin^2\theta \right]_{-\pi}^0 = \frac{2}{3}.$$

For the last part (c), the integrals are straightforward giving the answers $1/2, 2/3$ and $-1/2$ for the straight lines connecting $(-1, 0)$ to $(-1, 1), (-1, 1)$ to $(1, 1)$ and $(1, 1)$ to $(0, 1)$, respectively. Summing to $2/3$, the results imply that the integral is independent of path, and this is confirmed by taking the curl of $\mathbf{F}$. The vector field $\mathbf{F}$ is conservative.

8.3   With $\mathbf{F} = (z - 1)\mathbf{i}$, all three calculations are straightforward and all give the answer -2. Calculating $\nabla \times \mathbf{F}$ does not give $\mathbf{0}$ but $\mathbf{j}$. However, the normal to the domain is $\mathbf{k}$, and this is at right angles to $\nabla \times \mathbf{F}$; and this means no flux through any closed curve in the x, y plane. Hence, there is still no work done even though the field is not conservative. Stokes' theorem in Chapter 11 is relevant here. If the integrals all took place on the plane $z = 1$, they would all be zero as $\mathbf{F} = \mathbf{0}$ there. Again, $\nabla \times \mathbf{F}$ is not zero, but still at right angles to the surface enclosed by all closed curves, so the flux is zero. It is worth mentioning a different case. In applications, notably in fluid mechanics, there are important cases where despite the curl of the velocity being non-zero in general, it is zero on surfaces and the existence of a potential only on these surfaces is useful for both understanding and calculations.

8.4 The integral

$$\int_C \mathbf{F} \cdot d\mathbf{r}$$

is evaluated three times:

a) The straight line is parameterised by $x = y = t, z = 0$ with $0 \le t \le 2$, so $\mathbf{F} = x\mathbf{i} = t\mathbf{i}$ and

$$\int_C \mathbf{F} \cdot d\mathbf{r} = \int_0^2 t\, dt = \left[\frac{1}{2}t^2\right]_0^2 = 2.$$

b) With $\mathbf{F} = \mathbf{r}$, the helix is parameterised by $x = \sin t, y = \cos t$ and $z = t$; hence, $d\mathbf{r} = \cos t\, dt\mathbf{i} + \sin t\, dt\mathbf{j} + dt\mathbf{k}$ and

$$\mathbf{F} \cdot d\mathbf{r} = (\sin t \cos t - \cos t \sin t + t)dt = t\, dt$$

so

$$\int_C \mathbf{F} \cdot d\mathbf{r} = \int_0^{2\pi} t\, dt = \frac{1}{2}(2\pi)^2 = 2\pi^2.$$

c) With the same $\mathbf{F}$ as before, on $x = 0, y = 1$ and $0 \le z \le 2\pi$ gives $\mathbf{F} = \mathbf{j} + z\mathbf{k}$ so

$$\int_C \mathbf{F} \cdot d\mathbf{r} = \int_0^{2\pi} z\, dz = 2\pi^2,$$

as before.

The last two results are the same as $\mathbf{r} = \nabla \phi$ where $\phi = \frac{1}{2}r^2$ is the potential function. Therefore, the integral is only dependent on the value of ϕ at the end points and not on the path taken.

8.5 On $C, \mathbf{r} = t\mathbf{i} + \frac{1}{\sqrt{2}}t^2\mathbf{j} + \frac{1}{3}t^3\mathbf{k}$ so $ds = (1 + t^2)dt$ with

$$\mathbf{F} = t^2\mathbf{i} + \frac{1}{2}t^4\mathbf{j} + \frac{1}{9}t^6\mathbf{k},$$

and the limits are 0 and 1.

a) Therefore,

$$\int_C \mathbf{F} \, ds = \int_0^1 \left[\begin{array}{c} (t^2 + t^4)\mathbf{i} + \left(\frac{1}{2}t^4 + \frac{1}{2}t^6 \right)\mathbf{j} \\ + \left(\frac{1}{9}t^6 + \frac{1}{9}t^8 \right)\mathbf{k} \end{array} \right] dt,$$

which gives

$$\frac{8}{15}\mathbf{i} + \frac{6}{35}\mathbf{j} + \frac{16}{567}\mathbf{k}.$$

b)

$$\int_C \mathbf{F} \cdot d\mathbf{r} = \int_0^1 \left(t^2\mathbf{i} + \frac{1}{2}t^4\mathbf{j} + \frac{1}{9}t^6\mathbf{k} \right) \cdot (\mathbf{i} + \sqrt{2}\mathbf{j} + t^2\mathbf{k}) dt$$

that becomes

$$\int_0^1 t^2 + \frac{\sqrt{2}}{2}t^5 + \frac{1}{9}t^8 \, dt = \frac{28}{81} + \frac{1}{12}\sqrt{2}.$$

c) The third integral is

$$\int_C \mathbf{F} \times d\mathbf{r} = \int_0^1 \left(t^2\mathbf{i} + \frac{1}{2}t^4\mathbf{j} + \frac{1}{9}t^6\mathbf{k} \right) \times (\mathbf{i} + \sqrt{2}\mathbf{j} + t^2\mathbf{k}) dt$$

that is a messy calculation. The answer is

$$\int_C \mathbf{F} \times d\mathbf{r} = \mathbf{i} \left(\frac{34 - 7\sqrt{2}}{504} \right) - \mathbf{j}\frac{58}{315} + \mathbf{k} \left(\frac{5\sqrt{2} - 2}{20} \right).$$

8.6 The constant vector $\mathbf{U} = \nabla(\mathbf{U} \cdot \mathbf{r})$ so the given integral is solely dependent on the end points and not the path, hence for a closed path the integral is zero.

8.7 The work done is given by the integral

$$\int_C \mathbf{F} \cdot d\mathbf{r},$$

however,

$$\hat{\mathbf{T}} = \frac{d\mathbf{r}/dt}{|d\mathbf{r}/dt|} = \frac{d\mathbf{r}/dt}{ds/dt} = \frac{d\mathbf{r}}{ds}$$

so the above integral is

$$\int_C \mathbf{F} \cdot \hat{\mathbf{T}} ds.$$

a) With $\mathbf{F} = \hat{\mathbf{T}}\sqrt{x^2 + y^2}$ and on $C, x = (1 - \cos\theta), y = \sin\theta$ so $\sqrt{x^2 + y^2} = \sqrt{(1 - \cos\theta)^2 + \sin^2\theta}$, which is $2\sin\frac{1}{2}\theta$. In addition, $ds = \sqrt{(dx)^2 + (dy)^2} = d\theta$ on C. Work done is therefore

$$\int_C \mathbf{F} \cdot \hat{\mathbf{T}} ds = \int_0^{2\pi} 2\sin\frac{1}{2}\theta\, d\theta = 4.$$

b) For this problem,

$$d\mathbf{r} = f_1'(t)dt\mathbf{i} + f_2'(t)\mathbf{j} + \frac{t\, dt}{\sqrt{1+t^2}}\mathbf{k}$$

so working with the unit tangent has no advantage, the work done is

$$\int_C \mathbf{F} \cdot d\mathbf{r} = \int_0^1 \sqrt{1+t^2}\frac{t\, dt}{\sqrt{1+t^2}} = \frac{1}{2}.$$

[The functions $f_1(t)$ and $f_2(t)$ are a red herring; they do not contribute to the work done as they describe components of the path perpendicular to the force.]

c) Here, proceed as in (b)

$$d\mathbf{r} = \frac{1}{\sqrt{1+t^2}}\mathbf{j} + \frac{t}{\sqrt{1+t^2}}\mathbf{k}$$

and on $C, \mathbf{F} = \mathbf{i} + \mathbf{j} + t\mathbf{k}$, so

$$\int_C \mathbf{F} \cdot d\mathbf{r} = \int_0^1 \left[\frac{1}{\sqrt{1+t^2}} + \frac{t^2}{\sqrt{1+t^2}}\right] dt = \int_0^1 \sqrt{1+t^2}dt.$$

The integral can either be done using software or by substituting $t = \sinh\theta$ the result is

$$\frac{1}{2}\sinh^{-1}(1).$$

[Note: This answer is the same as $\frac{1}{2}\ln(1 + \sqrt{2})$ an answer that can be obtained using a tangent substitution.]

d) With $\mathbf{F} = mg\mathbf{k}$, this represents the force due to Earth's gravity. The curve is a *catenary* $x = 1, y = \sinh^{-1}t, z = \sqrt{1+t^2}$ (it's $z = \cosh(y)$ in the plane $x = 1$). Hence, the problem is finding the work done by a frictionless bead sliding down a catenary from $t = 0$ to $t = 1$ or from $(1, \sinh^{-1}1, \sqrt{2})$ to $(1, 0, 1)$. The calculation is

$$\int_C \mathbf{F} \cdot d\mathbf{r} = \int_0^1 \frac{mgt}{\sqrt{1+t^2}}dt = mg[\sqrt{1+t^2}]_0^1 = mg(\sqrt{2} - 1)$$

the expected answer, mg times the vertical distance, as gravity is a conservative force and the work done must only depend on the end points, specifically z in this case.

8.8 The work done in a anticlockwise direction will be positive (θ increasing) and equal to

$$2 \int_C \mathbf{F} \cdot d\mathbf{r}.$$

The ellipse is parameterised by $x = 3\cos\theta, y = 2\cos\theta$ and $\mathbf{F} = (3\cos\theta - 6\sin\theta)\mathbf{i} + (2\sin\theta - 6\cos\theta)\mathbf{j}$ on this ellipse. Hence

$$\mathbf{F} \cdot d\mathbf{r} = -3\sin\theta(3\cos\theta - 6\sin\theta)d\theta$$
$$+ 2\cos\theta(2\sin\theta - 6\cos\theta)d\theta$$

so evaluating the integral is routine, with answer 12π.

8.9 Since the curl of $-k\mathbf{r}/|\mathbf{r}|^3$ is zero, the path has no affect on the result, just the end points. The work done is

$$\int_C \mathbf{F} \cdot d\mathbf{r} = \int_{r=a}^{r=b} -\frac{k\mathbf{r}}{r^3} d\mathbf{r} = -k \int_{r=a}^{r=b} \frac{dr}{r^2} = \left[\frac{k}{r}\right]_a^b = \frac{k}{ab}(a-b).$$

8.10 Starting from the Biot–Savart law,

$$\int_C \frac{d\mathbf{s} \times \mathbf{p}}{|\mathbf{p}|^3}$$

with $\mathbf{p} = -z_0\mathbf{k} + (\mathbf{R} \cdot \mathbf{r}_0)\mathbf{R}$ and $d\mathbf{s} = a\, d\theta\hat{\theta}$ the result follows at once upon taking the cross product.

12.9 Chapter 9

9.1 These problems involve reasonably standard integration techniques but applied to double integrals.
a)

$$I_1 = \int_{y=0}^{y=3} \int_{x=1}^{x=2} (3x^2 + 2xy + 4y)dx\, dy.$$

This is

$$I_1 = \int_{y=0}^{y=3} [x^3 + x^2y + 4xy]_{x=1}^{x=2} dy,$$

$$= \int_{y=0}^{y=3} (8 + 4y + 8y - 1 - y - 4y)dy = \int_{y=0}^{y=3} (7 + 7y)dy,$$

$$= \left[7y + \frac{7}{2}y^2\right]_0^3 = 21 + \frac{63}{2} = \frac{105}{2}.$$

b)

$$I_2 = \int_{x=1}^{x=2} \int_{y=0}^{y=3} (3x^2 + 2xy + 4y)dy\ dx,$$

$$= \int_{x=1}^{x=2} [3x^2 y + xy^2 + 2y^2]_{y=0}^{y=3} dx$$

$$= \int_{x=1}^{x=2} (9x^2 + 9x + 18)dx$$

$$= \left[3x^3 + \frac{9}{2}x^2 + 18x\right]_{x=1}^{x=2}$$

$$= (24 + 18 + 36) - \left(3 + \frac{9}{2} + 18\right) = 57 - \frac{9}{2} = \frac{105}{2},$$

which of course is the same as the answer to part (a).

c)

$$I_3 = \int_{x=0}^{x=2} \int_{y=0}^{y=3} y \sin x\ dx\ dy,$$

$$= \int_{x=0}^{x=2} \left[\frac{1}{2}y^2 \sin x\right]_{y=0}^{y=3} dx$$

$$= \int_{x=0}^{x=2} \frac{9}{2} \sin x\ dx = \frac{9}{2}(1 - \cos 2).$$

d)

$$I_4 = \int_0^1 \int_0^1 \frac{x}{(xy + 1)^2} dy\ dx$$

$$= \int_0^1 \left[-\frac{x}{(xy + 1)} \cdot \frac{1}{x}\right]_0^1 dx$$

$$= -\int_0^1 \left[\frac{1}{1 + x} - 1\right] dx$$

$$= -[\ln(x + 1) - x]_0^1$$

$$= 1 - \ln 2.$$

e)

$$I_5 = \int_{\pi/2}^{\pi} \int_1^2 x \cos(xy)dy\ dx$$

$$= \int_{\pi/2}^{\pi} \left[\frac{1}{x}x \sin(xy)\right]_1^2 dx$$

$$= \int_{\pi/2}^{\pi} (\sin 2x - \sin x)dx$$

$$= \left[-\frac{1}{2}\cos 2x + \cos x\right]_{\pi/2}^{\pi}$$

$$= \left[-\frac{1}{2} - 1 - \left(\frac{1}{2} + 0\right)\right] = -2.$$

9.2 Here variable limits are more complex.

a)
$$I_1 = \int_0^1 \int_{-x}^{x^2} y^2 x \, dy \, dx$$
$$= \int_0^1 \left[\frac{1}{3} y^3 x \right]_{-x}^{x^2} dx$$
$$= \int_0^1 \left(\frac{1}{3} x^7 + \frac{1}{3} x^4 \right) dx$$
$$= \left[\frac{1}{24} x^8 + \frac{1}{15} x^5 \right]_0^1 = \frac{13}{120}.$$

b)
$$I_2 = \int_0^{\pi/3} \int_0^{\cos y} x \sin(y) dx \, dy$$
$$= \int_0^{\pi/3} \left[\frac{1}{2} x^2 \sin(y) \right]_0^{\cos y} dy$$
$$= \int_0^{\pi/3} \left(\frac{1}{2} \cos^2 y \sin(y) \right) dy$$
$$= \left[-\frac{1}{6} \cos^3 y \right]_0^{\pi/3}$$
$$= -\frac{1}{48} + \frac{1}{6} = \frac{7}{48}.$$

9.3 This problem involves recognising limits through graphs

a) The limits are filled in using the triangle shown in Figure 12.8

$$\int_{x=0}^{x=2} \int_{y=0}^{y=4-2x} x \, dy \, dx,$$

$$\int_{y=0}^{y=4} \int_{x=0}^{x=2-\frac{1}{2}y} x \, dx \, dy,$$

each gives the answer 8/3.

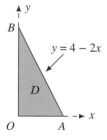

Figure 12.8 Triangle for Exercise 9.3(a).

Figure 12.9 Domain for Exercise 9.3(b).

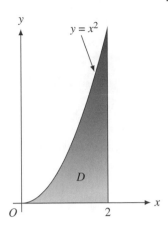

b) See Figure 12.9:

$$\int_{x=0}^{x=2} \int_{y=0}^{y=x^2} (x+y) dy\, dx,$$

$$\int_{y=0}^{y=4} \int_{x=\sqrt{y}}^{x=2} (x+y) dx\, dy,$$

choosing either the answer is 36/5; the first is much easier.

c) See Figure 12.10:

$$\int_{x=-2}^{x=2} \int_{y=x^2}^{y=4} x^2\, dy\, dx,$$

$$\int_{y=0}^{y=4} \int_{x=-\sqrt{y}}^{x=\sqrt{y}} x^2\, dx\, dy,$$

either way round gives the answer 128/15; the second is easier.

d) See Figure 12.11:

$$\int_{x=-2}^{x=2} \int_{y=0}^{y=\sqrt{4-x^2}} y\, dy\, dx,$$

$$\int_{y=0}^{y=2} \int_{x=-\sqrt{4-y^2}}^{x=\sqrt{4-y^2}} y\, dx\, dy,$$

the answer in both cases is 16/3; the first is marginally simpler.

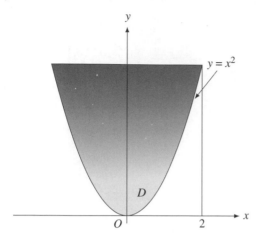

Figure 12.10 Domain for Exercise 9.3(c).

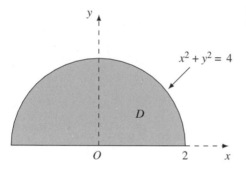

Figure 12.11 The semi-circular domain for Exercise 9.3(d).

e) See Figure 12.12:

$$\int_{y=0}^{y=6} \int_{x=y-6}^{x=2-\frac{y}{3}} dx \, dy$$

$$\int_{x=-6}^{x=0} \int_{y=0}^{y=x+6} dy \, dx + \int_{x=0}^{x=2} \int_{y=0}^{y=6-3x} dy \, dx$$

and both give the answer 24. As the integrand is unity, this is the area of the triangle (half the base times the height).

9.4 The domain is shown as marked D in Figure 12.13

$$\iint_D \cos x \, dD = \int_0^6 \int_0^{\frac{1}{2}y} \cos x \, dx \, dy + \int_6^{15} \int_0^{\sqrt{15-y}} \cos x \, dx \, dy$$

Figure 12.12 The triangular domain, shaded, for Exercise 9.3(e).

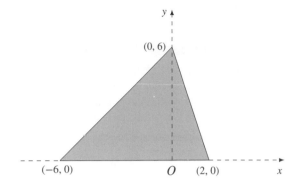

Figure 12.13 The domain *D* is shaded.

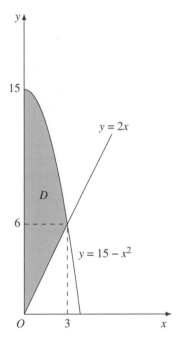

or

$$\iint_D \cos x \, dD = \int_0^3 \int_{2x}^{15-x^2} \cos x \, dy \, dx.$$

Using the second single expression double integral, the answer is

$$2 - 8 \cos 3 + 2 \sin 3.$$

9.5 (a) 1/12 domain is a semi-circle; (b) $3\pi/4$ domain is the top half of a cardioid.

9.6 (a) in polar co-ordinates the integral is:-

$$\int_0^\pi \int_0^1 r^3 \cdot r \, dr \, d\theta = \frac{\pi}{5};$$

(b) $9\pi/2$.

9.7 (a) $\pi/8$, the domain is a quarter circle, first quadrant radius 1; (b) $16/9$, the domain is a semi-circle above the diameter $(0,0), (2,0)$ and the integral is

$$\int_0^{\pi/2} \int_0^{2\cos\theta} r^2 \, dr \, d\theta;$$

(c) $\frac{\pi}{4}(\sqrt{5}-1)$, the domain is one-eighth of a circle and the integral is

$$\int_0^{\pi/4} \int_0^2 \frac{r}{\sqrt{1+r^2}} dr \, d\theta.$$

9.8 Here, we need to change the order of integration.

a) The integral becomes

$$\int_0^1 \int_0^{y^2} \sin\left(\frac{y^3+1}{2}\right) dx \, dy,$$

which in turn becomes

$$\int_0^1 y^2 \sin\left(\frac{y^3+1}{2}\right) dy = \frac{2}{3}\left[\cos\frac{1}{2} - \cos 1\right].$$

b) Reversing the order the double integral becomes

$$\int_0^1 \int_0^{\sqrt{y}} \frac{x^3}{\sqrt{x^4+y^2}} dx \, dy.$$

Substitute $u = x^4 + y^2$ and the answer is $\frac{1}{4}(\sqrt{2}-1)$.

c) Reversing the order gives

$$\int_0^{\pi/2} \int_0^{\cos x} e^{\sin x} \, dy \, dx = e - 1.$$

d) The integral becomes

$$\int_0^1 \int_0^y x^2 e^{y^2} \, dx \, dy,$$

which is

$$\int_0^1 \frac{1}{3} y^3 e^{y^4} \, dy.$$

Substituting $u = y^4$, this evaluates easily to $\frac{1}{12}(e-1)$.

Figure 12.14 The intersection of the plane $x + y + z = 2$ and the parabolic cylinder $y = x^2$.

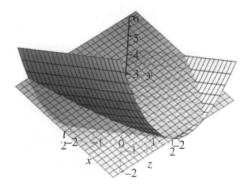

9.9 Using the shaded Figure 9.35, the equation of the circle is $r = 2a \sin \theta$ and the integral is

$$\int_0^\phi \int_0^{2a \sin \theta} r \, dr \, d\theta.$$

This gives the result in the question.

9.10 These integrals are reasonably straightforward: the answers are (a) 8/3; (b) 53/105.

9.11 The plane $x + y + z = 2$ cuts each axis at 2. In Figure 12.14, the y-axis points up, the z-axis points along the cylinder and the x-axis can be seen pointing to the left downwards. The required volume is that part of the parabolic cylinder cut off by these axes and this plane. Along the axis of the cylinder, the limits are 0 and the plane $z = 2 - x - y$, this line is then taken from the curved surface at the bottom $y = x^2$ to the slant surface at the top $y = 2 - x$ (remember z has been integrated out and the line already runs through all values of z). Finally, the plane is taken along the x direction between the two extremes that lie on the coordinate plane $z = 0$. That is where $y = x^2 = 2 - x$, i.e. $x = -2, 1$. The triple integral

$$\int_0^1 \int_{x^2}^{2-x} \int_0^{2-x-y} dz \, dy \, dx$$

is evaluated to be 81/20.

9.12 The triple integral, as shown in Figure 12.15, is

$$\int_0^2 \int_0^{\sqrt{4-x^2}} \int_0^{6-xy} dz \, dy \, dx.$$

A substitution $x = 2 \sin \theta$ is required for the final (x) integration and the answer is $6\pi - 2$.

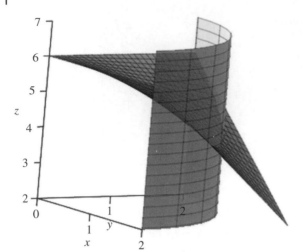

Figure 12.15 The solid inside the cylinder $x^2 + y^2 = 4$ above the x, y plane, with $x, y \geq 0$ but under the surface $z = 6 - xy$.

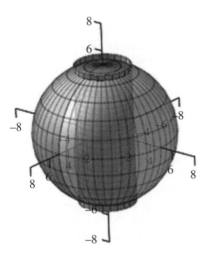

Figure 12.16 The sphere and piercing cylinder.

9.13 Figure 12.16 shows the volume that is given by the triple integral

$$\int_{-\sqrt{21}}^{\sqrt{21}} \int_{2}^{5} \int_{0}^{2\pi} R\, dR\, d\theta\, dz = 28\pi\sqrt{21}.$$

9.14 These integrals are reasonably straightforward. r changed to R here as we are using cylindrical polars.

a)

$$\int_0^{2\pi} \int_0^1 \int_0^{\sqrt{1-R^2}} zR \, dz \, dR \, d\theta = \int_0^{2\pi} \int_0^1 R \left[\frac{1}{2}z^2\right]_0^{\sqrt{1-R^2}} dR \, d\theta$$

$$= \frac{1}{2} \int_0^{2\pi} \left[\frac{1}{2}R^2 - \frac{1}{4}R^4\right]_0^1 d\theta$$

$$= \frac{1}{2} \cdot 2\pi \cdot \frac{1}{4} = \frac{\pi}{4}.$$

b)

$$\int_0^{\pi/2} \int_0^{\cos\theta} \int_0^{R^2} R \sin\theta \, dz \, dR \, d\theta = \int_0^{\pi/2} \int_0^{\cos\theta} R^3 \sin\theta \, dR \, d\theta$$

$$= \int_0^{\pi/2} \frac{1}{4}\cos^4\theta \sin\theta \, d\theta$$

$$= -\left[\frac{1}{4} \cdot \frac{1}{5}\cos^5\theta\right]_0^{\pi/2} = \frac{1}{20}.$$

9.15 These three-dimensional problems benefit from computer algebra to draw diagrams.

a) Figure 12.17 shows the paraboloid and the plane. The volume is

$$\int_0^9 \int_0^{2\pi} \int_0^{\sqrt{z}} R \, dR \, d\theta \, dz,$$

Figure 12.17 The plane $z = 9$ and the paraboloid $z = x^2 + y^2$.

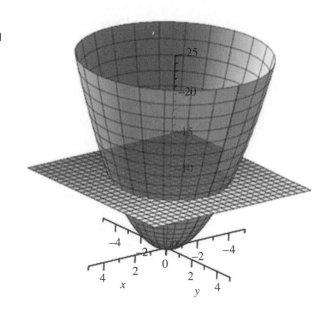

which integrates to

$$2\pi \int_0^9 \left[\frac{1}{2}R^2\right]_0^{\sqrt{z}} dz = \pi \int_0^9 z\, dz = \frac{81}{2}\pi.$$

b) Volume is simply, using a figure similar to Figure 12.16

$$\int_0^{2\pi} \int_0^2 \int_{-\sqrt{9-R^2}}^{\sqrt{9-R^2}} R\, dz\, dR\, d\theta,$$

that is

$$2\pi \int_0^2 2R\sqrt{9-R^2}\, dR = 2\pi \int_5^9 u^{1/2}\, du$$

writing $u = 9 - R^2$. Thus, the volume is

$$\frac{4\pi}{3}(27 - 5\sqrt{5}).$$

9.16 Here are some more.

a) This triple integral

$$\int_0^{\pi/2} \int_0^{\pi/2} \int_0^1 r^3 \sin\phi \cos\phi\, dr\, d\phi\, d\theta$$

is reasonably straightforward:

$$= \int_0^{\pi/2} \int_0^{\pi/2} \frac{1}{4} \sin\phi \cos\phi\, d\phi\, d\theta$$

$$= \frac{\pi}{2} \cdot \frac{1}{4} \left[\frac{1}{2}\sin^2\phi\right]_0^{\pi/2}$$

$$= \frac{\pi}{16}.$$

b) Similarly, this triple integral

$$\int_0^{2\pi} \int_0^{\pi/4} \int_0^{a\sec\phi} r^2 \sin\phi\, dr\, d\phi\, d\theta$$

$$= \int_0^{2\pi} \int_0^{\pi/4} \frac{1}{3}a^3 \sec^3\phi \sin\phi\, d\phi\, d\theta$$

$$= \frac{1}{3} \int_0^{2\pi} a^3 \int_0^{\pi/4} \frac{\sin\phi}{\cos^3\phi}\, d\phi\, d\theta = \frac{1}{3}a^3 \cdot \frac{2\pi}{2} \left[\frac{1}{\cos^2\phi}\right]_0^{\pi/4}$$

$$= \frac{a^3\pi}{3}[2-1] = \frac{a^3\pi}{3}.$$

9.17 Here's more spherical polar and cylindrical polar co-ordinate problems

a) This is the ice-cream cone-shaped object, with the top (sphere) the ice cream and the bottom (cone) the wafer. It is the volume common to the cone and sphere shown in Figure 12.18. The volume will be

$$
\begin{aligned}
\text{Vol} &= \int_0^{\pi/3} \int_0^{2\pi} \int_0^4 r^2 \sin\phi \, dr \, d\theta \, d\phi \\
&= \int_0^{\pi/3} \int_0^{2\pi} \left[\frac{1}{3} r^3\right]_0^4 \sin\phi \, d\theta \, d\phi \\
&= \frac{64}{3} \cdot 2\pi \cdot [-\cos\phi]_0^{\pi/3} \\
&= \frac{128}{3}\pi \left(1 - \frac{1}{2}\right) = \frac{64\pi}{3}.
\end{aligned}
$$

b) This volume is that between a cut through a diameter of a sphere and a parallel cut a quarter of the way along the radius perpendicular to the plane of the first cut; Figure 12.19 displays the situation. This time,

Figure 12.18 The sphere intersects the cone producing an ice-cream cornet shape.

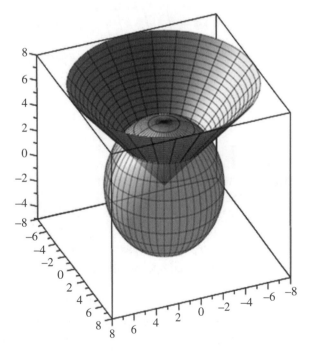

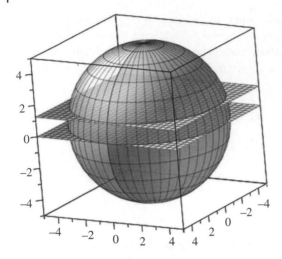

Figure 12.19 The sphere intersecting two planes.

cylindrical polars are used.

$$Vol = \int_0^a \int_0^{2\pi} \int_0^{\sqrt{4a^2-z^2}} R \, dR \, d\theta \, dz$$

$$= \int_0^a \int_0^{2\pi} \frac{1}{2}(4a^2 - z^2)d\theta \, dz$$

$$= 2\pi \cdot \frac{1}{2}\left[4a^2 z - \frac{1}{3}z^3\right]_0^a$$

$$= \pi\left(4a^3 - \frac{1}{3}a^3\right) = \frac{11}{3}\pi a^3.$$

9.18 This involves conversion to polars
a) The triple integral

$$\int_0^a \int_0^{\sqrt{a^2-x^2}} \int_0^{a^2-x^2-y^2} x^2 \, dz \, dy \, dx$$

denotes a z limit from 0 to $z = a^2 - x^2 - y^2$, which is from the x, y plane to a paraboloid. The other limits complete the integration over the nose cone shaped paraboloid (see Figure 12.17). Converting to cylindrical polars gives

$$\int_0^a \int_0^{\pi/2} \int_0^{a^2-R^2} R^2 \cos^2\theta \, dz \, d\theta \, dR.$$

Integrating them gives

$$
= \int_0^a \int_0^{\pi/2} (a^2 - R^2) R^3 \cos^2\theta \; d\theta \; dR
$$

$$
= \int_0^a (a^2 - R^2) R^3 \left[\frac{1}{2}\theta + \frac{1}{4}\sin 2\theta \right]_0^{\pi/2} dR
$$

$$
= \frac{\pi}{4} \left[\frac{1}{4}a^6 - \frac{1}{6}a^6 \right] = \frac{\pi a^6}{48}.
$$

b) The triple integral

$$
\int_{-2}^{2} \int_{-\sqrt{4-x^2}}^{\sqrt{4-x^2}} \int_0^{\sqrt{4-x^2-y^2}} z^2 \sqrt{x^2 + y^2 + z^2} \, dz \; dy \; dx
$$

indicates integration over a hemispherical domain, so we convert to spherical polars:

$$
\int_0^a \int_0^{2\pi} \int_0^{\pi/2} r^2 \cos^2\phi \cdot r \cdot r^2 \sin\phi \; d\phi \; d\theta \; dr
$$

$$
= \int_0^a \int_0^{2\pi} r^5 \left[-\frac{1}{3}\cos^3\phi \right]_0^{\pi/2} d\theta \; dr
$$

$$
= 2\pi \int_0^a \frac{1}{3} r^5 \; dr = \frac{\pi a^6}{9}.
$$

9.19 This is a challenging problem: first find the volume in the first octant; in fact, find the volume in half this and multiply by 16. The answer is $8(2 - \sqrt{2})a^3$, and the solid is called the Steinmetz solid. The detailed solution can be found on the Internet, for example the Wolfram Math-World site.

9.20 Using spherical polar co-ordinates

$$
V = \int_0^{2\pi} \int_0^{\pi/4} \int_0^{a \sec\phi} r^2 \sin\phi \; dr \; d\phi \; d\theta.
$$

The upper limit for r is the (general) distance between the origin and a point on the base of the cone. Careful evaluation yields the answer $\frac{\pi}{3}a^3$. The centre of mass is at the point

$$
\bar{z} = \frac{1}{V} \int_0^{2\pi} \int_0^{\pi/4} \int_0^{a \sec\phi} r^3 \cos\phi \sin\phi \; dr \; d\phi \; d\theta
$$

and yields the result $\bar{z} = \frac{3}{4}a$. Remember the cone is upside down, so the answer is the standard 'centre of mass is a quarter of the way up the axis of symmetry line from the centre of the base'.

9.21 The easiest way to do this is to consider a ring radius R that has moment of inertia

$$MR^2 = 2\pi \rho R^3 \, dR \, dz.$$

The solid of Figure 12.20 will then be an integral over R and over z. The limits are as follows: using cylindrical polar co-ordinates (R, θ, z) the cone has equation $R = z$ and the hemisphere on the top equation $(z - a)^2 + R^2 = a^2$. Therefore, the total moment of inertia will be

$$I = 2\pi \left[\int_0^a \int_0^{R=z} R^3 \, dR \, dz + \int_a^{2a} \int_0^{R=\sqrt{a^2+(z-a)^2}} R^3 \, dR \, dz \right],$$

which becomes

$$I = \frac{\pi \rho}{2} \left[\int_0^a z^4 \, dz + \int_a^{2a} (a^2 - (z - a)^2)^2 dz \right]$$

upon integrating once. Therefore, expanding and evaluating gives

$$I = \frac{\pi a^5 \rho}{2} \left[\frac{1}{5} + \frac{32}{3} - 16 + \frac{32}{5} - \frac{4}{3} + 1 - \frac{1}{5} \right]$$

so

$$I = \frac{11}{30} \pi \rho a^5$$

and since the mass of the hemisphere plus cone is M, where

$$M = \frac{2}{3} \pi a^2 \rho + \frac{1}{3} \pi a^3 = \pi a^3 \rho,$$

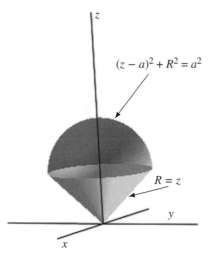

Figure 12.20 The hemisphere $(z - a)^2 + R^2 = a^2$ on the cone $R = z$.

$(z - a)^2 + R^2 = a^2$

$R = z$

the answer is

$$I = \frac{11}{30} Ma^2.$$

9.22 Let the probability distributions for the tube, train and on foot be $p_1(x), p_2(y)$ and $p_3(z)$, respectively, where x, y and z are the times taken by tube, train and walking, also respectively. Assuming that they are Poisson distributions:

$$p_1(x) = \begin{cases} \dfrac{1}{10} e^{-x/10} & x \geq 0, \\ 0 & x < 0, \end{cases}$$

$$p_2(y) = \begin{cases} \dfrac{1}{40} e^{-y/40} & y \geq 0, \\ 0 & y < 0 \end{cases}$$

$$p_3(z) = \begin{cases} \dfrac{1}{5} e^{-z/10} & z \geq 0, \\ 0 & z < 0. \end{cases}$$

The total time has to be less than an hour so $x + y + z \leq 60$ and of course $x, y, z > 0$. The total probability for the journey time is the product of the three probabilities as they are independent events. The three-dimensional distribution $p_1(x)p_2(y)p_3(z)$ thus represents the probability density, and summing over the space $x + y + z \leq 60$ with $x, y, z > 0$ will give the probability that the journey time is less than 1 h, hence give the probability of arriving home on time. Call this P, so

$$P = \frac{1}{2000} \int_0^{60} \int_0^{60-x} \int_0^{60-x-y} e^{-x/10} e^{-y/40} e^{-z/5} \, dz \, dy \, dx.$$

The calculation is messy but routine; ideal for computer algebra software. The answer is $P = 0.66164$ thus the probability of being late is $1 - P = 0.33836$.

12.10 Chapter 10

10.1 Parameterisation of the circle radius 2 is $x = 2\cos\theta, y = 2\sin\theta$ and the integral becomes

$$\oint_C (x^2 + y^2)dx + 3xy^2 \, dy = \int_0^{2\pi} -8\sin\theta + 48\cos^2\theta\sin^2\theta \, d\theta$$

and this is 12π.

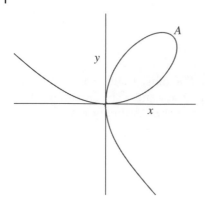

Figure 12.21 The Folium of Descartes $x^3 + y^3 = 3axy$.

10.2 Starting from

$$\frac{1}{2}\int_C (x\, dy - y\, dx)$$

use $x = R\cos\theta, y = R\sin\theta$ so that $dx = -R\sin\theta, dy = R\cos\theta$ and

$$\frac{1}{2}\int_C R^2(\cos^2\theta + \sin^2\theta)d\theta = \frac{1}{2}\int_C R^2\, d\theta.$$

With $R = a(1 + \cos\theta)$, this integral is

$$A = \frac{1}{2}a^2\int_0^{2\pi}(1 + 2\cos\theta + \cos^2\theta)d\theta = \frac{3}{2}a^2\pi.$$

10.3 The Folium of Descartes shown in Figure 12.21 has the line $y = x$ as an axis of symmetry. This line cuts the graph where

$$2x^3 = 3ax^2 \quad \text{so} \quad x = 0 \quad \text{or} \quad \frac{3}{2}a$$

so the point $(3a/2, 3a/2)$ is the point A in Figure 12.21. The folium has the polar equation

$$R^3 = 2R^2\frac{3a\sin\theta\cos\theta}{\cos^3\theta + \sin^3\theta},$$

so dividing by R^2, the area of the folium loop is

$$\frac{1}{2}\int_C R^2\, d\theta = \frac{1}{2}9a^2\int_0^{\pi/2}\frac{\sec^2\theta\tan^2\theta}{(1 + \tan^3\theta)^2}d\theta.$$

With the substitution $u = 1 + \tan^3\theta$, this integral reduces to

$$\frac{3}{2}a^2\int_1^\infty \frac{du}{u^2} = \frac{3}{2}a^2.$$

9.4 The tangent vectors are

$$\mathbf{r}_u = (\sin v, \cos v, 2u),$$
$$\mathbf{r}_v = (-u \sin v, u \cos v, 0),$$

so

$$|\mathbf{r}_u \times \mathbf{r}_v| = \begin{vmatrix} \mathbf{i} & \mathbf{j} & \mathbf{k} \\ \sin v & \cos v & 2u \\ -u \sin v & u \cos v & 0 \end{vmatrix} = (4u^4 + u^2)^{1/2}.$$

Hence,

$$A = \int_0^{2\pi} \int_0^1 (4u^4 + u^2)^{1/2} du \, dv = 2\pi \int_0^1 u(4u^2 + 1)^{1/2} du,$$

and using the substitution $t = 4u^2 + 1$, this integrates by elementary means to

$$\frac{\pi}{12}(\sqrt{125} - 1).$$

9.5 With

$$\mathbf{r} = (a + b\cos\phi)\cos\theta\mathbf{i} + (a + b\cos\phi)\sin\theta\mathbf{j} + b\sin\phi\mathbf{k},$$
$$a > b > 0,$$

we have

$$\mathbf{r}_\theta = (-(a + b\cos\phi)\sin\theta, (a + b\cos\phi)\cos\theta, 0),$$
$$\mathbf{r}_\phi = (-b\sin\phi\cos\theta, b\sin\phi\sin\theta, b\cos\phi),$$

and hence

$$|\mathbf{r}_\theta \times \mathbf{r}_\phi| = \begin{vmatrix} \mathbf{i} & \mathbf{j} & \mathbf{k} \\ -(a + b\cos\phi)\sin\theta & (a + b\cos\phi)\cos\theta & 0 \\ -b\sin\phi\cos\theta & b\sin\phi\sin\theta & b\cos\phi \end{vmatrix},$$

which gives $|\mathbf{r}_\theta \times \mathbf{r}_\phi| = b(a + b\cos\phi)$. Hence,

$$A = \int_0^{2\pi} \int_0^{2\pi} b(a + b\cos\phi)d\theta \, d\phi = 2\pi \int_0^{2\pi} b(a + b\cos\phi)d\phi$$
$$= 4\pi^2 ab.$$

9.6 Here the projection method is used, so the normal is needed. Taking the gradient yields

$$\hat{\mathbf{n}} = \frac{x\mathbf{i} + y\mathbf{j} - 2z\mathbf{k}}{\sqrt{x^2 + y^2 + 4z^2}}.$$

Projecting on to the x, y plane gives

$$\hat{\mathbf{n}} \cdot \mathbf{k} = \frac{-2z}{\sqrt{3(x^2 + y^2)}}$$

as on the cone, $2z^2 = x^2 + y^2$, but the minus sign can be ignored as $-\hat{\mathbf{n}}$ is also a unit normal to the cone. Therefore, the projection of the integral on the x, y plane is

$$\iint_D \frac{3(x^2 + y^2)(\sqrt{2}(x^2 + y^2)^{1/2})}{\sqrt{3(x^2 + y^2)}} dx \, dy,$$

where D is an annulus between the two circles $x^2 + y^2 = 2$ and $x^2 + y^2 = 18$. Converting to polar co-ordinates (R, θ) yields

$$\int_0^{2\pi} \int_{\sqrt{2}}^{3\sqrt{2}} \frac{3\sqrt{2}}{\sqrt{3}} R^3 \, dR \, d\theta$$

giving the result $40\pi\sqrt{6}$.

9.7 The key to evaluating a surface integral is parameterisation, so if $z = f(x, y)$ then $x = x, y = y$ and $z = f(x, y)$ is a parameterisation (Monge patch, see Chapter 5). Therefore, $\mathbf{r}_x = (1, 0, f_x), \mathbf{r}_y = (0, 1, f_y)$ and $\mathbf{r}_x \times \mathbf{r}_y = (-f_x, -f_y, 1)$ giving

$$A = \iint_S \sqrt{1 + f_x^2 + f_y^2} \, dx \, dy$$

and using $z_x = f_x, z_y = f_y$ as an alternative notation gives the result. For the second part where the surface is given by $F(x, y, z) = 0$ if the surface is parameterised with parameters u, v then $F(x(u, v), y(u, v), z(u, v)) = 0$ and the chain rule gives

$$F_x x_u + F_y y_u + F_z z_u = 0 \quad \text{or} \quad \nabla F \cdot \mathbf{r}_u = 0$$

and

$$F_x x_v + F_y y_v + F_z z_v = 0 \quad \text{or} \quad \nabla F \cdot \mathbf{r}_v = 0.$$

Thus, ∇F is parallel to $\mathbf{r}_u \times \mathbf{r}_v$ and the textbook result involving the double integral of $|\mathbf{r}_u \times \mathbf{r}_v|$ is regained. To get an expression involving ∇F note that

$$\hat{\mathbf{n}} = \nabla F / |\nabla F|$$

so that the projection

$$\hat{\mathbf{n}} \cdot \mathbf{k} = \frac{|F_z|}{|\nabla F|}$$

yielding the projection result:

$$\iint_D \frac{|\nabla F|}{|F_z|} dx \, dy = \iint_D \frac{\sqrt{F_x^2 + F_y^2 + F_z^2}}{|F_z|} dx \, dy.$$

9.8 Using parameterisation, this is $x = r \cos \theta \sin \phi, y = r \sin \theta \sin \phi,$ $z = r \cos \phi$ with $r = 1$ on the sphere so the integral becomes

$$\int_0^{\pi/2} \int_0^{\pi/2} (\cos \theta \sin \phi + \sin \theta \sin \phi + \cos \phi) \sin \phi \, d\phi \, d\theta$$

that is routine integration with answer $3\pi/4$.
Projection means projection on to the x, y plane. The normal to S is $\hat{\mathbf{r}}$
so that on the x, y plane

$$\hat{\mathbf{r}} \cdot \mathbf{k} = \cos \phi = \sqrt{1 - R^2}$$

writing R, θ as the plane polar co-ordinates. The integrand $x + y + z$ is
written as $R \cos \theta + R \sin \theta + \sqrt{1 - R^2}$. It is important to notice that we
are projecting on to the x, y plane here *not* putting $z = 0$. The integral
is thus

$$\iint_S (x + y + z) dS = \int_0^1 \int_0^{\pi/2} \frac{R \cos \theta + R \sin \theta + \sqrt{1 - R^2}}{\sqrt{1 - R^2}}$$

$$R \, d\theta \, dR,$$

which does give the same answer $3\pi/4$ but parameterisation is by far
the easier method.

9.9 The surface is as in the last question, but this time the result is a vector.
Here are the details:

$$\iint_S \mathbf{F} \times d\mathbf{S} = \iint_D \mathbf{F} \times \hat{\mathbf{n}} \frac{dD}{|\hat{\mathbf{n}} \cdot \mathbf{k}|}.$$

With $\mathbf{F} = z\mathbf{k}$ and as in the last exercise, $\hat{\mathbf{n}} = \cos \theta \sin \phi \mathbf{i} + \sin \theta \sin \phi \mathbf{j} +$ $\cos \phi \mathbf{k}$ and so $\mathbf{F} \times \hat{\mathbf{n}} = z(\cos \theta \sin \phi \mathbf{j} - \sin \theta \sin \phi \mathbf{i})$ and $|\hat{\mathbf{n}} \cdot \mathbf{k}| = \cos \phi,$
which is z of course. Therefore, the z cancels thus saving the complica-
tion of evaluating it on the surface. Hence, as on D, $\sin \phi = R$ we have

$$\iint_S \mathbf{F} \times d\mathbf{S} = \int_0^1 \int_0^{\pi/2} (R \cos \theta \mathbf{j} - R \sin \theta \mathbf{i}) R \, d\theta \, dR$$

so the integration is not difficult and the answer is

$$\frac{1}{3}(\mathbf{j} - \mathbf{i}).$$

9.10 The difficulty with this exercise is that it is devoid of notation. Let the sphere be centred at $\mathbf{r}_0$ so it has equation $|\mathbf{r} - \mathbf{r}_0| = a$, where a is its radius. The temperature field is $T = bz$ so that the flux through the sphere is

$$\iint_S \nabla T \cdot d\mathbf{S} = b \iint_S \mathbf{k} \cdot d\mathbf{S}$$

and using spherical polar co-ordinates centred on the sphere this is, using parameterisation

$$b \int_0^\pi \int_0^{2\pi} a \cos\theta \sin\phi \, d\theta \, d\phi$$

and this is immediately zero through either integration.

9.11 This problem will be revisited in the next chapter, here the integrals are evaluated directly. With $\mathbf{E} = (\mu \hat{\mathbf{r}})/r^2$:

a) For a sphere, direct parametrisation gives

$$\int_S \mathbf{E} \cdot d\mathbf{S} = \mu \int_0^{2\pi} \int_0^\pi \frac{\hat{\mathbf{r}} \cdot \hat{\mathbf{r}}}{a^2} \cdot a^2 \sin\phi \, d\phi \, d\theta = 4\pi\mu.$$

b) For the curved surface of the cylinder, the integration in R, θ, z co-ordinates gives

$$\int_S \mathbf{E} \cdot d\mathbf{S} = 2 \int_0^a \int_0^{2\pi} \mu a \frac{\sin\phi}{r^2} dz \, d\theta$$

with $\sin\phi = a/r$, ϕ being the angle between $\hat{\mathbf{r}}$ and the vertical, and $r^2 = a^2 + z^2$. After integration with respect to θ, this becomes

$$4\pi a\mu \int_0^a \frac{a}{(z^2 + a^2)^{3/2}} dz.$$

The integral succumbs to the substitution $u = a \tan u$ and gives the answer $2\pi\mu\sqrt{2}$.

For the flat ends, $\cos\phi$ is involved, and the integral is

$$\int_S \mathbf{E} \cdot d\mathbf{S} = 2\pi\mu a \int_0^a \frac{R}{(R^2 + a^2)^{3/2}} dz.$$

This integral succumbs to the substitution $u = R^2 + a^2$ and gives the answer $2\pi\mu(2 - \sqrt{2})$. The sum of these parts is $4\pi\mu$ the same as for the sphere in part (a).

In the next chapter, it is found that this integral over any closed surface that includes the origin would give the same result.

9.11   Chapter 11

11.1   In two dimensions, $\mathbf{F} = F_1\mathbf{i} + F_2\mathbf{j}$ so demanding that $\nabla \times \mathbf{F} = 0$ gives $(F_2)_x = (F_1)_y$. This in turn leads to the existence of a ϕ such that $F_1 = \phi_x$ and $F_2 = \phi_y$. In vectors, this is $\mathbf{F} = \nabla\phi$. Starting with $\mathbf{F} = \nabla\phi$ retraces these steps.

11.2   Call the intersection of the two cylinders C, then on C, $x = a\cos\theta$, $y = z = a\sin\theta$. Calculate $\mathbf{F} \cdot d\mathbf{r}$ then use Stokes' theorem to give

$$\iint_S \nabla \times \mathbf{F} \cdot d\mathbf{S} = \oint_C \mathbf{F} \cdot d\mathbf{r} = \int_0^{2\pi} f(\theta)d\theta,$$

where

$$f(\theta) = 2a^3\sin^3\theta - (a^2\cos^2\theta + 3a^2\sin\theta\cos\theta - 2a\cos\theta)$$
$$+ a^3\sin^3\theta + a\cos\theta\sin\theta$$

and only the $\cos^2\theta$ term contributes, giving the answer $-\pi a^2$.

11.3   Start from the identity $\nabla \times (\phi\mathbf{a}) = \nabla\phi \times \mathbf{a}$ as $\mathbf{a}$ is a constant. Now consider the integral

$$\iint_S \nabla \times (\phi\mathbf{a}) \cdot d\mathbf{S} = \iint_S \nabla\phi \times \mathbf{a} \cdot d\mathbf{S} = \oint_C a\phi \, d\mathbf{r}$$

using Stokes' theorem. The right-hand integral is

$$\mathbf{a} \cdot \oint_C \phi \, d\mathbf{r},$$

whereas the triple scalar product $\nabla \times \mathbf{F} \cdot d\mathbf{S} = d\mathbf{S} \times \nabla\phi \cdot \mathbf{a}$ leads to the left-hand side

$$\mathbf{a} \cdot \left[\iint_S d\mathbf{S} \times \nabla\phi \right].$$

Hence,

$$\mathbf{a} \cdot \left[\iint_S d\mathbf{S} \times \nabla\phi \right] = \mathbf{a} \cdot \oint_C \phi \, d\mathbf{r},$$

so

$$\mathbf{a} \cdot \left[\iint_S d\mathbf{S} \times \nabla\phi - \oint_C \phi \, d\mathbf{r} \right] = 0$$

and the result follows as $\mathbf{a}$ is non-zero.

11.4 Using Gauss' flux theorem and $\nabla \cdot \mathbf{r} = 3$ gives the answer $3V$, where V is the volume enclosed by the closed surface $\mathbf{S}$.

11.5 This follows from Stokes' theorem together with $\nabla \cdot \nabla \times \mathbf{A} = 0$.

11.6 Noting that $\nabla \cdot (\mathbf{a} \times \mathbf{r}) = \mathbf{r} \cdot \nabla \times \mathbf{a} - \mathbf{a} \cdot \nabla \times \mathbf{r}$ but both curls are zero, therefore $\nabla \cdot (\mathbf{a} \times \mathbf{r}) = 0$. Gauss' flux theorem therefore implies, for a closed surface S

$$\iint_S (\mathbf{a} \times \mathbf{r}) \cdot d\mathbf{S} = 0.$$

Hence,

$$\mathbf{a} \cdot \iint_S \mathbf{r} \times d\mathbf{S} = 0$$

but as $\mathbf{a}$ is arbitrary we are forced to conclude that

$$\iint_S \mathbf{r} \times d\mathbf{S} = 0$$

for any closed surface S.

11.7 Noting that

$$\nabla \times (\mathbf{a} \times \mathbf{r}) = (\mathbf{r} \cdot \nabla)\mathbf{a} + \mathbf{a}(\nabla \cdot \mathbf{r}) - \mathbf{r}(\nabla \cdot \mathbf{a}) + (\mathbf{a} \cdot \nabla)\mathbf{r} = 2\mathbf{a}$$

the result follows from Stokes' theorem:

$$\iint_S \nabla \times (\mathbf{a} \times \mathbf{r}) \cdot d\mathbf{S} = 2 \iint_S \mathbf{a} \cdot d\mathbf{S} = \oint_C (\mathbf{a} \times \mathbf{r}) \cdot d\mathbf{r}.$$

11.8 Using

$$\nabla \times (\phi \nabla \psi) = \phi \nabla \times \nabla \psi + \nabla \phi \times \nabla \psi$$

the curl of grad is zero, so Stokes' theorem gives

$$\iint_S \nabla \times (\phi \nabla \psi) d\mathbf{S} = \iint_S \nabla \phi \times \nabla \psi \, d\mathbf{S} = \oint_C \phi \nabla \psi \, d\mathbf{r}$$

as required.

11.9 Gauss flux theorem on the field $\phi \nabla \psi$ yields

$$\iiint_V \nabla \cdot (\phi \nabla \psi) dV = \oiint_S (\phi \nabla \psi) \cdot d\mathbf{S}.$$

Expanding

$$\nabla \cdot (\phi \nabla \psi) = \phi \nabla^2 \phi + \nabla \phi \cdot \nabla \psi$$

gives the result required:

$$\oiint_S (\phi \nabla \psi) \cdot d\mathbf{S} = \iiint_V \phi \nabla^2 \phi + \nabla \phi \cdot \nabla \psi \, dV,$$

Green's first theorem.

11.10 This exercise examines the one case not covered in the chapter, namely when $r = 0$ *on* S. In this case, surround the point by a small hemispherical bubble of radius ϵ, and integrate as in the case when $r = 0$ inside V. The integration is now:

$$\iint_{S_\epsilon} \mathbf{E} \cdot d\mathbf{S} = \mu \int_0^{2\pi} \int_0^{\pi/2} \frac{\epsilon^3 \sin \theta}{\epsilon^3} \, d\theta \, d\phi = 2\mu\pi.$$

The second top limit for a whole sphere was π, whereas for this hemisphere it is $\pi/2$. The answer is $2\mu\pi$.

11.11 This exercise demands some knowledge of electromagnetism. The Poynting vector is $\mathbf{P} = \mathbf{E} \times \mathbf{H}$. The divergence of the Poynting vector is interesting:

$$\nabla \cdot \mathbf{P} = \nabla \cdot (\mathbf{E} \times \mathbf{H}) = \mathbf{H} \cdot (\nabla \times \mathbf{E}) - \mathbf{E} \cdot (\nabla \times \mathbf{H}).$$

Using Maxwell's equations, the two curls on the right can be substituted as follows:

$$\nabla \cdot \mathbf{P} = -\mathbf{H} \cdot \frac{\partial \mathbf{B}}{\partial t} - \mathbf{E} \cdot \frac{\partial \mathbf{D}}{\partial t}$$

provided that there is no current ($\mathbf{J}$). With $\mathbf{B} = \mu \mathbf{H}$ and $\mathbf{D} = \epsilon \mathbf{E}$, this can be written

$$\nabla \cdot \mathbf{P} + \mu \mathbf{H} \cdot \frac{\partial \mathbf{H}}{\partial t} + \epsilon \mathbf{E} \cdot \frac{\partial \mathbf{E}}{\partial t} = 0$$

or

$$\nabla \cdot \mathbf{P} + \frac{\partial}{\partial t} \left(\frac{1}{2} [\epsilon E^2 + \mu H^2] \right) = 0.$$

This is the same as

$$\nabla \cdot \mathbf{P} = -\frac{\partial \mathcal{E}}{\partial t},$$

where $\mathcal{E}$ is the energy density $\frac{1}{2}[\epsilon E^2 + \mu H^2]$. This last equation can be integrated over the surface S and Gauss' flux theorem then gives

$$\iint_S \mathbf{P} \cdot d\mathbf{S} = -\frac{\partial}{\partial t} \iiint_V E \, dV$$

so the flux of the Poynting vector is the rate of change of the energy density, called the *power density*. This is the same relationship as the fluid velocity has to density in a compressible fluid.

11.12 The derivative of the circulation is

$$\frac{d}{dt} \oint_C \mathbf{u} \cdot d\mathbf{r} = \oint_C \frac{d\mathbf{u}}{dt} \cdot d\mathbf{r} + \oint_C \mathbf{u} \cdot \frac{d}{dt}\left(\frac{d\mathbf{r}}{ds}\right) ds,$$

where s is the arc length, a parameter for the path $\mathbf{r} = \mathbf{r}(s, t)$ taken by the fluid. The second term on the right is

$$\oint_C \mathbf{u} \frac{d\mathbf{u}}{ds} ds,$$

since the order of differentiation can be swapped and $\mathbf{u} = d\mathbf{r}/dt$. This integrates to $[\frac{1}{2}\mathbf{u}^2]_C$, which is zero as C is closed and the kinetic energy does not change when evaluated around a closed curve. Using the equation in the question,

$$\frac{d}{dt} \oint_C \mathbf{u} \cdot d\mathbf{r} = \iint_S \nabla \times \left(-\frac{1}{\rho}\nabla p + \nabla \phi\right) \cdot d\mathbf{S}.$$

As curl of grad is zero, only the inverse of density term gives a non-zero factor, which leads to

$$\frac{d}{dt} \oint_C \mathbf{u} \cdot d\mathbf{r} = \iint_S (\nabla \rho \times \nabla p) \cdot d\mathbf{S}.$$

This is certainly zero for constant density, and is also zero wherever these vectors are parallel. This is called barotropic flow. Therefore, the circulation is constant in a fluid that is barotropic; this is Kelvin's circulation theorem.. Using Stokes' theorem,

$$\oint_C \mathbf{u} \cdot d\mathbf{r} = \iint_S \nabla \times \mathbf{u} \cdot d\mathbf{S} = 0,$$

where $\mathbf{S}$ is the area enclosed by the closed curve C. As $\nabla \times \mathbf{u}$ is the vorticity, this shows that the flux of vorticity through the area enclosed by the closed curve C is also zero.

11.13 This equation expresses the rate of change of energy. The hint is $\rho\left(\frac{1}{2}v_iv_i + e\right)$, which is kinetic energy plus internal energy. This is summed over the volume V and differentiated, hence dissipation of energy. The term ρb_iv_i must represent the body force, and in the surface integral T_{ji} is the stress tensor with q_i representing the heat flux out of V through the surface S. In this expression, both i and j are summed over; there is no free index thus it is a scalar equation.

References

1 Gillespie, R.P. (1951) *Partial Differentiation*, Oliver and Boyd, p. 105. Comment: short but contains material not in many books now. Good for chapter 2 here.

2 Eisenhart, L.P. (1909) *A Treatise on the Differential Geometry of Curves and Surfaces*, reprinted by Dover Publications in 1960, p. 474. Comment: a classic text on differential geometry completely without vectors. Probably now only of historic interest.

3 O'Neill, B. (1966) *Elementary Differential Geometry*, Academic Press, p. 411. Comment: still the go-to book for differential geometry. Manages to walk the line between mathematical precision and readability.

4 Arnol'd, V.I. (1988) *The Mathematical Methods of Classical Mechanics*, Springer Graduate Texts in Mathematics, 2nd edn, Springer-Verlag, p. 505. Comment: this is the classical text for mechanics, a very advanced approach for the mathematicians amongst you.

5 Phillips, E.G. (1962) *A Course of Analysis*, Cambridge University Press, p. 372. Comment: classic real analysis, but readable.

6 Acheson, D.J. (1990) *Elementary Fluid Dynamics*, Oxford University Press, p. 397. Comment: there are many books on fluid mechanics, this is the best written in my view.

7 Marsden, J.E. and Hughes, T.J.R. (1983) *The Mathematical Foundations of Elasticity*, Prentice Hall, Englewood Cliffs, NJ. reproduced by Dover in 2003, p. 556. Comment: now downloadable, a classic elasticity text.

8 Lang, S. (1987) *Linear Algebra*, Springer Undergraduate Texts in Mathematics, 3rd edn, p. 287. Comment: a classic text on linear algebra by a leading author.

9 Dyke, P. (1998) *Advanced Calculus*, Macmillan Work Out Series, p. 193. Comment: this is a series of worked examples with more or less similar coverage to the present text.

10 Dyke, P. and Whitworth, R. (2001) *A Guide to Mechanics*, 2nd edn, Macmillan Publishers, p. 348. Comment: a contrast to Arnold, an

elementary text accessible by those with only high school or sixth form mathematics.

11 Marsden, J.E. and Tromba, A. (2011) *Vector Calculus*, 6th edn, W. H. Freeman and Sons, p. 545. Comment: a good vector calculus text, well written but not to everyone's taste.

12 Spiegel, M.R. (1959) *Vector Analysis, Theory and Problems*, Schaum Series, McGraw-Hill, p. 225. Comment: just lots of solved problems, very useful for self study.

Index

Two and Three Dimensional Calculus: With Applications in Science and Engineering, First Edition. Phil Dyke.
© 2018 John Wiley & Sons Ltd. Published 2018 by John Wiley & Sons Ltd.